MW01181184

Nutrient Management of Food Animals to Enhance and Protect the Environment

Edited by
E.T. Kornegay

Acquiring Editor: Joel Stein
Project Editor:: Les Kaplan
Marketing Manager: Greg Daurelle
Direct Marketing Manager: Arline Massey
Cover Designer: Dawn Boyd
PrePress: Kevin Luong
Manufacturing Asistant: Sherri Schwartz

Library of Congress Cataloging-in-Publication Data

Nutrient management of food animals to enhance and protect the environment / edited by E. T. Kornegay.
p. cm.
Includes bibliographical references and index.
ISBN 1-56670-199-6 (alk. paper)
1. Animal nutrition--Congresses. 2. Food animals--Nutrition--Congresses. 3. Feeds--Congresses. I. Kornegay, E. T.
SF94.6.N87 1996
6368.08′52—dc20 95-47437
CIP

Lewis Publishers is an imprint of CRC Press

International Standard Book Number 1-56670-199-6
Library of Congress Card Number 95-47437
Printed in the United States of America 2 3 4 5 6 7 8 9 0
Printed on acid-free paper

PREFACE

Nutrient management has always been an important entity in feeding livestock and poultry in competitive animal production systems. A heightened awareness of the environmental impacts associated with animal production and the increase in size and intensity of modern production units are causing animal nutritionists to refocus their thinking, practices, and expectations regarding the form and level of nutrients that are supplied to animals.

This book is based on the Proceedings of the John Lee Pratt International Symposium on Nutrient Management of Food Animals to Enhance and Protect the Environment held on June 4–7 at Virginia Polytechnic Institute and State University. The symposium began with a discussion of environmental challenges faced by the livestock and poultry industries. Then new and future technologies for enhancing amino acid and nitrogen utilization and reduction of excretion were discussed by internationally recognized scientists. Next, new feeding strategies to reduce environmental pollution were addressed by recognized experts. Also, new and future technologies for improving nutrient utilization and reduction of excretion (primarily phosphorus and trace minerals) were presented by scientists who are actively engaged in this research. Finally, a discussion of the environmental concerns and experiences of using animal manure in the Netherlands was discussed by a Dutch expert.

The environmental impact of modern livestock and poultry production is of major concern and interest in many countries. Professionals representing industry, government, universities, and private practice who share an interest in nutrient management of food animals to enhance and protect the environment should find this book to be of great interest and value.

The organizing committee thanks the John Lee Pratt Animal Nutrition Program for financial support of the symposium and for generous support of research conducted by the organizing committee. The committee also thanks guest speakers who have been so generous with their time and ideas.

The assistance of Mrs. Cindy Hixon in the preparation of the proceedings for publication is acknowledged.

Organizing Committee

E. T. Kornegay
K. E. Webb, Jr.
J. P. Fontenot
C. E. Polan
J. H. Herbein
D. Blodgett
C. Thatcher

Contributors

V. G. Allen, Ph.D.
Department of Plant and Soil Science
Texas Tech University
Lubbock , Texas

G. A. Ayangbile, Ph.D.
Agri-King, Inc.
Fulton, Illinois

David H. Baker, Ph.D.
Department of Animal Sciences and
Division of Nutritional Sciences
University of Illinois
Urbana, Illinois

D. J. Blodgett, Ph.D.
Departments of Biomedical Sciences
and Pathobiology
Virginia-Maryland College of
Veterinary Medicine
Blacksburg, Virginia

Paul T. Chandler, Ph.D.
Chandler & Associates, Inc.
Dresden, Tennessee

M. Terry Coffey, Ph.D.
Vice President, Operations
Murphy Family Farms
Rose Hill, North Carolina

R. W. Engel, Ph.D.
Emeritus Professor of Biochemistry
and Nutrition
Virginia Polytechnic Institute and
State University
Blacksburg, Virginia

D. E. Eversole, Ph.D.
Department of Animal and Poultry
Sciences
Virginia Polytechnic Institute and
State University
Blacksburg, Virginia

G. C. Fahey, Jr., Ph.D.
Department of Animal Sciences
University of Illinois
Urbana, Illinois

Joseph D. Fontenot, Ph.D.
Department of Animal and Poultry
Science
Virginia Polytechnic Institute and
State University
Blacksburg, Virginia

Keith R. Hansen, Ph.D.
Nutrition Service Associates
Hereford, Texas

Chris H. Henkens, Ph.D.
Agricultural Advisory Service for
Matters Relating to Soil and
Manuring
Wageningen, The Netherlands

Joseph H. Herbein, Ph.D.
Dairy Science Department
Virginia Polytechnic Institute and
State University
Blacksburg, Virginia

H. S. Hussein, Ph.D.
Department of Animal Sciences
University of Illinois
Urbana, Illinois

Donald E. Johnson, Ph.D.
Department of Animal Sciences
Colorado State University
Fort Collins, Colorado

Age W. Jongbloed, Ph.D.
Department of Nutrition of Pigs and
Poultry
Institute of Animal Science and
Health (ID-DLO)
Lelystad, The Netherlands

K. Karunanandaa, Ph.D.
Department of Dairy and Animal Science
Pennsylvania State University
University Park, Pennsylvania

Roger A. Kleese, Ph.D.
Director of Market Development
Pioneer Hi-Bred International, Inc.
Des Moines, Iowa

E. T. Kornegay, Ph.D.
Department of Animal and Poultry Sciences
Virginia Polytechnic Institute and State University
Blacksburg, Virginia

M. D. Lindemann, Ph.D.
Department of Animal Sciences
University of Kentucky
Lexington, Kentucky

J. C. MacRae, Ph.D.
Rowett Research Institute
Bucksburn, Aberdeen, Scotland

J. C. Matthews, Ph.D.
Department of Animal and Poultry Sciences
Virginia Polytechnic Institute and State University
Blacksburg, Virginia

M. Q. McCollum, B.S.
Department of Animal and Poultry Sciences
Virginia Polytechnic Institute and State University
Blacksburg, Virginia

Mark Morrison, Ph.D.
Department of Animal Science
School of Biological Sciences and Center for Biotechnology
University of Nebraska
Lincoln, Nebraska

Y. L. Pan, Ph.D.
Department of Animal and Poultry Sciences
Virginia Polytechnic Institute and State University
Blacksburg, Virginia

Carl E. Polan, Ph.D.
Department of Dairy Science
Virginia Polytechnic Institute and State University
Blacksburg, Virginia

Jon J. Ramsey, Ph.D.
Metabolic Laboratory
Department of Animal Sciences
Colorado State University
Fort Collins, Colorado

R. L. Rice, DVM, Ph.D.
Departments of Biomedical Sciences and Pathobiology
Virginia Polytechnic Institute and State University
Blacksburg, Virginia

Keith E. Rinehart, Ph.D.
Perdue Farms Incorporated
Salisbury, Maryland

G. G. Schurig, DVM, Ph.D.
Departments of Biomedical Sciences and Pathobiology
Virginia-Maryland College of Veterinary Medicine
Blacksburg, Virginia

Martin N. Sillence, Ph.D.
School of Agriculture
Charles Sturt University
Wagga Wagga, New South Wales, Australia

Jerry W. Spears, Ph.D.
Department of Animal Science
North Carolina State University
Raleigh, North Carolina

H. M. Stahr, Ph.D.
Veterinary Diagnostic Laboratory
College of Veterinary Medicine
Iowa State University
Ames, Iowa

W. S. Swecker, DMV, Ph.D.
Large Animal Clinical Sciences
Virginia-Maryland College of Veterinary Medicine
Blacksburg, Virginia

L. A. Swiger, Ph.D.
College of Agricultural and Life Sciences
Virginia Polytechnical Institute and State University
Blacksburg, Virginia

C. D. Thatcher, DMV, Ph.D.
Large Animal Clinical Sciences
Virginia-Maryland College of Veterinary Medicine
Blacksburg, Virginia

G. A. Varga, Ph.D.
Department of Dairy and Animal Science
Pennsylvania State University
University Park, Pennsylvania

S. Wang, Ph.D.
Department of Animal and Poultry Sciences
Virginia Polytechnic Institute and State University
Blacksburg, Virginia

Gerald M. Ward, Ph.D.
Department of Animal Sciences
Colorado State University
Fort Collins, Colorado

K. E. Webb, Jr., Ph.D.
Department of Animal and Poultry Sciences
Virginia Polytechnic Institute and State University
Blacksburg, Virginia

Contents

INTRODUCTION 1

Reflections on John Lee Pratt

R. W. Engel

Mr. John Lee Pratt entered my life in a somewhat unexpected manner. I was quite contented with a professorial appointment in the Animal Nutrition Laboratory at Auburn University when I was first contacted to consider an offer to organize a Department of Biochemistry and Nutrition at Virginia Tech, a process that culminated in my acceptance of the offer in 1952. However, the story actually began earlier. Dr. Harold Young, Director of the Virginia Agricultural Experiment Station, assembled members of his staff for a meeting with Mr. Pratt, who had come to the campus for advice on a disease problem he had with his beef cattle. Present at the meeting were Dr. I. D. Wilson, Head of the Biology Department, and Dr. Wilson Bell, Professor of Veterinary Science in the same department. They had suggested to Mr. Pratt that they were quite certain that the cattle were suffering from a noninfectious, perhaps nutritional, disease condition, but they were not certain of the cause.

An outcome of the meeting with Mr. Pratt was a decision by Director Young to seek advice from the University of Wisconsin, Cornell, Penn State, and possibly other land grant universities with strong programs in nutritional biochemistry to see how he might improve his existing Department of Agricultural Chemistry so that Mr. Pratt's questions could be addressed more adequately in the future. The consultants were also asked to recommend candidates for the headship of the new or improved department. It was through this process that I became head of the new Department of Biochemistry and Nutrition at Virginia Tech in 1952, a process unknown to me at the time of my appointment, but which gradually was revealed as I settled into my new duties.

Mr. Pratt wanted additional advice on the role of the rumen. Dr. Ken King of our staff was interested in studying the bacteria in the rumen that were capable of breaking down complex carbohydrates, the celluloses. This fascinated Mr. Pratt and he wanted papers on this subject. He also was very interested in the work that Dr. Ed Bunce had under way. He was feeding Brazilian beans that produced blind offspring when fed to pregnant rats. This was an outcome of our participation in a U.S. Public Health Service nutrition survey in Brazil. Mr. Pratt's interest may have stemmed from his own experiences there. When Mr. Pratt was a student in engineering at the University of Virginia, and for some years after he graduated, he spent some time in Brazil as a construction engineer on the railroad. He used to reflect on his early interest in developing countries and of the encounters with hostile natives as the railhead extended into the vast reaches of western Brazil and the Amazon Jungle. His main interest in nutrition had always been the trace elements.

In 1957, when Wilson Bell became associate director of the experiment station, he informed me that it was time I met Mr. Pratt and that he would arrange for me to see him at his estate, a mansion on the bluffs of the Rappahannock River overlooking Fredricksburg, VA. By way of preparation for the meeting, Wilson thought it would be appropriate for me to review for Mr. Pratt the steps that had been taken to improve the capability of the department and the biochemical-nutritional work that had been initiated. I met with Mr. Pratt in his study. The meeting of about 2 hours mostly involved my acquainting him with the activities under way in the department. He thanked me for briefing him and added that he would like to see me about once a year for further talks. "Have your secretary call me here to set up the next meeting about a year from now" is the way he put it. About a week after this visit, Mr. Pratt sent a check to the Virginia Tech Treasurer for $100,000 to be used to enhance the effectiveness of faculty and graduate students in biochemistry and nutrition.

I saw Mr. Pratt once a year for the next 9 years. The format was always the same. It seems I was teaching a 2-hour course that met once a year, had no problems with student attention or interest, and paid a handsome tuition fee. What did I actually teach in this unusual course? Early on we worked jointly with the veterinary staff; they maintained some steers for us that had surgically implanted glass windows (cannula) in their rumens. This facilitated determining whether or not the problem with his cattle might not also involve nutrition.

When Mr. Pratt sent the check in 1957 for $100,000 he had specified that it was to be used over a period of about 10 years. When I met with him in 1965, I reminded him that the fund was approaching exhaustion. One week after that meeting, Virginia Tech received a check for $150,000 to continue support. The department had doubled, and the administration at Virginia Tech had continued strong support, particularly in the new graduate training program.

Incidentally, while the new building was under construction for the Biochemistry and Nutrition Department in 1960, Mr. Pratt told me he was not interested in paying for buildings. He did, however, transfer shares of General Motors stock, valued at about $50,000, to Virginia Tech to finance the purchase of an amino acid analyzer for the new facility,

Mr. Pratt never wanted a written record of how his money was spent. He did not want copies of publications except for a time or two when he became particularly interested in a subject under discussion. As in the beginning of our relationship, he continued to remind me that he wanted his name kept out of anything we did with his contributions.

To my knowledge the McCollum-Pratt Institute at Johns Hopkins University is the only case in which John Lee Pratt identified himself with his benevolence in supporting scientific research and education during his lifetime. He was adamant in insisting that his assistance be treated anonymously.

In closing, I must confess that when I accepted the appointment at Virginia Tech in 1952 I had no notion that I would be involved with a retired General

Motors executive in a shared learning experience for the majority of years of my service as a department head. Incidentally, Mr. Pratt never mentioned any problems with his beef cattle, nor did I ever see them. Nevertheless, the fruitful outcome of my connections with him must be credited to the vision of Director Harold Young, and particularly the input of Dr. I. D. Wilson and Dr. Wilson Bell, which gave me the opportunity to get to know this great humanitarian of our time. He confided to me that he assisted untold numbers of young black students with their education — anonymously, of course.

For Mr. Pratt's financial support during my time, I should also credit all of the faculty and students in the Department of Biochemistry and Nutrition for they were the producers of the subject matter for my sessions with him.

In conclusion, I wish you a successful symposium, made possible in part by Mr. Pratt's benevolence.

INTRODUCTION 2

The Role of the John Lee Pratt Animal Nutrition Program in Supporting Nutrient Management of Food Animals to Enhance the Environment

L. A. Swiger

It is indeed a pleasure for Virginia Tech and the College of Agriculture and Life Sciences to serve as your host for this important symposium sponsored by the John Lee Pratt Animal Nutrition Program. The Pratt program represents a unique opportunity for our scientists, teachers, and students working in the area of animal nutrition. Initial funds were made available in 1977. I became associated with the program when I came to Virginia Tech as Head of Animal Science in 1980, and I have administered the program since 1986 as the Associate Dean for Research. When I became Acting Dean on January 1, 1992, I tried to get out of this job. The faculty would not let me off the hook, most likely because they found me easy to control. But, let me go back to the origins of the program and tell you of its financial and programmatic history.

Let me read to you the preamble of the master plan for the program written in 1976.

> Early this year (1976), the University and the public received news of the bequest of the late Mr. John Lee Pratt of Stafford County. The generous bequest for "promoting the study of animal nutrition, . . . and to provide equipment and material for experiments in feeding and the preparation of feeds for livestock and poultry, and to publish and disseminate the practical results thereby obtained . . ." affords opportunities and presents challenges which are without parallel in the history of the College of Agriculture and Life Sciences at VPI&SU.
>
> Judicious use of these funds will provide support for the development of a premier center of learning in animal nutrition at VPI&SU which should rank first in the nation. We are also accepting with this bequest the concomitant obligation to become a world leader in deriving and disseminating "practical information in feeding and preparation of feed for livestock and poultry."
>
> Present programs in livestock nutrition at VPI&SU have grown in scope and quality from past efforts at this institution. Traditional funding has not, however, allowed these programs to reach full potential. Infusion of the Pratt monies will stimulate the initiation of new programs as well as appropriate expansion of present programs for the benefit of the livestock industry and the consuming public.

No academic or research program can rise above the excellence and expertise of the faculty it embraces. Among the benefits of the bequest to this University will be the opportunity to attract new and exceptional scientists in addition to multiplying the efforts of the fine faculty already here. These funds will also provide for the continuous and vigorous search for outstanding young talent through scholarships, fellowships, graduate stipends, and post-doctoral fellowships.

The funds will also permit us to purchase modern laboratory equipment resulting in greater efficiency of researchers and better instruction, particularly at the graduate level. They may also be used to purchase major equipment for animal nutrition in the proposed new Animal Sciences Building.

In accord with Mr. Pratt's wishes, livestock producers will be furnished a wealth of reliable nutritional information resulting from this research. Such information will make for a more efficient animal agriculture in Virginia for the ultimate benefit of all consumers. Continuing education programs and institutional services (e.g. analytical and computer) can be provided which should greatly accelerate the rate of progress made in Virginia and this region. Livestock producers and those allied segments of the livestock and feed industry will be the recipient of the best programs available anywhere. They should, therefore, become the pacesetters for the entire region and nation. Thus, the reputation of VPI&SU will be further enhanced consistent with our Land-Grant University mission of service to the people.

Mr. Pratt, after taking care of family and employees, left his legacy to several universities — Johns Hopkins, University of Virginia (to be transferred to Washington and Lee if they misused it) and, similarly, to Virginia Tech with the same qualifier. Seven other institutions were named as benefactors, including Washington and Lee. It is not known to me why the University of Virginia and Virginia Tech were singled out in this way. I quote from Mr. Pratt's last will and testament.

I give, devise and bequeath twenty (20) parts thereof to the Virginia Polytechnic Institute, of Blacksburg, Virginia, ten (10) parts to be used by it for promoting the study of animal nutrition, for supplementing salaries in order to procure and adequately compensate proficient personnel, and to provide equipment and materials for experiments in feeding and the preparation of feeds for livestock and poultry, and to publish and disseminate the practical results thereby obtained as recognized by the President, and approved by the Board of Visitors, or their successors, of the said Institute. The other ten (10) parts shall be used for the support of research and scholarships in the School of Engineering.

This bequest and gift is subject to the following provision, namely: that it shall not be used, directly or indirectly, at any time, as a substitute for any appropriations by the State of Virginia, for the said Institute for the aforesaid purposes, but shall be supplementary thereto; and in the event that the State of Virginia, or the Board of Visitors, or their successors, of the Institute will not permit these funds and property to be used as a supplement to the appropriations made by the State of Virginia for the Institute for the aforesaid purposes, then this bequest shall

> terminate and cease to belong to the said institute, and by it shall forthwith be delivered and given to the Washington and Lee University, Lexington, Virginia.

Our half amounted to about $5.5 million.

It is my understanding that Mr. Pratt's benevolence was due to his pleasure with and respect of the work of Dr. R. W. Engel, for whom our biochemistry building is named.

Initially, the program concentrated primarily in two areas, research projects that were spread widely throughout the college and represented a very broad spectrum of research and the placing of postdoctoral personnel in the laboratories of a few key researchers.

In about 1983, we stopped funding projects and concentrated on graduate support only. This program did not work well, and for several years the Pratt fund was underspent. Then, in 1986, we assembled the animal nutrition faculty from the departments of Animal Science and Poultry Science (since merged), Biochemistry and Nutrition and Anaerobic Microbiology (since merged), Dairy Science, and Fisheries and Wildlife Sciences. We discussed, at length, where we wanted the program to go and prioritized several possibilities. Clearly, the faculty, and in keeping with the will, wanted these funds to be used to enhance other funds, not to replace them. We desired to really make a difference in our teaching, research, and extension programs in animal nutrition. I will tell you that the last 5 years have strained, but not broken, our resolve in that regard. We agreed to put about half of our resources into an elite Ph.D. Fellowship program that featured an attractive stipend plus an operating allotment. We have 15 of these. We support a handful of M.S. assistantships, usually four or five at most. About 65% of the funds go for direct graduate student support. Additionally, 15% goes to award 40 $1000 freshman scholarships in departments engaged in animal nutrition, and, a program we are very proud of, in which six Pratt senior animal nutrition research scholars are chosen annually and receive scholarships and research support. As you might expect, most of these students wind up in graduate school in nutrition.

The remaining funds enhance student programs with laboratory and livestock equipment purchases, usually serving several faculty and department programs, and by bringing in visiting professors and seminar speakers who engage the Pratt program with the entire university.

The chart that follows shows expenditures by years:

The 1995 income was estimated to be $612,323, and the budget includes an additional $193,275 of carryover funds. We are nearing the $10,000,000 mark.

The accomplishments, at this point, include over 200 refereed journal articles and over 200 abstracts, research reports, and extension publications. Nearly 100 graduate students have been supported. Many of them hold key positions around the world. We have no direct assessment of the important discoveries and impact of these students and faculty, but I judge it to be considerable.

Because of Virginia's concern with the Chesapeake Bay and the great rivers flowing to it, Virginia Tech and our College of Agriculture and Life Sciences is heavily involved in water quality programs. Virtually all of our projects have an

PRATT EXPENDITURES

Year	Regular ($)	Nutrient mgt. ($)
1977	1,200	
1978	89,466	
1979	363,298	
1980	453,112	
1981	1,123,979	
1982	402,515	
1983	408,270	
1984	312,823	
1985	238,141	
1986	206,935	
1987	419,239	
1988	359,614	
1989	609,061	
1990	552,418	
1991	615,732	
1992	628,545	101,537
1993	587,074	227,386
1994	537,765	263,388
Total	7,909,187	592,311
Grand Total		8,501,498

environmental protection component. Phosphorus and nitrogen are of immense concern for ground and surface water in Virginia. Under the leadership of our Department of Crop and Soil Environmental Sciences (formerly Agronomy), our undergraduate major in Environmental Sciences has grown in three years to an enrollment of 300. We are out front in this important area and proud of it.

We favor preventative medicine over surgery. We would rather keep those nutrients from being used excessively as fertilizers and pesticides and keep those nutrients from being dealt with in animal waste if we can take care of things on the front end. We found we had about $800,000 of accumulated funds from those low expenditure years that you saw earlier. I asked the faculty to propose the best use of these funds. The faculty acted quickly, and with unanimity, in proposing a three-year project in nutrient management involving our food animals to culminate in this symposium. I quote from the project:

> The concept of reducing N and P excretion through improved nutritional regimes is the method which has the potential to achieve the greatest balance of retained production efficiencies and reduced nutrient excretion at the least cost. The methods currently thought to have the greatest potential are: reduction of dietary P through use of microbial phytases which liberate unavailable P (phytate P) and make it utilizable, reduction of dietary protein content and supplementation with synthetic amino acids to make a more ideal amino acid pattern and limit N excesses, and the use of chelated minerals to improve trace mineral absorption and decrease the excess mineral that is competing with other minerals for absorption sites. These methods have been known for quite some time but have not been commercially feasible. But now with the purification and concentration of some of the phytases, with the development of commercial technology to produce larger

quantities of several amino acids, and with the recent availability of numerous mineral chelates, the potential to utilize dietary means to reduce N and P excretion, as well as several trace minerals, on a commercial basis is here.

The original gift was a tribute to the dedicated work and accomplishments of our faculty and students, the ability to take resources and their own abilities and serve science and serve students. This Pratt project, this topic of nutrient management, this symposium, reflect the dedication of faculty and students throughout the world in the pursuit of knowledge in service to animal agriculture and in service to man. It is in this spirit of able scientists and teachers pursuing their work to further their professions and solve industry problems in applied agriculture that we look forward to sharing our knowledge in this symposium.

CHAPTER 1

Environmental Challenges as Related to Animal Agriculture — Beef

Keith R. Hansen

INTRODUCTION

The charge: To address needs for future research, focusing on those needs and/or voids of our present knowledge of nutrient requirements and management systems of beef animal nutrition, as these relate to nutrients of particular environmental concern.

It is difficult to address future needs without at least some reference to the past. Like many issues, the environmental concerns were voiced early on with little more than rampant speculation based on conjecture and, in the research community, the hope of generating research funds to augment the declining resources available for research. In an effort to stimulate this resource, many extrapolated numbers were used, and worst-case scenarios were represented as undeniable fact. This generally mobilized the government regulatory groups to begin legislation to "clean up" and save the environment from agricultural waste and contamination with a minimal amount of real data and a universal approach, with little regard for regional differences for the potential of pollution.

As unfounded as much of the information was and as often misleading as the interpretations were, the effort seems to have identified some very logical considerations, such as the need to restrict manure from running into surface waters used for public water supplies.

And, recently, we are seeing an effort by regulatory groups to recognize and address regional differences in potential pollution with differential guidelines on waste application rates.

When we talk about the environment and its relation to beef production, we are essentially addressing the topic of how do we manage the manure, solid and liquid waste products, and particulate matter in the air. I am not, in this chapter, going to give space to the subject of gases of digestion and their effect on the ozone layer, greenhouse effect, or any other consideration. This problem of managing the waste products into the environment must be heavily influenced by consideration for the most efficient requirements of production.

1-56670-199-6/96/$0.00+$.50

THE CHALLENGE

There is little data to support a concern for environmental pollution from grazing animals. In fact, the available information supports the case that grazing animals do not alter the nongrazing background values for pollutants. The concentration is obviously not adequate to overwhelm the natural system.

The obvious challenge of animal agriculture to the environment then becomes disposal of waste products from highly concentrated feedlot areas to areas of plant production where controlled application can enhance production of plant products, grains, and forages, to be used in the production of beef. The primary nutrients for plant production to be derived from manure are nitrogen (N), phosphorus (P), and potassium (K). Also, trace elements are present in manure for utilization in plant growth.

There is a defined need for these nutrients provided by animal waste to go back into the production of feedstuffs. We will need some interdisciplinary cooperation to address the problems and concerns for this very obvious and efficient recycling of nutrients.

The questions of proper ratio, availability, consequence of excess, and economics of use all become important in identifying the solutions posed by the challenge of intense concentration.

Dr. John Sweeten from Texas A&M University in College Station published a paper on "Heavy Metals in Cattle Feedlot Manure" (Sweeten, 1993). Using the U.S. Environmental Protection Agency lifetime loading rates and average heavy metal content of manure, the addition of 10 to 12 tons of manure per acre per year on an as-received basis at 30 to 40% moisture would not exceed the limits for cadmium in 169 years. Zinc would require 388 years, copper would require 660 years, and it would require more than 40,000 years to reach the limit for lead.

In 1975, there was an International Seminar on Animal Wastes held in Bratislava, Czechoslovakia (Day, 1975). At that time, much of the discussion was based on handling waste for recycling as animal feed. This approach has not been found to be economically or biologically palatable. Hansen et al. (1969), feeding two levels of dry feedlot solid waste, demonstrated increased intake, reduced gains, and poorer feed efficiency when compared to comparable levels of alfalfa hay in the diet or the all concentrate diet.

Past

In the last 20 years, we have researched and discarded many potential solutions to potential waste problems. Now maybe we can turn our efforts toward the goal of identifying nutrient requirements based on biological value and establish definitive requirements. We have worked with very general mineral and protein requirements in beef nutrition to the point that there are wide variations in ration nutrient values depending primarily upon undocumented experience.

There is more than a 25% difference in protein levels routinely used throughout the beef industry. Finishing rations for beef cattle are formulated from 10 to

14% for protein on a dry matter basis. Amino acid metabolism is being addressed in ruminants by programs such as the Cornell protein and carbohydrate system (O'Connor et al., 1993). This is a good start, but the database to support this system needs to grow and expand in order to validate the system for definitive use in practice.

When we look at actual mineral use, it is not uncommon to see 30 to 100% variation in macromineral recommendations and, where trace minerals are concerned, as much as fourfold ranges are common.

There is an opportunity here for the development of a "WIN:WIN" situation. It is possible by establishing bioavailability values of the various minerals and, based on bioavailability, to establish more definitive requirements. Then we could see that what is good for the animal is also good for the environment.

Present

We have several researchers already engaged in an effort to identify the interrelationships between trace mineral bioavailability and health and production systems.

Nockels et al. (1993) studied stress induction effects of copper and zinc balance in calves fed organic and inorganic copper and zinc sources. Research at North Carolina State University has addressed the issue of bioavailability of various sources of trace minerals on weaning weights and feedlot production (Spears and Kegley, 1991). Kerley (1992) measured intake, absorption, and retention of different sources of trace minerals. Hutcheson (1992) in Texas developed a model to study trace mineral source and its effect on the immune system response.

The surface may not have even been scratched when you consider recent data showing definite breed differences for nutrient requirements (Ward et al., 1995). These may relate to physiological maturity, which provides an area of research that, although addressed, has not been studied at the levels of cellular efficiency.

Specifically, we need repeatable data on accurately defined mineral products that express availability to the cellular level of production. We need accurate measurement methods to assess the nutritional status of the animal. We need data to reflect the effects of supplementation at the different stages of physiological maturity. This is particularly true when addressing the needs of the light-weight calf destined to 200 to 300 d in the feedyard.

Future

In the beef production industry, we are faced with the economics of competing with other meat animals and poultry. As we address nutrient management of the animal, we will also address nutrient impact on the environment. But we need research. Today there is much too much arbitrary determination of potentially excess levels, disregarding source and leaving our industry open to criticism from an environmental perspective. The end result may be legislated mineral levels.

There is a real need for interdisciplinary cooperation between animal scientists and chemists to identify, characterize, and provide an explanation of mode of action of mineral sources at the cellular level of absorption and utilization. Research over the last 10 years has begun to explain and demonstrate the potential benefits of chelated and complexed trace mineral supplementation.

Nutrition Service Associates have been involved in establishing methods to characterize chelated and complexed trace minerals (Holwerda et al., 1995). The object of this research is to differentiate these products and possibly help explain why variable research results have been obtained.

From this research, we need to develop basic data without regard for economics. This research should further our basic knowledge of mineral sources. Practicality and economics should not drive the research engine. Industry will decide those issues.

We are fortunate in the United States to have vast land areas of productive agriculture to help us manage our environmental pollutants. But as populations grow almost exponentially, we will certainly see increased pressures to monitor and justify the environmental effects of all waste materials placed back into the environment.

How do we reduce phosphates in the solid waste? How do we reduce the nitrates excreted by the animal? Does sodium excretion need to be a consideration in the disposal of waste and thus impact site location?

We have seen studies over the past 25 years that demonstrate that feedlot cattle will perform with little or no added sodium to the ration (Klett et al., 1972). However, most of these studies have focused on feeding periods of less than 165 d. That is a good average feeding period, but we have many cattle in this country fed for 200 plus days and up to 1 year.

All animals have an amazing ability to conserve nutrients and increase the efficiency of utilization as the supplemental nutrients are reduced in the diet. There are even sparing interactions that exist. All too often our research focuses on a single variable without consideration of those interactions. The “most limiting nutrient” concept needs to be alive and well in the thought process of discovery to insure that nutrient limitations do not result in long-term deficiencies and poor economic performance.

We have advanced grain processing to a science that can extract starch utilization to whatever level we can support with balanced nutrition and appropriate management. Percentages of nutrients in diets are no longer adequate definitions of requirements. In beef cattle nutrition, we need to express these mineral, protein, and vitamin needs relative to daily requirements based on energy intake and the composition of body growth.

We expect to see some of these concepts presented in the revised Recommended Diet Allowances for Beef Cattle, National Academy of Science, Washington, D.C. However, there is still a large gap in the knowledge base to support these concepts relative to implant strategies, physiological maturity of various breeds and crosses, and ruminal effects of feed additives on the requirements and efficiency of utilization.

SUMMARY

The beef industry wants to be sensitive and responsive to environmental questions. We want to be ahead of the knowledge curve so we can deal with actual problems and not perceptions. The universal concern for the environment continues to grow as we are told that population numbers will double early in the 21st century. Animal agriculture anticipates increased production spurred by the increase in consumers and demand. With increased demand will come increased production and increased pressures on the environment from the by-products and waste products of digestion.

Total cattle inventory has been projected to reach 103.3 million head in 1995 (USDA, 1995). Twenty-five million of those cattle will be fed to slaughter in 1995, consuming more than 41 million tons (37.2 million metric tons [mt]) of dry matter feedstuffs. Those feedstuffs represent 1.3 billion bushels (45.8 million m^3) of grain; over 6 million tons (5.45 million mt) of roughage dry matter; and 4 million tons (3.64 million mt) of fats, molasses products, minerals, and protein meals. Their level of consumption will produce 5 million tons (4.55 million mt) of dry matter waste.

We have to be excited about the potential for animal agriculture to meet the future demand for animal protein. The beef industry is particularly exciting because of its ability to utilize a wide range of feedstuffs. But we need a concentrated research effort to define accurately from the vantage point of bioavailability nutrient requirements relative to energy intake and physiological maturity.

Research models are needed to detect small differences. Consider that a currently undetectable 1% advantage in efficiency could produce a return of $66 million annually to our industry.

We have a willingness and ability to manage our environment. What we need is quality, defined research in macro mineral, trace mineral, and protein utilization to improve efficiency that will in turn be favorable at the environmental level.

REFERENCES

Day, D. L., Utilization of Livestock Wastes as Feed and Other Dietary Products, Paper presented at the International Seminar on Animal Wastes, 28 September to 5 October, 1975, Bratislava, Czechoslovakia, 1975.

Hansen, K. R., Furr, R. D., and Sherrod, L. B., Paper and Feedlot Solid-Waste in Finishing Rations, Texas Tech University, Lubbock, TX, ICASALS Sp. Rpt. No. 18, 1969.

Holwerda, R. A., Albin, R. C., and Madsen, F. C., A Chemical Perspective on Chelation Effectiveness in Zinc Proteinates, Department of Chemistry and Biochemistry, Texas Tech University, Lubbock, TX, personal communication, 1995.

Hutcheson, D. P., Zinc Bioavailability and Immune Response in Cattle, NFIA Nutrition Institute — Minerals, 1992.

Kerley, M. S., Practical Applications for Chelated Minerals in Cattle, NFIA Nutrition Institute — Minerals, 1992.

Klett, R. H., Hansen, K. R., and Sherrod, L. B., Sodium Levels in Beef Cattle Finishing Rations as Related to Performance and Concentration in Feedlot Solid-Waste, Texas Tech University, Lubbock, TX, ICASALS Rpt. No. 104, 1972.

Nockels, C. F., DeBonis, J., and Torrent, J., Stress induction affects copper and zinc balance in calves fed organic and inorganic copper and zinc sources, *J. Anim. Sci.,* 71, 2539–2545, 1993.

O'Connor, J. D., Sniffen, C. J., Fox, D. G., and Chalupa, W., A net carbohydrate and protein system for evaluating cattle diets: IV. Predicting amino acid adequacy, *J. Anim. Sci.,* 71, 1298–1311, 1993.

Spears, J. W. and Kegley, E. B., Effect of zinc and manganese metmonial on performance of beef cows and calves, *J. Anim. Sci.,* 69 (Suppl. 1), 59, 1991 (Abstract).

Sweeten, J. M., Heavy Metals in Cattle Feedlot Manure, Extension publication, Texas Agricultural Extension Service, Texas A & M University System, College Station, TX, 1993.

U.S. Department of Agriculture (USDA), Statistical Reporting Service, Washington, DC, 1995.

Ward, J. D., Spears, J. W, and Gengelbach, G. P., Differences in copper status and copper metabolism among Angus, Simmental, and Charolais cattle, *J. Anim. Sci.,* 73, 571–577, 1995.

CHAPTER 2

Environmental Challenges as Related to Animal Agriculture — Dairy

Paul T. Chandler

INTRODUCTION

The dairy industry today faces economic and environmental challenges. The goal must be to structure a dairy operation on a land base that is adequate to operate within the rules and regulations of environmental control, but at the same time this operation must produce adequate returns to satisfy the needs of ownership and management. Profits have not been excessive within this industry over the past 10 to 15 years, and today we find many operations struggling to meet economic goals. The additional challenge of meeting environmental goals will likely become the factor that results in liquidation for a large number of operations.

Considering the status of the economy today and the trends that have occurred over the past 10 years, greater emphasis is placed on increased production efficiency. There are really no factors, apparent now or expected in the immediate future, that will change these trends. Protected regulations are being removed with respect to milk marketing orders, and no longer can a dairyman expect premium prices with respect to distance from Eau Claire, WI, other than what is associated with trucking freight. This means that a dairy operation, in any area, must position itself to compete on a national scale, even with the mega dairy units of the Southwest or any other area where they may emerge.

ECONOMIC CHALLENGE

Much discussion and debate have revolved around selecting criteria for measuring the success of a dairy operation. Is it milk sold per cow per year, milk produced per labor unit, or some other factors or combinations thereof? Obviously, the final and most important measurement is profit as related to dollar investment. Some very sound answers in this area are provided by studying analysis produced by Rogers (1993).

Increased production alone does not assure profit, as illustrated in Figure 1. Rogers (1993) selected the 183 most profitable dairy farms (top 25%) from the 731 farm samples and grouped them into five fairly distinct operating styles, which are summarized in Table 1. Dairy farm operations achieved their ranking

1-56670-199-6/96/$0.00+$.50

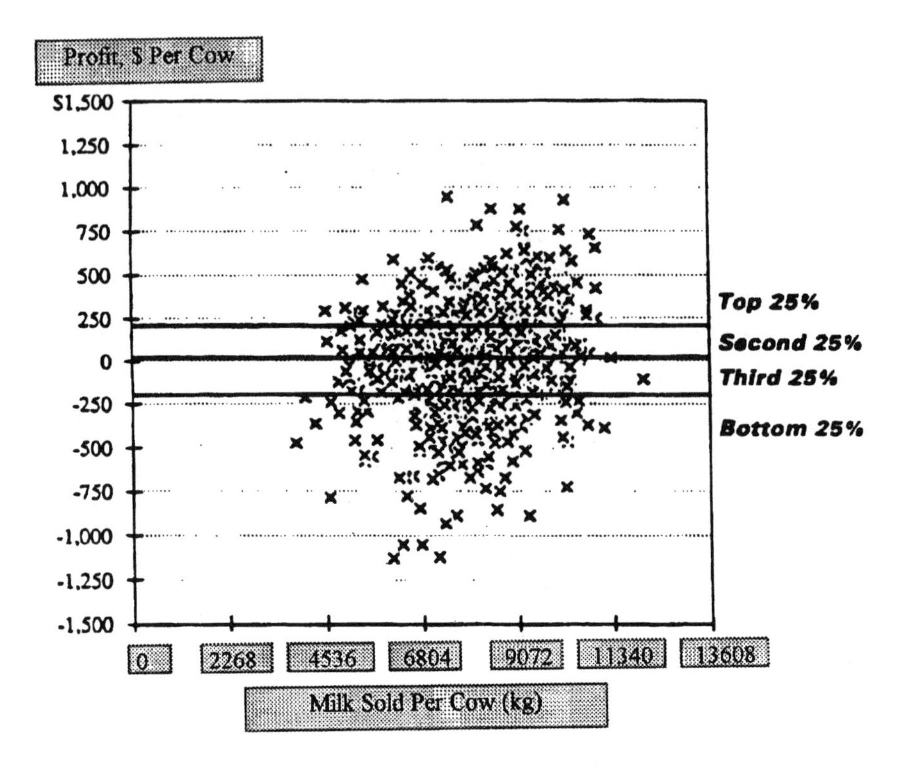

Figure 1 Profit vs. milk sold per cow for 731 farms in the northeast dairy region. (From Rogers, D. R., Northeast Dairy Farm Summary — 1993, Farm Credit Bank, Springfield, MA, 1993. With permission.)

in the top 25% group by being (1) good with cows, (2) labor efficient, (3) superior in price, (4) very conservative spenders, and (5) utility infielders (strong in all traits mentioned previously). Several points are important with respect to this table. Clearly, there are different ways to achieve profits necessary for the top 25% ranking. Those dairymen who were good with their cows reached this goal through high annual production (9959 kg.). Other dairymen concentrated on labor efficiency, producing greater than 475,000 kg of milk per worker. One group was there because of superior milk price (9% higher than average), which could have been due to milk quality factors and/or a better milk market. The group that was most impressive was classified as "conservative spenders." This group achieved the greatest net earning, with per cow values exceeding $500. The "utility infielders" incorporated some or all of the specific traits mentioned. Even though this summary represents the top 25% of a 731-dairy farm summary for 1993, the return to equity was from 5.3 to 9.3%. The average return to equity for all the 731 farms in the analysis was only 0.9%.

Limited profit within the dairy industry has resulted in limited capital investment, causing many of the facilities that are in use today to be outdated by 15 to

Table 1 Top 25% of Dairy Farms Classified by Management Style[a]

	Good with cows	Labor efficient	Superior milk price	Very conservative spenders	Utility infielders
Average herd size	135	185	99	129	86
Net earnings[b]					
Per farm	$50,220	$60,060	$25,641	$65,145	$21,156
Per cow	372	330	259	505	246
Per kg	0.0373	0.0381	0.0331	0.0639	0.0311
Expense per cow					
Total adjusted operation	$2,896	$2,501	$2,672	$2,239	$2,340
Family living and taxes	238	184	298	185	373
Total	$3,134	$2,685	$2,970	$2,424	$2,713
Cost of production					
$$/kg	.2601	.2606	.2967	.2302	.2809
Kilograms					
Per cow	9959	8654	7830	7973	7944
Per worker	335,194	476,150	240,007	302,980	228,247
Per crop hectare	8781	8695	4992	8225	6496
Per $1,000 assets	1258	1267	454	1050	1024
Price of milk ($/kg)	0.2886	0.2912	0.3146	0.2875	0.2873
Return on equity	7.2%	7.5%	5.6%	9.3%	5.3%

[a] Northeast Dairy Farm Summary — 1993, the top 25% of 731 herds summarized (Rogers, 1993).

[b] This represents total earnings for the farms. Net earnings from milk sold was 76, 80, 54, 90, and 20%, respectively, for the five classifications.

20 years. As the industry has to tool up to face environmental challenges, a valid question to address is where will the industry get the funds.

ENVIRONMENTAL CHALLENGE

From an environmental point of view, there are two areas of concern. The first involves the potential pollution of surface water and groundwater from land application of excessive amounts of nitrogen, phosphorus, and other minerals that are produced by the dairy unit. Also associated with this aspect is the emission of methane into the air, which is of concern with respect to global warming or the "greenhouse effect." The second environmental concern is in the area of a nuisance. This does not involve federal control, but is controlled by numerous local and state regulations. In the second area, the problem is primarily odorous compounds that can be very difficult to quantify. Also included in the nuisance area is the perception that is presented based on a physical view of the dairy unit in the eyes of neighbors and other nonfarm people. This can involve the type of husbandry that is practiced on the farm, including the presence of muddy lots, calf units, and cow housing.

An ever-increasing number of people, who are several generations away from the farm, may not appreciate or understand the necessary components of a modern successful dairy farm. If they encounter significant odors from these operations, serious problems may result, leading to litigation and, in most cases, serious consequences for the dairy farm operation. The problem is that an increasing number of people expect cows to reside on the green pastures of rolling hills and do not understand the economic situation that dairy farms operate within. The trends of the industry have forced and will continue to force more intensive dairy operations, concentrating more animal numbers per operational unit. This increases the possibility of single point pollution of streams and, at the same time, makes the units more visible, causing the likelihood of a nuisance complaint.

DAIRY WASTE MANAGEMENT

Dairy waste management is by far the most significant factor to consider with respect to the environmental challenge. Waste would include feces, urine, parlor effluent, silo seepage, and other sources of water and materials that become mixed together. Historically, cows and dairy manure were strong assets for man. These components enhanced the standard of living. The heat generated by the cows helped keep the family warm during the cold times, and the manure produced provided fertilizer for the fields to grow crops as food for the family as well as the livestock. Such is not true today, as manure and other waste products have become a significant liability to the dairy farm business.

MANURE HANDLING

Presently, management systems can be classified as follows:

1. None
2. Scrape and haul
3. Modifications of scrape and haul
4. Water flush

None

There are numerous dairy units, especially in the moderate climates such as the southern section of the country, that basically operate without any manure and waste management system. Cows are maintained outside in small lots and pastures on a year-round basis. There may be several feeding areas, and cows are moved frequently to prevent accumulation of waste materials. Generally, the holding pen and parlor have the waste material scraped and hauled or washed onto the adjoining area. These units are small in size and do not create the awareness from the public as do some of the larger units. But on a pollution index expressed as units per animal or units per food product sold, these units may be contributors to environmental problems. Their production is frequently at or below the average for the area, and the manure produced is certainly not distributed evenly over land areas.

Scrape and Haul

The scrape and haul system of manure and waste management is by far the most common within the industry today. The system involves scraping manure from various areas around the dairy unit and spreading it on a daily basis or holding the semisolid mixture in piles or pits before being moved into spreaders for field application. This is a relatively low-cost system, but generally requires daily attention and can result in applying manure to fields at times when conditions are not ideal.

Modified Scrape and Haul

Several modifications of the scrape and haul system have been introduced to make the system more labor efficient and to provide a system that can be managed more effectively within the constraints of environmental regulations. One approach is to install some type of filtration or sedimentation system, allowing separation of some liquid from the solid material. The liquid material is diverted to a lagoon with the solids held for disposal on fields. This reduces the frequency of hauling, allowing for planned application consistent with field preparation for crops. Another approach is construction of lagoons or other waste holding facilities, above or below ground, with a storage capacity of 6 to 8 months. This requires a rather large capital investment in both the holding areas as well as equipment for stirring and mixing, pumping, and field application. The dairyman is still faced with the problem of assurance that field application does not result in nutrient loss in runoff or that groundwater contamination via leaching does not occur. This method of waste management is perhaps most likely to result in nuisance claims, especially if field application is close to urban areas.

Water Flush

The movement of semisolid manure and other waste with water is a very effective way of cleaning alleys, holding pens, and feeding areas. This requires minimum labor, but produces excessive water. The primary problem of this system is the management of the excess water produced. It is recognized that solids separation with water recycling is a requirement. Solids separate very effectively through mechanical screening or through the use of sedimentation basins. The effluent can proceed to a primary lagoon and then be diverted to a secondary lagoon. From the secondary lagoon, the water will achieve sufficient purity to be reused for flushing.

There are problems with the water flush approach that must be considered. All bedding materials do not work the same in these systems. Plain sand is perhaps the most ideal bedding for cows in free stall facilities, but it can be very detrimental to mechanical pumping systems that are necessary in these systems. Herd health problems, especially mastitis, can result from recycled water, especially if herd management does not coordinate the flushing of alleys at times when cows are removed. Disposal of the separated solid material can present problems if sufficient storage areas are not provided. Use of this material for rebedding the free stalls has been suggested, but practical experience does not support its use.

With the water flush approach, adjacent land area is essential to apply surplus water via irrigation. Also, the separated solids must be applied on this land area or removed from the dairy farm. Forage crop production on this irrigated land area can be significant, benefiting from the nutrients supplied in the irrigation water and the water itself.

The water flush approach seems to be the system being used by a large number of new dairy facilities that are emerging around the country. But it is very difficult to introduce this concept in many of the existing dairy units simply because of location, facility design, or available water.

Nutrients In Dairy Waste

Nitrogen, Phosphorus, and Potassium

Significant efforts by the academic community have been directed toward quantitative measures of waste nutrients from dairy operations. Van Horn et al. (1994) provided an excellent review of the quantitation of nutrient waste. They note that total excretion of numerous waste nutrients can be obtained by subtracting the amounts produced in milk from the amounts consumed. There will be considerable variation from farm to farm based on dry matter intake and the level of feed nutrients contained within the dry matter.

Data from Van Horn et al. (1994) was used to develop Figures 2, 3, and 4. Observe that the predicted excretion of nitrogen (N) (Figure 2), phosphorus (P) (Figure 3), and potassium (K) (Figure 4) decreased per unit of milk as the level of milk production increased. Even though dairies are generally perceived by some groups as being major contributors to environmental pollution, their incorporation of feed nutrients into food nutrients (milk) is relatively high among the species involved in animal agriculture. High-producing dairy herds approach a 30% conversion of feed N (Figure 2) and exceed a 30% conversion of feed P (Figure 3) into milk. The conversion of feed K into milk K is lower, being 15 to 18% (Figure 4).

Also, increased levels of production or more intensely managed herds do not increase pollution, but in fact result in a considerable decrease in waste nutrient excretion per unit of milk.

Nitrogen excretion demands special consideration because of the loss from dairy units by volatilization prior to and during the application to crop growing areas. These losses by volatilization can reach 50 to 75% (Van Horn et al., 1994). As noted by Van Horn et al. (1994), the gaseous ammonia losses have been considered to be an economic issue, relating to the fertilizer value of the waste rather than an environmental concern. Over 100 years of data relating to rainwater showed the composition, relative to nitrogen compounds, to remain relatively constant even with increased uses of fertilizers and increasing livestock numbers.

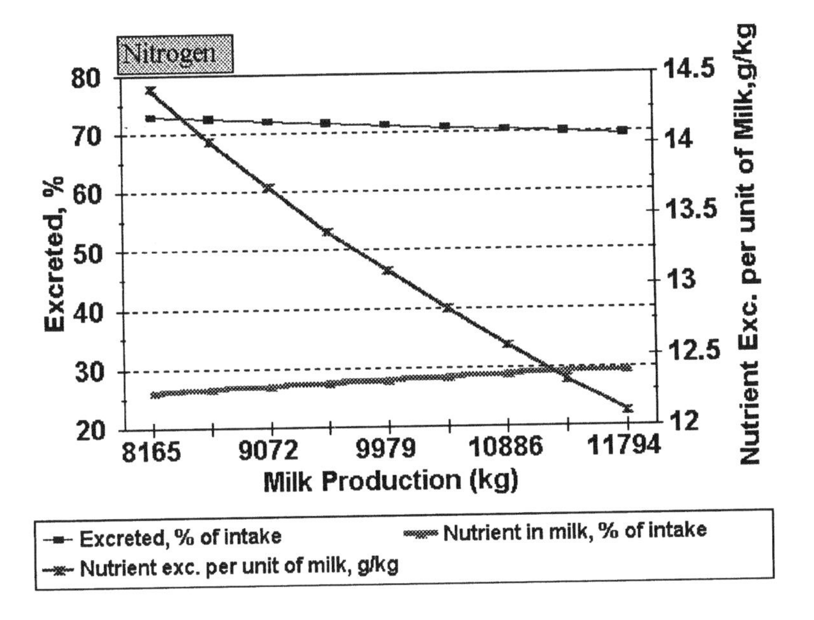

Figure 2 Effect of milk production on excretions of nitrogen. (From Van Horn, H. H., et al., *J. Dairy Sci.,* 77, 2008, 1994.)

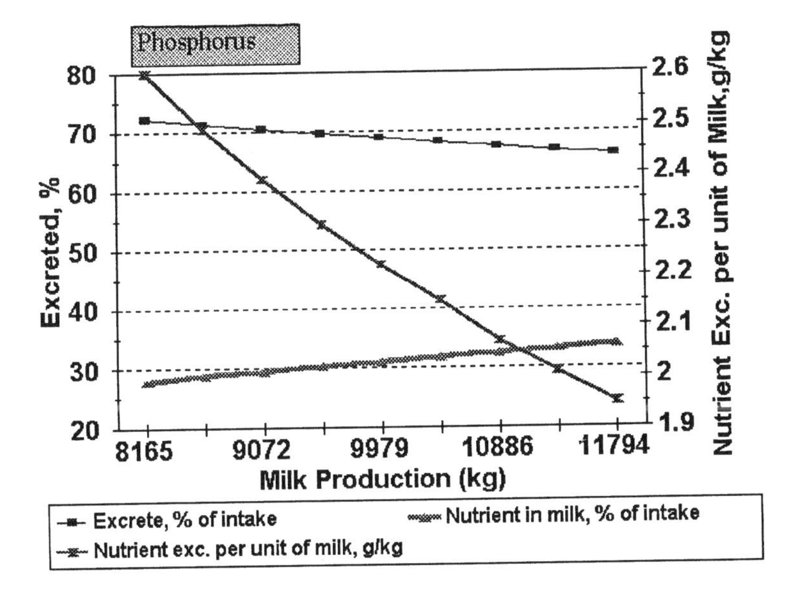

Figure 3 Effect of milk production on excretions of phosphorus. (From Van Horn, H. H., et al., *J. Dairy Sci.,* 77, 2008, 1994.)

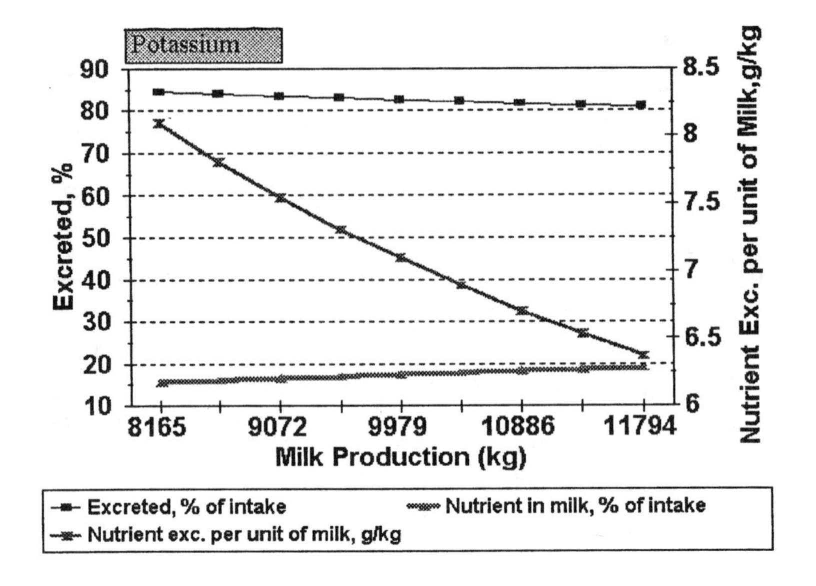

Figure 4 Effect of milk production on excretions of potassium. (From Van Horn, H. H., et al., *J. Dairy Sci.*, 77, 2008, 1994.)

It is important to note, however, that in European countries concern is directed toward N loss via volatilization and efforts are directed toward systems for reduction of N loss to the atmosphere (Van Horn et al., 1994).

Carbon and Energy

It is necessary to present the carbon and energy balance of a dairy cow to understand the environmental concern relating to methane production. Around 35% of the carbon and energy is contained in animal waste and provides the bulk of the solid's component. The major route for absorbed carbon and energy loss from the dairy cow is through the process of heat production, accounting for 40 to 44% of that consumed (Figure 5). This is dissipated to the atmosphere from the animal as carbon dioxide and should not be of environmental concern. At least 50% of the cow's diet can be corn silage or other forages, which by the photosynthetic cycle will effectively recycle the carbon back for animal feed. Milk captures some 18 to 20% of consumed energy and carbon, and the amount captured is directly related to production level.

Methane, which also is of environmental concern with respect to global warming, is produced at a rate of 5% with respect to carbon and energy consumed (Figure 5). A more realistic expression of dairy production in this aspect is to relate the values in terms of food products. In this respect, consumed carbon and energy appear as methane at a rate of 17.6 g and 171.8 kcal/kg of milk, respectively. As was demonstrated for the N, P, and K nutrients, these numbers are

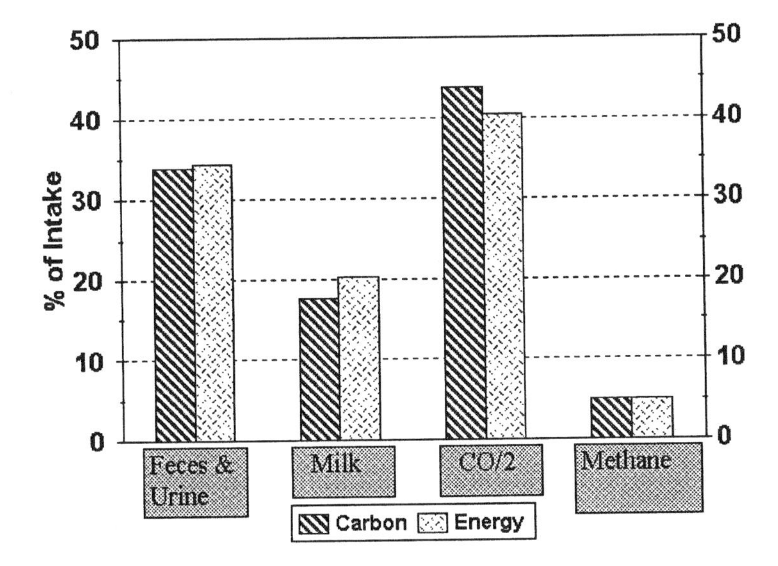

Figure 5 Excretion of carbon and energy for a dairy cow producing 22.7 kg of milk. (From Van Horn, H. H., et al., *J. Dairy Sci.*, 77, 2008, 1994.)

negatively correlated with production level, resulting in increased production and generating less pollution per unit of food nutrient produced.

FEEDING APPROACHES AND WASTE ALLOCATION

The economic environment under which dairy farms have operated over the past several years has forced a very critical evaluation of all feed inputs. Feed nutrients are not allotted unless positive responses are expected with respect to production and/or herd health. Two of the most monitored nutrients from an environmental viewpoint, N (protein) and P, are viewed as very important cost factors (Table 2). Protein is the most expensive organic component of the dairy diet when evaluated on a cost per unit basis. In mineral and vitamin premixes for dairy cattle, P accounts for more than 50% of the cost. Obviously, good dairy farm managers who strive to achieve profits under current economic conditions will not intentionally waste these very costly nutrients by feeding them in excess.

Table 2 Major Nutrient Cost for Dairy Ration

Nutrient	Cost ($/kg)
Protein	0.2366
Energy (TDN)	0.1265
Calcium	0.1508
Phosphorus	1.9321

The use of a total mixed ration in smaller herds, where single group feeding for the entire lactation string is necessary, does result in some overfeeding of protein for cows that are in later lactation. But as overall herd production increases, the degree of excessive protein intake and excretion is reduced. Also, industry trends resulting in increasing herd size, which soon leads to the opportunity for feeding multiple groups, allows for the allocation of protein to be aligned with productive needs.

Over the past few years as herd production levels have increased, there has been an ever-increasing use of more complex nutritional models for nutrient allocation. For example, with respect to protein, N sources are provided to supply the microbial needs for rumen fermentation followed by additional protein supplied in a form that is undegradable in the rumen and complements microbial protein production. This system views the proteins of feedstuffs based on component parts and allows their allocation to accomplish maximum efficiency for conversion of feed N into milk N. Table 3 provides an example of the nutritional guidelines provided by a typical feeding model.

A more complex form of nutrient allocation is achieved through the use of computerized models such as the one being developed by workers at Cornell University (O'Connor et al., 1993). At its current stage of development, this model evaluates the amino acid balance at the metabolic level, allowing the user to substitute different feed sources to achieve a more efficient utilization of dietary protein or to recommend the use of synthetic amino acids that can be provided in rumen protected form. Application of these new formulation technologies has allowed dairy farmers and their consultants to achieve greater overall efficiency of nutrient utilization for production of milk, but there are perhaps some serious problems concerning the efficient plant utilization of the nutrients from the waste by plants.

In many dairy operations, adequate attention has not been directed toward fertility levels of fields receiving large supplies of dairy waste nutrients. Commercial fertilizer applications continue at rates established for the farm, resulting in nutrient imbalance or excesses. This establishes conditions that are ideal for environmental pollution via runoff or leaching. Because of the waste management system in place, many times the crop areas involved are located close to the dairy unit.

Dairy farmers must apply the same technologies to their cropping areas as they do to the feeding of their herds. With adequate land area, it is possible to achieve a desirable nutrient balance for the entire operation. For those dairy operations residing on a small land base and purchasing most of their feed from off-farm sources, serious problems are likely to occur. Their alternative is to make arrangements with neighbors involved in crop production to utilize their excess waste or plan to relocate where the dairy unit can be sized to operate with sufficient land area. This factor, plus differential in land prices, is one of the reasons for movement of dairy operations from relatively heavy urban populated areas of the country to areas where urban pressure is less.

Table 3 Nutrient Model for Lactating Cows

Nutritional profile:	
NEl = 1.61 Mcal/kg	NDF intake
Fat = 5.00 percent	% BW = 1.1
Weight:	
Loss = 0.00 kg/day	BW(kg) = 599
Gain = 0.00 kg/day	Growth = No

	Milk production (kg/day), Fat % = 3.5				
Nutrient parameter:	**41**	**36**	**32**	**27**	**23**
Dry matter intake, kg	22.5	21.1	19.6	8.0	16.4
DMI, % of BW	3.8	3.5	3.3	3.0	2.7
NDF intake, kg	7.1	6.8	6.6	6.3	6.0
DIP, % of DM	11.3	11.0	10.6	10.3	9.9
SOL CP, % of DM	5.4	5.2	5.0	4.9	4.7
UIP, % of DM	7.0	6.7	6.3	5.8	5.3
Crude protein, % DM	18.3	17.6	16.9	16.1	15.2
	Protein fractions				
Total crude protein	100	100	100	100	100
DIP, % of CP	62	62	63	64	65
Sol CP, % of CP	29	30	30	30	31
UIP, % of CP	38	38	37	36	35
Fat, % of DM	5.0	5.0	5.0	5.0	5.0
NEl, Mcal/kg	1.68	1.64	1.61	1.58	1.55
NEl:CP(kcal/kg)	9.15	9.32	9.53	9.79	10.14
NEl at 3% fat	1.54	1.51	1.48	1.44	1.41 +
Gms DIP/Mcal NE1	73.3	72.7	72.0	71.2	70.2
NDF, % of DM	31.3	32.4	33.6	35.1	36.8
Forage NDF, % of DM	23.5	24.3	25.2	26.3	27.6
ADF, % of DM	21.2	21.9	22.8	23.7	24.9
Mineral (ash),% of DM	8.3	8.3	8.3	8.3	8.3
NSC, % of DM	37.1	36.7	36.2	35.5	34.7
DIP/NSC	0.30	0.30	0.29	0.29	0.29

Note: BW = body weight; DMI = dry matter intake; NDF = neutral detergent fiber; DIP = degradable intake protein; SOL CP = soluble crude protein; UIP = undegradable intake protein; CP = crude protein; NE1 = net energy lactation; NSC = nonstructural carbohydrate; ADF = acid detergent fiber.

ALTERNATIVE PRODUCTION SYSTEMS

It is appropriate to consider the use of production systems that differ from those in common use today. One approach that differs considerably from the conventional is a return to intensive pasture utilization. Such an approach is certainly not new for this country, as this was the common method of milk production 40 years ago.

Recently, there has been extensive discussion and research, along with some industry adoption, of a more substantial dairy farming approach, featuring maximum utilization of pasture forage, lower rates of concentrate feeding, and lower total equipment investment. In almost every state you can find examples of success stories from these types of operations. But when true land values are evaluated, along with the necessity of providing a continuous 12-month supply of milk, it does not seem that this dairying concept will become significant in meeting the nation's milk supply.

Mass movements for the industry in that direction would result in an increased cow population and more operating dairy units. Even though the presence of this type of dairy farming activity may not alarm the public from the urban sector, it must be recognized that overall national pollution index from dairy farming activity would increase. Small movements in this direction will likely continue and may in fact become dominant in certain areas of the country. It certainly offers a good avenue for young people with limited capital to enter the industry. It also allows older producers, who are faced with a need of significant capital investment to meet environmental regulations, the opportunity to scale back and achieve several additional years of income before retirement.

SUMMARY AND CONCLUSIONS

The dairy industry has been operating over the past several years under an economic challenge that has resulted in a significant improvement in overall productive efficiency. These conditions have forced a rather critical evaluation of all input costs, including feed nutrients. As a result of these conditions, many operations within the industry have lagged behind with respect to facility updates and now face major decisions with respect to new investments to face environmental challenges.

Sufficient work has been conducted to allow adequate estimations of nutrient waste excretion relative to herd production and diet composition. The challenge now must be directed toward deriving cost-effective procedures for collecting, holding, and recycling these waste nutrients to land areas used for cropping. Coupled with this process, the nutrient balance for crop production must follow the same effort that the dairy industry is applying to nutrient allocation to the dairy cows.

It would seem that some merit would be associated with the thinking directed toward pollution indexes expressed in terms of product or food produced. Today we find greater attention being directed toward some of the larger, more production-efficient units with almost no concern for the average and smaller dairy units. When pollution indexes are viewed relative to food produced, the current direction of movement within the dairy industry today seems to offer less opportunity for having negative environmental action. The nation's supply of milk and dairy products is being produced with a decreasing numbers of cows, and certainly the number of active dairy units is dropping very rapidly.

This movement brings with it the construction of new feeding and housing facilities for dairy units where problems of waste management can be properly addressed. This is certainly a more cost-effective way to approach the problem of waste disposal, rather than considering the renovation of many existing dairy units. After renovation, to satisfy environmental standards for waste management, there is little opportunity for increasing herd numbers, resulting in a significant increase in overhead cost per producing cow unit.

REFERENCES

O'Connor, J. D., Sniffen, C. J., Fox, D. G., and Chalupa, W., A net carbohydrate and protein system for evaluating cattle diets: IV. Predicting amino acid adequacy, *J. Anim. Sci.,* 71, 1298, 1993.

Rogers, D. R., Northeast Dairy Farm Summary — 1993, Farm Credit Bank, Springfield, MA, 1993.

Van Horn, H. H., Wilkie, A. C., Powers, W. J., and Nordstedt, R. A., Components of dairy management systems, *J. Dairy Sci.,* 77, 2008, 1994.

CHAPTER 3

Environmental Challenges as Related to Animal Agriculture — Poultry

Keith E. Rinehart

INTRODUCTION

Protection and enhancement of the environment is a well-recognized priority among populations of the developed world. Production agriculture is an important factor relative to the environment. Water pollution is the most potentially damaging and widespread concern in regard to production agriculture, including the poultry industry. Although poultry producers perceive themselves as friendly to the environment, there are recognized improvements in current practices that can be made and a further need for new processes to be developed that will enhance and/or slow degradation of the environment.

Trends in world population growth are well established and documented. The requirement for world food production to keep pace with population growth will dictate continuous improvement in efficiencies on a shrinking amount of land available for agricultural production. The debate will continue regarding the place of animal proteins in human nutrition as the poultry industry competes for resources and land area for production.

Data in 1994 (USDA-FAS) indicate that per capita meat and poultry consumption continues to increase, especially in developing nations (Table 1). This, coupled with world population growth, causes one to predict that growth of the poultry industry over the next few decades will be consistent with recent past patterns (Table 2). World broiler production may be expected to expand approximately 5% per year, turkey production between 3 to 4%, and egg production less than 3% (Tables 3, 4, and 5).

In looking back over the past several decades, we can see tremendous improvements in the efficiency of nutrient utilization for all classes of livestock and poultry. These improvements have come from all areas to include genetics, physiology, nutrition, and disease control and management. While the individual bird has become more efficient in the conversion of nutrients to meat or eggs, the large increase in animal units has led to an overall increase in environmental burden.

Challenges for researchers in agriculture, food, and environmental sciences are broad, and there remains considerable opportunity to improve nutrient utilization. These range from improved feedstuffs, to feeding and management of the

1-56670-199-6/96/$0.00+$.50

Table 1 Poultry, Meat, and Egg Consumption Per Capita (Selected Countries)

	1990			1992			1994		
	Total	Broken	Egg	Total	Broken	Egg	Total	Broken	Egg
	Poultry (kg), eggs (each)								
U.S.	36.5	27.7	186	30.3	39.2	181	41.5	32.5	177
Brazil	13.8	13.5	88	15.8	16.1	90	18.8	18.4	82
U.K.	24.0	14.9	199	17.4	27.4	196	27.9	18.3	194
Israel	37.0	25.6	360	28.2	41.1	390	41.2	28.3	387
China	2.8	1.5	136	1.7	3.8	163	4.6	2.1	175
Hong Kong	33.5	27.0	241	35.7	44.7	237	46.7	38.6	253
Japan	13.7	12.6	235	14.1	14.1	276	13.6	13.6	272
South Africa	14.4	12.5	—	15.5	18.3	—	17.1	14.5	—

Table 2 Assumptions Based on Trends

World population will double in 25–30 years.
U.S. population will double in 50–60 years.
U.S. broiler production will double in 10–11 years.
World broiler production will double in 13–14 years.
U.S. and world turkey production will double in 19–20 years.
World egg production will double in 25–30 years.
Poultry meat will continue to gain favor for human nutrition.

Table 3 Trends — Broiler Production[a]

	1990	1992	1994	% of world	% Charge 5-year average
North America	12,046	13,582	14,832	35.3	6.4
South & Central America	3,033	3,928	4,471	10.6	12.4
EEC	6,757	7,369	7,654	18.2	3.4
Former Soviet Union	3,287	2,807	2,518	6.0	(4.7)
Asia	6,327	7,890	9,078	21.6	11.0
Total	34,994	38,977	42,008	100.0	5.3

[a] 1000 metric tons.

Table 4 Trends — Turkey Production[a]

	1990	1992	1994	% of world	% Charge 5-year average
North America	2048	2167	2233	56	3.8
World	3596	3847	3979	100	3.8

[a] 1000 metric tons.

Table 5 Trends[a] — Egg Production[b]

	1990	1992	1994	% of world	% Charge 5-year average
North America	67,987	70,592	71,880	12.4	1.4
World	527,833	566,832	581,203	100.0	2.7

[a] Million.

[b] USDA-FAS, 1994.

bird, to waste handling. All these potential improvements have an associated cost as well as political implications.

Much of the following information and opportunities for improvement may be similar to that discussed by other authors in this book. When one considers the environmental challenges as related to poultry production and further narrows this to nutrient management relative to the environment, the issues are similar across species and geography. The United States will follow some of the developments that those in Europe have already faced relative to environmental pressures. de Lange (1993) reported on the situation in the Netherlands.

POULTRY-RELATED ISSUES

The major environmental concerns, as they relate to groundwater protection, are nitrogen, phosphorus, and trace minerals (e.g., copper, selenium, arsenic). Other environmental issues relative to poultry production are dust, ammonia, odors, and pathogens; however, these are not included in this chapter. Possible groupings that come to mind in being able to reduce environmental loading of nutrients from poultry production are refining nutrient requirements needed to optimize meat and egg production, improving the efficiency of utilization for those nutrients consumed, and improving methods for the disposal of waste products produced. There are factors that impact each of these groupings, as well as interactions between the various factors within and across groupings.

For the purpose of this discussion, we will limit our discussion principally to nitrogen, phosphorus, and trace minerals.

Cromwell (1994) noted that livestock and poultry excrete approximately 158 million tons of dry matter manure in the United States per year. Poultry accounts for over 800,000 tons (726 metric tons) of nitrogen and 250,000 tons (227 metric tons) of phosphorus from this total. Lyons (1994) stated that it may be conceivable that, at some future time, bioavailability will determine what mineral sources are allowed for use in environmentally sensitive areas.

Zublena et al. (1993) reported concentrations of nutrients in solid manure or litter sources (Table 6). Likewise, Sims and Wolfe (1994) indicated that litter application could build soil levels of phosphorus, zinc, and copper to levels that no longer required supplementation for crop production.

REDUCING NUTRIENT REQUIREMENTS FOR OPTIONAL PRODUCTION

There remains an opportunity to refine the bird's nutrient requirement that will promote optional economic returns while minimizing the effect on the environment. In times past, decisions on nutrient levels fed to birds were made on a combination of production response with an economic consideration with very little attention given to those nutrients that passed through the bird.

Table 6 Concentrations of Nutrients in Solid Manure and Litter Sources[a]

	Nitrogen	Phosphorus	Zinc	Copper
Manure				
Layer	1.9	2.8	0.02	0.002
Litter				
Broiler	3.6	3.9	0.03	0.020
Breeder	1.6	2.7	0.03	0.010
Turkey	2.9	3.6	0.03	0.030

[a] Zublena et al. (1993) as is basis.

Except for the most expensive nutrients in a formulation, it has been common practice to include a margin of safety or enough excess over the actual requirement to guard against a deficiency when targets were not met or other issues interfaced with utilization of nutrients. Trace elements were good examples of this, as well as phosphorus and amino acids to a lesser extent.

In many instances, nutrient levels for birds of older ages are extrapolations from research done in the very early life of chickens or turkeys. There is a need for refinement of requirements for amino acids, phosphorus, and trace elements in birds of all ages and especially for those beyond the starting period.

Most of today's commercial broiler genetic lines have the capacity for very rapid growth that is fueled by a large appetite. Perdue research data indicate the following relative changes in growth and performance between 1960 and 1995.

	1960	1995
Broiler age (days)	72	48
Average weight (kg)	1.80	2.20
Feed conversion (feed/meat)	2.27	1.95

Typical protein for the average broiler feed consumed has been reduced from 23 to 21.5% over this period, while available phosphorus content has been reduced by more on a percentage basis, 0.50 to 0.45%.

An example of a feeding program for today's broiler may contain three or four different feed products. These employ a change in nutrient level as the bird increases in age, thus attempting to match requirement to age. Programs for turkeys and egg-laying breeds may contain more products but for a similar reason.

This phase feeding could be greatly refined if the nutrient levels for each age, sex, and level of production were more precisely known. Other factors that limit this refinement are logistics of feed manufacture plus delivery and approved drug programs. One tends to feed above or below the absolute nutrient requirement curve due to timing of feed product changes (Figure 1). This causes impaired efficiency of performance and/or nutrients lost to the environment.

Baker (1993) has reported on the ideal protein concept for poultry as well as swine. He indicated that use of the ideal protein concept in feed formulation would minimize nitrogen excretion in waste products. The availability and use of synthetic amino acids has allowed feeds to be lower in total proteins, thus reducing

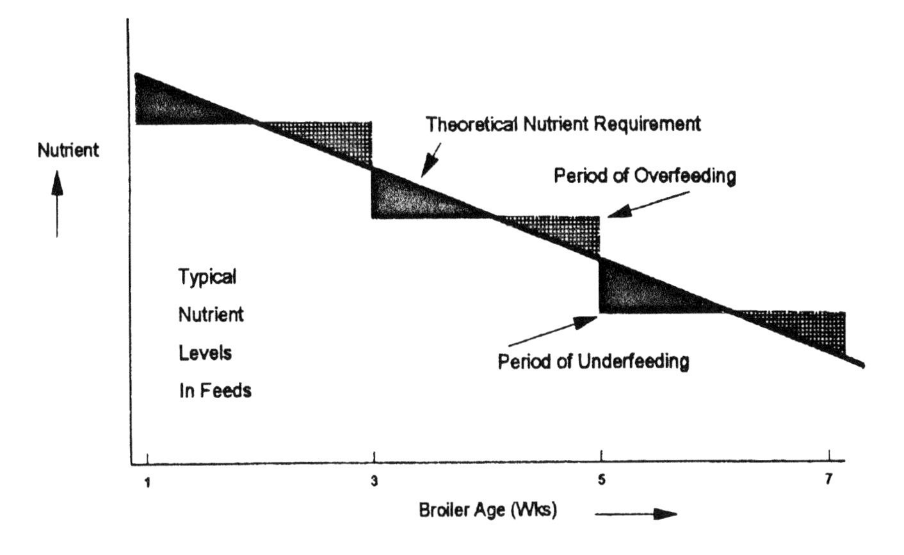

Figure 1 Nutrient intake (feed pattern vs. requirement).

excesses of nitrogen, which in turn reduces the environmental impact while still maintaining economical bird performance. This trend will continue as more economically useful amino acids are made available for animal feeding.

Apparent requirements are impacted by bird health or immune status, which also interact with factors leading to excretion of nutrients.

IMPROVED EFFICIENCY OF NUTRIENT UTILIZATION

In the past, there has been limited coordination between plant geneticists and animal or poultry nutritionists. Their concern was primarily one of maximizing salable yield per unit of land area with no recognition of biologically available nutrients for the bird. Bird-resistant milo was a perfect example of this. A grain was produced that had increased harvested yield, but it had drastically reduced biologically available nutrient content, which caused it to be uneconomical for poultry feeding.

A more recent trend toward enhancement of economically important nutrients in feed grains has been accelerated by some of the newer molecular genetic techniques. Research that alleviates critical nutrient levels, improves nutrient balances, reduces antinutrient factors, and improves resistance to fungus are examples. These factors can either individually or in combination improve nutrient utilization, which in turn reduces waste products and nutrient loading of soils. An example is the use of microbial phytase to increase the release of phosphorus from grain or vegetable meals that could dramatically reduce the need for inorganic phosphorus required for the monogastric animal and bird. This could have a major impact on lowering the phosphorus loading on soil from manure application. A broiler ration may contain 0.75% total phosphorus with only 0.45 being available for utilization. This means a tremendous amount of phosphorus has to be voided in the manure, not to mention the endogenous contribution.

Poultry geneticists continue to make progress in improving efficiency of feed utilization. This will average 0.5 to 1.0% per year. Selection for rapid growth is correlated with a large capacity to consume feed or appetite. More rapid growth to desired market weight reduces nutrients required for maintenance. Some recent, high-yielding lines have in fact demonstrated smaller appetites, better feed efficiency, greater lean muscle accretion, and lower fat. These factors improve efficiency and reduce waste to the environment. There may be an opportunity to lower critical nutrients passed to the environment through genetic selection.

Improvements in disease control have continually reduced mortality and morbidity. These relate both indirectly through nutrient utilization efficiency and directly through lower dead bird disposal, thus impacting nutrients returned to the environment. There remains a tremendous opportunity for work in nutrient–immune interactions to improve nutrient utilization efficiency. More effective use of antibiotics, probiotics, and related materials relate directly to animal or bird well being and efficiency that relate to reduction of waste products. Our industry could enjoy a different attitude from regulators in terms of more flexibility in usage of such materials.

Improvements in the management of poultry continue to enhance efficiency of nutrient utilization and play a part in lowering those nutrients that must be returned to the environment. There have been innovations in environmental controls such as temperature, air movement, and light, plus continuous discoveries and refinements over the past several decades that help to improve biological utilization of nutrients from feedstuffs. One way this works is to lower maintenance requirement. All these have allowed fewer excesses in formulation, which relates to lower amounts of nitrogen, phosphorus, and trace elements returned to the soil with their implication on groundwater quality.

Other examples are improved sources of feed phosphates, more biologically usable forms of trace elements, and work on biologically available amino acids by feed ingredient source. Work over the years on specific processing techniques has improved nutrient availability.

Several companies are working with specific anabolic agents that improve nutrient utilization by altering carcass composition. Recently, there has been an increase in work with direct fed enzymes, which greatly alter the ability of the bird to utilize nutrients from some ingredient sources.

Efficiency of nutrient utilization can be enhanced by certain types of feed intake restriction apart from phase feeding. This may relate to physical feed removal for certain periods of time. Restriction has also been done by various lighting regimes as well as ration density.

We continue to improve ingredient and/or feed processing with the objective of improving nutrient availability to the animal or bird. Different types of heating, pelleting, and more recently expansion have been effective. We should not forget that the animal or bird has a positive role to play in using materials as feedstuffs that might in themselves become an environmental burden. Many byproducts or coproducts of food processing for human use get used in feeds.

Development of more economically viable amino acids will allow further refinement of nutrient balance. This, coupled with a continued better understanding of requirements for all ages, sexes, and breeds, and improved mineral sources will continue to lessen the environmental impact of nutrient waste.

DISPOSAL OF WASTE PRODUCTS

Disposal of dead birds and manure has evolved to more environmentally friendly methods in the recent past. Older methods of pit disposal for dead birds have given way to composting. Considerable work has been done on lactic acid fermentation with a goal of returning the nutrients to the feed chain. Composting allows the final product to be used as regular manure or litter for crop production. Rendering offers yet another opportunity. Considerable research is ongoing on the effect of land application of manure on groundwater nutrient levels. There are continuous improvements in soil testing and precision in application for litter and manure. It has been reported that properly established buffer zones with grasses between fields can reduce the leaching of nutrients from a cultivated plot. Some have used settling pits and subsequent nutrient reclamation from manure. Other specialty fertilizer compounds are being made from certain waste materials such as manure, wastewater sludge, and hatchery waste.

SUMMARY

The poultry industry recognizes that it can and does have an effect on the protection of our environment. It considers issues of the environment to be high priority. Some of the areas where progress has been made, but for which there remains room for work, are listed here.

Trends in poultry production:

- Lower protein, better amino acid balance
- Lower phosphorus intake per unit of gain
- Better digestibility of feedstuffs
- Improved mineral balance
- Use of byproducts/coproducts
- Development of workable enzymes
- Improved environmental management
- Better disease control
- Nutrient management plans and research
- Spread of production units geographically
- Genetic improvements in poultry and crops

Areas of potential advancement:

- Crops with unique nutrient levels
- Crops that have unique nutrient requirements
- Systems to stabilize waste streams, feeding value
- Extractable, reusable litter sources
- Capturing waste nutrients for alternate uses
- Nutrition immune interactions
- Improved enzyme technology
- Refinement of optional nutrient requirements

REFERENCES

Baker, D. H., Digestible Amino Acids of Broilers Based Upon Ideal Protein Considerations, Arkansas Nutrition Conference Proceedings, 1993, 22.

Cromwell, G. L., Diet Formulation to Reduce the Nitrogen and Phosphorous in Pig Manure, Nutrient Management Symposium Proceedings, Chesapeake Bay Commission, Harrisburg, PA, December 1994.

de Lange, C. F. M., Diet Formulation to Minimize Contribution of Livestock to Environmental Pollution, Arkansas Nutrition Conference Proceedings, September 1993, 9.

Lyons, T. P., Improving bioavailability of minerals, *World Poultry — Misset,* Vol. 10, No. 4, 1994.

Sims, J. R. and Wolfe, D. C., Poultry waste management: Agricultural and environmental issues, *Adv. Agron.,* 52, 1, 1994.

USDA-FAS, Dairy, livestock, and poultry: world poultry situation, *Fl & P,* January 1994, 1.

Zublena, J. P., Barker, J. C., and Carter, T. A., Poultry Manure as a Fertilizer Source, Soil Facts, AG 439-5, North Carolina Cooperative Extension Service, 1993.

CHAPTER 4

Environmental Challenges as Related to Animal Agriculture — Swine

M. Terry Coffey

INTRODUCTION

The evolution of agriculture has been driven by competitive economic, social, and political forces toward increased specialization. Within the swine industry, specialization has resulted in a worldwide shift from extensive outdoor production to the use of confinement rearing systems.

The total number of hogs produced in the United States and worldwide has not changed dramatically in the past 50 years. In the United States, concentration of the production capacity into fewer producers has been under way since the early part of this century (Table 1). The industry has been generally located close to areas of efficient grain production (Table 2), but concentrated areas of pork production exist around the world that do not coincide with areas of grain production. The Netherlands, Denmark, parts of the United Kingdom, and the southeastern United States are a few examples. Improved production efficiency and economies of scale have continued to encourage concentration of swine production. Today farms are larger and are being constructed near key infrastructure facilities such as feed mills and packing plants to reduce transportation costs. In the United States, accelerated application of technology, specialization, favorable climate, closeness to consumers, and other factors have led to dramatic growth of the swine industry in the mid-Atlantic region, particularly in North Carolina.

In contrast with pasture or range production systems, modern housing systems are designed for improved animal care and safe management of nutrients in manure. From a global perspective, swine production has not been a source of increased manure nutrient production. In fact, dramatic improvements in utilization of nutrients in swine feeds have reduced the excretion of nutrients per pig. Larger and more intensive farms have resulted in a concentration of manure that has increased the focus on the industry and developed new challenges for nutrient management. As the volume of waste is concentrated, we must continue to assess the adequacy of existing technologies and management capabilities. Those circumstances will encourage development of new technology and the applications of technologies that are not currently used in agricultural systems to achieve more efficient and economical methods of safe and effective nutrient management.

1-56670-199-6/96/$0.00+$.50

Table 1 Production of Hogs in the United States[a]

Year	Number of hogs	Number of farms
1900	62,879,000	4,335,989
1920	59,350,000	4,852,430
1940	34,040,000	3,767,875
1950	55,720,000	3,013,549
1960	59,030,000	1,848,784
1965	50,519,000	1,057,570
1975	49,267,000	661,700
1985	52,314,000	388,570
1993	56,798,000	235,840

[a] Data from Brandt et al., 1985 and USDA, 1988–1994.

CURRENT SITUATION

From a nutrient management perspective, the swine industry is primarily concerned with the availability of systems to protect water quality. Modern pork production uses a wide variety of facilities. However, they have similar components or features to achieve effective nutrient management. Pigs are housed in buildings designed to ensure their comfort and encourage efficient production. Buildings have special characteristics, including pen types and ventilation capacity, that correspond to the size of the animal, stage of production, management activities, among other factors. In most modern farms, pen flooring is slatted or of a woven material designed to allow waste to fall through to remove it from the pigs' area. Space below the floor is used for manure storage in some systems. In newer farms, it is more common to provide a method for the frequent removal of manure from the building to a storage and processing facility prior to application as a fertilizer. Usually, waste is removed from the building by flushing with water recycled from the storage and processing facility or by a mechanical scraper system. Features of the storage and processing facility are influenced by a number of factors, including soil type, climate, and the crop production system that will

Table 2 Production of Hogs in the United States by Region[a]

	Percentage of total production	
Region	**1970**	**1994**
Eastern Corn Belt (OH, IN, IL, MI, WI)	28.6	23.3
Western Corn Belt (MN, IA, MO)	37.2	37.7
Northern Plains (ND, SD, NE, KS)	13.7	12.8
Southeast (AR, LA, KY, TN, MS, GA, FL, SC, NC, VA, AL)	14.4	19.3
Southwest (TX, OK, NM)	2.6	1.9
Other	3.5	5.0
48-State total	100.0	100.0

[a] Data from Agricultural Statistics Board, 1993.

Table 3 Nutrient Composition of Swine Manure[a]

Manure type	Total Nitrogen	Phosphorus	Potassium
		(kg/1000 l)	
Liquid slurry	3.72	2.64	2.04
Anaerobic lagoon sludge	2.64	5.87	0.84
Anaerobic lagoon liquid	0.60	0.24	0.60
Aerobic lagoon liquid	0.31	0.12	0.24

[a] Data from NC Cooperative Extension Service, 1991.

utilize the nutrients. Material from the storage facility is periodically applied to the land as organic fertilizer. Frequency and volume of application must coincide with crop production and harvesting to achieve effective nutrient removal from the soil and to avoid surface runoff. Excessive soil nutrient accumulation must be avoided to prevent contamination of groundwater.

The design of storage/processing facilities has a dramatic impact on the concentration of nutrients in the effluent material (Table 3). Factors having the greatest effect are volume of the system and aerobic vs. anaerobic status of the facility. Larger-capacity facilities allow for nutrient dilution and aerobic treatment of the larger volume. Aerobic microorganisms in lagoons more effectively utilize and reduce concentration of nutrients compared with liquid from anaerobic lagoons or slurry facilities (Table 3).

The type of swine production facility affects the nutrient concentration in effluent. Data in Table 4 illustrate variations in plant-available nitrogen content of aerobic lagoon liquid from various types of farms. Nutrient content of stored liquid or slurry is analyzed prior to application as fertilizer. The concentration of nutrients in aerobic lagoon liquid varies, depending upon the depth at which the sample was obtained, and increases dramatically below 3.5 m (Table 5). Obviously, appropriate land application of effluent will be affected by the depth from which the material is obtained.

Adequate knowledge of land application sites is required to complete nutrient management plans. Soil types, existing nutrient concentrations, topography, erosion potential, crop productivity potential, and crop systems must be considered to determine appropriate application rates. Timing of application should coincide with nutrient utilization by the crop. Soil sampling schemes are developed based on soil type and topography (Figure 1). Results of soil analysis and crop plans form the basis for application recommendations (Table 6). As described here, the technology is available for concentrated production of swine and management of nutrients in manure to protect water quality.

Table 4 Plant-Available Nitrogen (PAN) Averages in Lagoon Effluent for the Period 1992–1994[a]

	Type of swine farm		
	Sow	Nursery	Finisher
Plant-available nitrogen	19.02	30.93	24.10

[a] From unpublished data of Murphy Family Farms research.

Table 5 Nutrient Concentration of Lagoon Effluent at Various Depths[a]

	Depth of sample (m)				
Nutrient (kg/1000 l)	0.5	1.5	2.5	3.5	5
Nitrogen	0.32	0.31	0.31	0.33	1.86
Potassium	0.28	0.28	0.28	0.28	0.50
Phosphorus	0.11	0.12	0.10	0.13	5.20

[a] From unpublished data of Murphy Family Farms research.

SIGNIFICANT CHALLENGES

A recent report to the North Carolina General Assembly identified nitrogen and phosphorus as nutrients with the greatest potential to adversely affect water quality as a result of swine production operations (Swine Odor Task Force, 1995b). Nitrate nitrogen is mobile in soils. If concentrations reach high levels at depths below the root zone of plants, nitrate may leach into groundwater. Surface water contamination by nitrate nitrogen can occur with runoff from overapplication. The solubility and therefore mobility of phosphorus in soils is much less than nitrate nitrogen. The threat to groundwater from phosphorus is minimal, but transport to surface water can occur from direct runoff or erosion of soils. Operation of nutrient management systems based on nitrogen-removal requirements may result in a buildup of soil phosphorus over time and increase the potential of this threat.

The availability of technology does not imply that all existing facilities employ adequate systems or that there have not been instances of poorly managed or improperly operated systems. Inadequate systems are more common in older facilities, and in general, these systems require increased management. In my opinion, the greatest potential for failure of systems to prevent nutrients from reaching water sources is because of poor management. The most extreme event would be overflow of the storage facility. More subtle danger comes during land applications. Overapplication, improper timing, failure to maintain crop systems, and application while raining can result in runoff.

In some locations, a key environmental concern is fresh water requirements to operate large-scale production facilities. Water is obviously a necessary nutrient for hogs. Also, substantial volumes of water are used in cleaning the facilities.

Seepage can possibly occur from old storage facilities constructed in coarse, sandy soil without liners of clay or other material (Swine Odor Task Force, 1995b). Those same studies confirmed the safety of current design standards and, particularly, the effectiveness of clay liners. Therefore, the potential for contamination of water from lagoon seepage has been greatly reduced in new facilities, but is of concern for certain older farms.

In a broader context, we need to evolve toward managing nutrients within entire watersheds, not just within small areas associated with an agriculture enterprise such as a hog farm. Many factors influence the concentration of nutrients in a watershed. Understanding the dynamics of a watershed, including

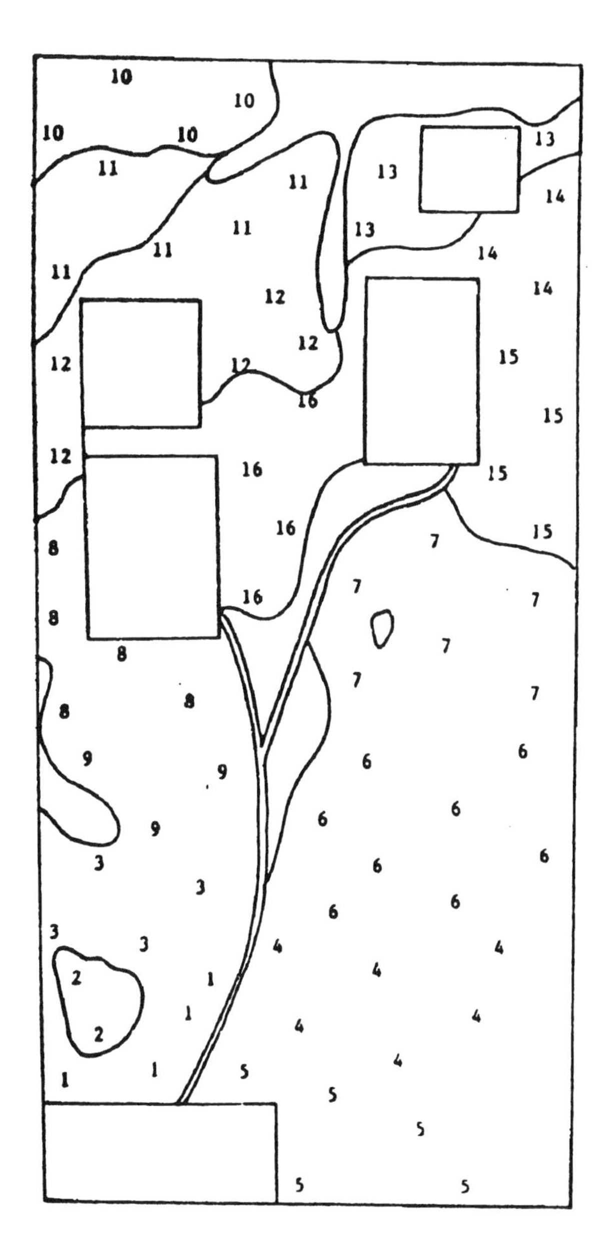

Figure 1 Example of a soil sampling map from a commercial sow production facility. Numbers 1 to 16 indicate location from which the sample was obtained. Samples from locations with the same numbers were combined to provide a composite. (Unpublished data from Murphy Family Farms research.)

discharge points such as municipal systems, sources of runoff, and other factors, will be important for adequate planning.

Today's technology is effective, but has limited flexibility because of fairly specific requirements for the appropriate combination of factors related to soil

Table 6 Example of Recommended Application of Aerobic Swine Lagoon Effluent[a]

	Locations									
	1	2	3	4	5	6	7	8	9	10
Applications/year	2	1	1	1	1	1	2	3	2	2
Volume (cm/ha)	1.3	1.3	1.9	1.3	1.3	.65	1.3	1.3	1.3	1.3
Nutrients applied										
Nitrogen (kg/ha/year)	257.6	128.8	192.6	128.8	128.8	65.0	257.6	386.4	257.6	257.6
Phosphorus (kg/ha/year)	51.5	25.8	38.0	25.8	25.8	19.0	51.5	77.3	51.5	51.5
Potassium (kg/ha/year)	25.8	13.4	19.0	13.4	13.4	17.9	25.8	39.2	25.8	25.8

[a] From unpublished data of Murphy Family Farms research.

type, proximity to people, climate, amount of land required, and cropping systems. Swine manure contains high levels of water, and the nutrient content is extremely diluted compared to concentrated commercial fertilizers. High moisture content and low nutrient concentration of material removed from storage facilities make it expensive to store, transport, and distribute on the land. Because of this, most land application is in close proximity to the swine production unit. The need for nearby land for application of wastes often limits expansion and modernization of existing facilities. This will limit the ability of the industry to lower production costs in order to compete in the marketplace. This is particularly true for older operations, which tend to be smaller.

Perhaps the most critical issue facing the industry is odor. Odor has always been a product of livestock enterprises, but larger farms, new development, and migration of people from urban to rural residences have combined to intensify the problem. The production of odor, source of odor, and human response to odor are extremely site specific (Swine Odor Task Force, 1995a). Inability to measure odor, quantify thresholds, or clearly define the problem make it difficult to develop strategies toward solutions. Public objections to odors from swine operations generally focus on property values and quality of life issues (Swine Odor Task Force, 1995a).

WHAT IS THE EFFECT OF INTENSIVE SWINE PRODUCTION ON WATER QUALITY?

There are a number of ways to approach understanding the impact of swine production on water resources. The dynamic nature of the industry makes it dangerous to generalize. The move toward intensive production has increased the concentration of animals, but construction of new facilities has incorporated more modern designs to manage nutrients. In outdoor systems there was potential for runoff with every rainfall. Old, outdated, or improperly designed and constructed confinement operations represent a greater environmental threat than modern units.

With 140,000 sows and over 2 million market hogs, Sampson County, NC, is the largest hog-producing county in the United States (NC Cooperative Extension Service, 1994). In addition, there is a large poultry industry in the county. Most of the swine production facilities have been constructed in the last 10 years, incorporating waste management technology that is relatively modern.

Facts are available that represent the status of groundwater and surface waters in that area and are useful to examine, in a broad way, the relationship of modern animal agriculture to water quality (NC Cooperative Extension Service, 1994). In 1991, the North Carolina Cooperative Extension Service assessed groundwater quality by testing 317 wells providing drinking water. Fourteen of the wells had concentrations of nitrate nitrogen above the 10 ppm standard. Contamination of these wells could not be traced to swine or poultry production facilities. Contamination was associated with shallow depth, faulty construction, or, in some cases, proximity to home septic systems. Surface water quality is reflected by the status of the Black and South rivers. In 1990, both rivers were classified as "High-Quality Waters" based on chemical and biological data. In 1994, the rivers were reclassified as "Outstanding Resource Waters," indicating that, in addition to excellent water, the rivers met additional criteria, including special ecological or scientific significance. These facts do not eliminate our need to be concerned about previously mentioned issues, but do indicate that intensive swine production can be conducted while maintaining and even improving the quality of water resources.

ALTERNATIVES FOR THE FUTURE

The goal of pork producers is to operate in a sustainable manner. Components of sustainability will include, among others, requirements for environmental soundness, social acceptability, and profitability. More specifically, our nutrient management systems should strive for no net nutrient buildup.

One of the most obvious strategies is to reduce excretion of nutrients of greatest concern. Nitrogen excretion in swine waste can be substantially reduced by a number of strategies. Improved productivity is the most obvious. As the efficiency of feed utilization is improved, there is a reduction in excretion of nitrogen. An improvement in feed conversion of 0.25 units would reduce nitrogen excretion by 5 to 10%. Over the past 20 years, the feed efficiency of pigs growing from 25 to 110 kg has decreased from approximately 4.0 to less than 2.85 in top-producing herds. Nutrition and genetics are the disciplines that will be most important to future improvements.

Formulation of diets with synthetic amino acids to provide the appropriate level of essential amino acids while avoiding large excess is another approach. Feedstuffs are combined to meet the pigs' needs for the most limiting amino acid. This results in higher-than-required protein content of the diet because of the presence of excess amino acids. The use of synthetic amino acids will reduce total nitrogen while still meeting the pigs' amino acid needs. Koch (1990) estimated the impact of lowering the protein content of the diet of finishing pigs growing from 55 to 100 kg on nitrogen excretion. Assuming a feed-to-gain ratio of 2.9, nitrogen requirements for maintenance at 6 g/d and nitrogen retention at 24.5 g/d, then crude protein in the diet could be reduced by 2.5%, and nitrogen excretion in urine would be decreased by 29% (1527 to 1083 g) over the period when amino acids were used.

Synthetic amino acids are commonly added to swine diets. L-Lysine-HCL is the most commonly used, and DL-methionine is used in some diets. Synthetic L-threonine and L-tryptophan have recently become commercially available. Our ability to efficiently utilize competitively priced synthetic amino acids is limited by our knowledge of amino acid requirements of pigs and of biological availability of amino acids in feed ingredients.

Table 7 The Effect of Phase Feeding on Nitrogen Excretion[a]

		Two feeds		
Item	**One feed**	**Grower**	**Finisher**	**Overall**
Protein (%)	16	16.5	14	—
Feed/gain	3.0	2.5	3.3	—
Feed intake (kg)	210	75	132	207
N intake (kg)	5.38	1.98	2.95	4.93
N excreted (kg)	3.48	1.16	1.86	3.02
N excreted (%)	65	58	63	61
N retention (kg)	1.9	0.82	1.09	1.91

[a] Data from Lenis, 1989.

If diets are precisely formulated to meet the protein/amino acid requirements of pigs, nitrogen excretion will also be reduced. Lenis (1989) calculated the reduction in nitrogen excretion that would result from changing from one feed system that is common in Europe to a two-phase system (Table 7). He estimated that meeting the nitrogen needs more precisely would reduce nitrogen in waste by 13%. When diets are formulated more nearly to the requirements of pigs, nutrient excretion is reduced because of decreased dietary excess and improved utilization of nutrients.

Knowledge about the bioavailability of phosphorus in feedstuffs, the available phosphorus requirements of pigs, and the effectiveness of phytase to increase the availability of plant phosphorus will give the nutritionists tools to reduce phosphorus in swine waste without affecting performance. Cromwell (1990) illustrated the effect of formulating grower diets for pigs on either a total or available phosphorus basis (Table 8). When the corn-soy diet was formulated to meet the total phosphorus requirement (0.5%), the available phosphorus level was at the requirement of 0.23% (Diet 1). The substitution of wheat for corn resulted in excess available phosphorus when the diet was formulated to contain 0.5% total phosphorus (Diet 2). When the diet containing wheat was formulated to meet the available

Table 8 Diets for Growing Pigs Formulated on an Available or Total Phosphorus Basis[a,b]

		Diet	
Ingredient (%)	**1**	**2**	**3**
Corn	79.79	—	—
Wheat	—	83.24	83.35
Soybean meal (48)	17.76	14.53	14.51
Dicalcium phosphate	0.96	0.57	0.32
Limestone	0.84	1.01	1.17
Salt	0.35	0.35	0.35
Premixes	0.30	0.30	0.30
Calculated analysis			
Total phosphorus (%)	0.50	0.50	0.45
Available phosphorus (%)	0.23	0.28	0.23

[a] Diets formulated to either 0.50% total phosphorus or 0.23% available phosphorus.

[b] Data from Cromwell, 1990.

phosphorus requirement of 0.23%, total phosphorus was reduced by 10% (Diet 3). This not only lowered costs, but reduced the phosphorus content in swine waste. Similarly, increasing the availability of plant phosphorus through the use of the enzyme phytase will reduce total dietary phosphorus and phosphorus excretion in feces.

Training, assessment, and certification programs for operators of swine nutrient management systems should be considered. Nutrient management systems being utilized today incorporate numerous technologies from a variety of disciplines. Proper maintenance and operation of housing, manure transport, and storage/treatment and distribution components are necessary for proper function. It is illogical to expect effective management of these systems without adequate training.

We should encourage refinement and improvement of systems to reduce the potential for runoff. Utilization of natural and constructed riparian buffers, including vegetative strips and constructed wetlands, offers potential.

In contrast to other environmental challenges, technologies for elimination of odor from swine production facilities are not readily apparent. Sources of odor have been identified as the animal facilities, manure storage facilities, and land application procedures. In a report titled "Options for Managing Odor," the North Carolina Swine Odor Task Force listed reduced nitrogen content of feed, use of odor control additives such as enzymes, and biogas or methane generation as promising topics that could be explored to reduce odor production. Practical approaches that are available to producers today include proper cleanliness, site selection, adequate waste management capabilities and improved practices for land application (Swine Odor Task Force, 1995a).

There is a need for innovative alternatives to the basic technologies employed in our current system. This is especially true if we are to make significant progress to reduce odor production. In its purest form, specialization of swine production involves the transport of nutrients from areas of the world well suited to intensive feed production to locations where pork is produced. Research should focus on a systems approach to the development of technologies that will allow the efficient concentration of byproduct nutrients from swine production facilities into a form that could be economically returned as fertilizer to crop production areas. If this were accomplished, complete closure of the recycling loop could be achieved. An efficient system would focus on nutrient retention in the final product to maximize fertilizer value. The swine farm of the future could incorporate solid separation, filtration, or other strategies to produce concentrated organic fertilizer.

A key will be to redesign our production systems to reduce water input into the waste stream. This will serve to reduce the total farm water requirements and reduce the requirement for removal of water in order to achieve efficient concentration of nutrients. Progress in this area could completely eliminate the need for large-capacity, on-farm facilities for manure storage and treatment; drastically reduce on-site land requirements; and result in truly effective options for odor reduction. Without a fundamental shift in waste management capability, we will begin to see the removal of production capacity because of limited land availability in some locations. The Netherlands is an example of where this is already occurring.

OUR FUTURE

Without question, the swine industry must achieve the goal of sustainability. Many forces will shape our future. Environmental concerns will certainly influence the evolution of our industry, and economic implications will increase. Advancement in nutrient management methods could help us solve problems with existing facilities or hasten their demise in favor of new units, which justify the incorporation of more elaborate processes into the swine enterprise. In order to deal with odor concerns, production units might be moved to sparsely populated regions and assembled into large, geographically self-contained units.

The only thing that we can be certain of is continued change. The future will need to be addressed with open minds and a commitment to the environment. The swine industry will continue to adapt to the requirements necessary to protect our planet, and we must insist that change is driven by facts rather than emotions. Over the past decade, pork production has been the most dynamic segment of American agriculture. We eagerly anticipate the next 10 years.

REFERENCES

Agricultural Statistics Board, USDA, *Hogs and Pigs,* December, 1993.

Brandt, J. A., Deiter, R. E., Hayenga, M., and Rhodes, J. V., *The U.S. Pork Sector: Changing Structure and Organization,* Iowa State University Press, Ames, 1985.

Cromwell, G. L., Application of Phosphorus Availability Data to Practical Diet Formulation, Proceedings of the Carolina Swine Nutrition Conference, Raleigh, NC, Carolina Feed Industry Association and the North Carolina Cooperative Extension Service, 1990.

Koch, F., Amino Acids Formulation to Improve Carcass Quality and Limit Nitrogen Load in Waste, Proceedings of the Carolina Swine Nutrition Conference, Raleigh, NC, Carolina Feed Industry Association and the North Carolina Cooperative Extension Service, 1990.

Lenis, N. P., Lower nitrogen excretion in pig husbandry by feeding: current and future possibilities, *J. Agric. Sci.,* 37, 61–70, 1989.

NC Cooperative Extension Service, North Carolina State University College of Agriculture & Life Sciences, Soil Facts: Swine Manure as a Fertilizer Source, March 1991.

NC Cooperative Extension Service, North Carolina State University College of Agriculture & Life Sciences, Economic Importance of Agriculture and Water Quality Studies of Sampson County Surface Water Quality in Sampson County, North Carolina, 1994.

Swine Odor Task Force, Options for Managing Odors, North Carolina Agricultural Research Service, North Carolina State University, Raleigh, 1995a.

Swine Odor Task Force, Water Quality and the North Carolina Swine Industry, North Carolina Agricultural Research Service, North Carolina State University, Raleigh, 1995b.

USDA, Number of operations with hogs, inventory, value per head, and total value, *Hogs and Pigs,* December issues, 1988–1994.

CHAPTER 5

Advances in Amino Acid Nutrition and Metabolism of Swine and Poultry

David H. Baker

INTRODUCTION

Nitrogen and phosphorus pollution of the environment will become more important in the future, and animal production will be called upon to do its part to minimize excessive nitrogen and phosphorus output in animal excreta. Geneticists have done their part in providing a new generation of pigs and poultry that are capable of producing protein gain at greater efficiencies than ever before. It now behooves the nutritionist to design feeding programs that can take advantage of the "new" genetics to produce a palatable product with a high lean-to-fat ratio and, in the process, produce this product in a way that maximizes use of amino acids for protein accretion and minimizes excretion of nitrogenous waste products.

Animals used to produce food for humans can be likened to machines whose purpose is to convert plant-source foodstuffs and byproducts to highly nutritious and palatable animal-source protein. Meat and eggs are rich in high-quality protein and in several important micronutrients, including zinc, iron, and B vitamins. Unfortunately, these foods are also rich in saturated fat, and attempts are being made to reduce the amount and kind of fat in these products to make them more consumer friendly. Also, animal nutritionists are working toward the goal of making the (animal) machine run as efficiently as possible while, at the same time, minimizing its efflux waste products.

Protein is a costly item in pig and poultry diets, so maximizing the efficiency of protein and amino acid utilization is very important. How can one maximize lean meat production with the absolute minimum intake of amino acids? Clearly, diets containing amino acids at minimally required levels (for maximal lean growth) with minimal excesses is a critically important factor. Using chemically defined diets containing amino acids as the sole source of dietary nitrogen, we have reported that, with a near perfect amino acid balance, a 15-kg pig is capable of converting 87% of its absorbed nitrogen above maintenance to carcass protein (Chung and Baker, 1992a). This does not mean, however, that each of the 23 amino acids found in dietary protein are utilized at 87% efficiency for protein accretion (Chung and Baker, 1992b). Indeed, some amino acids are used more efficiently than others, and understanding the rationale for this "differential" efficiency is important in applying the concept of an ideal protein to practical pig and poultry production (Baker and Chung, 1992). It is important to remember that

1-56670-199-6/96/$0.00+$.50

of the amino acids used in protein synthesis, at least 60% are derived from body protein degradation. Thus the nutritionist must design dietary amino acid profiles that will complement the profile being provided by protein turnover.

IDEAL PROTEIN FOR PIGS

Theoretically, an ideal pattern of amino acids should exist for each physiological function, but clearly the ideal pattern will be different for each function, i.e., maintenance, protein accretion, reproduction, and lactation. For meat production, amino acid requirements can be separated into those required for protein accretion and those required for maintenance. Moughan (1989) and Fuller (1991) described maintenance as comprising (1) urinary excretion of unmodified amino acids, (2) use of amino acids as precursors for other essential body metabolites (e.g., creatine from arginine and glycine; taurine and glutathione from cysteine, or indirectly from methionine; thyroxin, melanin, and catecholamines from tyrosine, or indirectly from phenylalanine; serotonin from tryptophan; nucleic acids and choline from glycine or serine; carnitine from trimethyl lysine; nitrous oxide and polyamines from arginine; carnosine from histidine), (3) amino acids lost from integuments and epidermal structures, (4) obligatory oxidation of amino acids, and (5) amino acids lost from gastrointestinal epithelia (mucous, mucosal cells, digestive enzymes).

Mitchell (1964) discussed the concept of an ideal protein or perfect amino acid balance in 1964. Thus he recognized that chemical scoring (Mitchell and Block, 1946) using egg protein as the ideal standard was flawed. Indeed, egg protein is *too* rich in essential amino acids, with isoleucine, for example, being present at twice the level that most animals require. Cole (1978) proposed that diet formulation for pigs could be done on the basis of ideal amino acid ratios, and the ARC (1981) subsequently described ideal ratios of amino acids for pigs. The ideal pattern received considerable attention over the next 12 years (Wang and Fuller, 1989, 1990; Fuller et al., 1989; Chung and Baker, 1992a), and in 1992, Baker and Chung (1992) proposed ideal amino ratios for pigs in three different weight classes (Table 1). Based upon additional calculations as well as recent empirical research (Baker, 1993, 1994; Hahn and Baker, 1995), our original ratios for later growth phases have been modified somewhat. Because the maintenance requirement for lysine is low relative to maintenance requirements for certain other amino acids (e.g., threonine, cystine), ideal ratios of certain amino acids for young pigs cannot be applied without adjustment to older market-type pigs.

Of the total requirement for a given amino acid, protein accretion comprises well over 90% of the need for pigs weighing 10 kg, but as pigs approach slaughter weight, maintenance assumes greater prominence in the total requirement for an amino acid (Fuller et al., 1989; Black and Davies, 1991; Chung and Baker, 1992b; Baker, 1994). Thus, while the ideal ratio of sulfur amino acids (SAA, i.e., methionine + cystine) to lysine is 60% for 10-kg pigs (Chung and Baker, 1992a),

Table 1 Ideal Amino Acid Patterns for Pigs in Three Weight Categories[a]

	Ideal patterns (% of lysine)		
Amino acid	**5–20 kg**	**20–50 kg**	**50–110 kg**
Lysine	100	100	100
Threonine	65	67	70
Tryptophan	17	18	19
Methionine	30	30	30
Cystine	30	32	35
Methionine + cystine	60	62	65
Isoleucine	60	60	60
Valine	68	68	68
Leucine	100	100	100
Phenylalanine + tyrosine	95	95	95
Arginine	42	36	30
Histidine	32	32	32

[a] From Chung and Baker (1992), Baker and Chung (1992), and Baker (1993, 1994); ratios are expressed on a true digestible basis.

the ideal ratio for the maintenance component of that requirement is much higher (Fuller et al., 1989; Fuller, 1994). Indeed, in every species where requirements for maintenance have been specifically studied, including studies with humans, the SAA requirement has exceeded the lysine requirement (Baker and Han, 1993). With an increasing contribution for maintenance as an animal grows toward slaughter weight, the ratio of SAA to lysine must increase, probably in a straight-line fashion, as a growing pig advances from 10 to 110 kg. Gut losses of SAA, particularly cystine, account for less than half of the maintenance need for SAA, but gut losses of threonine account for 75% of the maintenance need for threonine (Fuller, 1994). Estimates of the threonine requirement for swine maintenance are considerably higher than estimates of the maintenance need for lysine (Baker et al., 1966; Fuller, 1994), with the threonine to lysine ratio being about 150%. Also, the ideal tryptophan to lysine ratio for maintenance per se is greater than the ideal ratio for protein accretion per se (Fuller et al., 1989; Fuller, 1994).

Least-cost diet formulation on an ideal protein basis can be influenced significantly by the ratios used. Work by Burgoon et al. (1992) and our work with 15-kg pigs (Han et al., 1993) could be interpreted as suggesting that tryptophan at a level of 17% of the true digestible lysine is just as effective for young pigs as higher ratios. The point being made here is that whether one uses tryptophan at a level representing 17 → 18 → 19% of lysine (starting → growing → finishing) or, instead, higher ratios (Fuller, 1991) can have profound effects on diet formulation and, in particular, on quantities as well as sources of both lysine and tryptophan that go into finished pig feeds. Interestingly, the situation could be quite different for barley-wheat-soybean meal diets such as those used outside the United States. Threonine and lysine are the important amino acids in these diets, and threonine unlike tryptophan is not abundant in soybean meal (nor in barley or wheat, either). Thus the threonine to lysine ratio in soybean meal is 64%, which is lower than that prescribed in ideal protein.

Use of Ideal Protein to Formulate Pig Diets

Whether using total or true digestible lysine requirement data, one can accurately formulate pig diets. Clearly, controversy surrounds the topic of digestibility, i.e., whether to use apparent, true, or "real" digestibility values or whether "bioavailability" (growth assay) values should be used for some ingredients like cottonseed meal. With corn-soybean meal diets, *true* digestibility values for the key amino acids (lysine, threonine, tryptophan, and SAA) are about the same. Thus, using either true digestible lysine or total lysine as a base, similar ratios and therefore similar predicted requirements are obtained.

If diet formulation is done using *apparent* digestibility data (i.e., for requirements as well as for feed ingredients), problems are encountered. The apparent digestibilities of threonine and tryptophan are lower than the apparent digestibility of lysine, and this is true for virtually all types of diets. Also, apparent amino acid digestibility values for cereal grains are underestimated, because at lower protein levels endogenous losses contribute heavily to total ileal losses of amino acids. Thus one can calculate that the ideal ratios for threonine on an apparently digestible basis are 58 → 60 → 62% (of apparently digestible lysine) for starting, growing, and finishing pigs, respectively, instead of the 65 → 67 → 70% values (Table 1) that apply when formulating diets on a total or true amino acid digestibility basis. The other amino acid whose ideal ratio changes when formulating on an apparent digestibility basis is tryptophan. Apparently digestible tryptophan as a percent of apparently digestible lysine calculates to be 15 → 16 → 17% instead of 17 → 18 → 19% (true or total amino acid basis) for starting, growing, and finishing pigs, respectively.

Another factor that can cause overestimation of dietary need for amino acids other than lysine is use of unrealistic requirement values for lysine. There has been a tendency for some to assume lysine requirement values for finishing pigs that are higher than they really are. Thus, with either cheap lysine or cheap soybean meal it is not uncommon for dietary levels of lysine to increase. If excess (or "safety factor") lysine levels are fed, one should ratio threonine, SAA, and tryptophan to the "real" lysine requirement, not to the excess dietary lysine level. If ratios for threonine, SAA, and tryptophan fall below 70, 65, and 19%, respectively, for finishing pigs (Table 1), then supplementation with one or more of these amino acids (or with more soybean meal) is necessary.

Realistic Lysine Requirements for Finishing Pigs

Hahn et al. (1995) recently reported apparently digestible and total lysine requirements of finishing barrows and gilts during two growth periods, 50 to 90 kg and 90 to 110 kg. The pigs used were from Pig Improvement Company (PIC) line 26 × Camborough 15 breedings. At a kill weight of 110 kg, these pigs typically have tenth-rib loin-eye areas of 40 (barrows) and 43 cm^2 (gilts) with tenth-rib fat depths of 2.5 (barrows) and 2.1 cm (gilts). Using a determined lysine digestibility of 85% in a corn-soybean meal diet, we have translated our determined

Table 2 Requirements (% of Diet) for Key Amino Acids in Finishing Pigs[a]

Amino acid	Ideal ratio	50–90 kg Barrow	50–90 kg Gilt	90–110 kg Barrow	90–110 kg Gilt
Lysine	100	0.68	0.75	0.58	0.61
Threonine	70	0.48	0.53	0.41	0.43
Methionine + Cystine	65	0.44	0.49	0.38	0.40
Tryptophan	19	0.13	0.14	0.11	0.12

[a] For PIC pigs consuming corn-soybean meal diets containing 3450 kcal ME/kg.

apparent digestible lysine requirements to total lysine requirements (Table 2). Requirements for threonine, methionine + cystine, and tryptophan were estimated by multiplying the total lysine requirement by 70, 65, and 19%, respectively. The requirements for these amino acids are generally met when a 15.2% crude protein (CP) corn-soybean meal diet is fed during early finishing and when a 13.2% CP diet is fed during late finishing (Table 3). Moreover, within a given genetic line, stress-related decreases in voluntary feed intake (e.g., crowding, disease, heat stress) lower lean growth potential and thus are not likely to cause increased requirements expressed in terms of diet concentration (percent of diet or percent of calories).

Genotypes that have higher lean-to-fat ratios than the PIC pigs used in our research (Hahn et al., 1995; Hahn and Baker, 1995) will have higher amino acid requirements and will therefore require higher protein levels in corn-soybean meal diets than those shown in Tables 2 and 3. On the other hand, pigs with less potential for lean growth will have lower requirements than those presented in Tables 2 and 3.

In designing diets that will meet ideal levels of lysine, threonine, SAA, and tryptophan with minimal excesses of any of these amino acids, crystalline lysine is necessary. Moreover, if economical to do so, use of crystalline threonine, methionine, and tryptophan (all available in feed-grade form) would minimize dietary excess amino acids even more and, as a result, would reduce nitrogen output in both feces and urine. Careful calculations reveal, however, that if using lysine supplementation alone, the upper limit of use is about 0.1% (equivalent to 2.5 lb of feed-grade lysine per ton of complete feed) for corn-soybean meal diets. Thus, if one uses more than 0.10% of added lysine, ideal ratio calculations reveal that threonine becomes limiting first, followed closely by SAA and tryptophan.

Table 3 Corn-Soybean Meal Diets for PIC Finishing Pigs (Mixed-Sex Feeding)[a]

Body weight (kg)	Option I	Option II[b]
50–90 kg	15.2% CP	13.8% CP + 0.1% L-lysine
90–110 kg	13.2% CP	11.8% CP + 0.1% L-lysine

[a] Assumes 2% fat in diet.

[b] 0.1% L-lysine provided by 0.127% feed-grade lysineŸHCl (78.8% L-lysine activity).

Modeling a Lysine Requirement for Finishing Pigs

Can one calculate what the lysine requirement should be for a given genotype fed a given type of diet in a given environment? Theoretically, this is possible, but there are problems in doing this just as there are problems in arriving at lysine requirements from empirical research. The problems focus on (1) characterizing the lean growth potential of the pig, (2) estimating the maintenance lysine requirement, (3) estimating the digestion efficiency and the efficiency of converting absorbed lysine to lysine in whole-body protein, and (4) converting daily lysine intake requirements to needed concentrations in the diet — at variable feed intakes.

In our study (Hahn and Baker, 1995) on lysine accretion, an average of 17.8 g/d of total lysine consumed (14.9 g/d of apparently digestible lysine) were required to average 10 g/d of lysine accretion in PIC gilts during the finishing period 56 to 111 kg (Table 4). One can calculate from this data set (where the lysine levels fed were slightly below requirements for maximal lean gain) that 56% of the lysine fed was recovered in whole-body protein of the pig. This efficiency value includes the lysine need for maintenance as well as the indigestible lysine in the diets fed. The efficiency of using apparently digestible lysine for maintenance and protein accretion is calculated to be 67%, and the efficiency above maintenance of using apparently digestible lysine is calculated to be 71%. This value is slightly lower than the 74% value calculated by Bicker et al. (1994), and it is considerably lower than the 86% value determined by Batterham et al. (1990), who fed graded levels of lysine that were distinctly below the lysine requirement. Their straight-line plot of lysine retained as a function of ileal digestible lysine intake had a slope of 0.86. Hence the range covered in their assay did not include what we have termed "protein accretion II," i.e., the portion of the accretion curve that is less efficient by virtue of having increasing numbers of pigs whose lysine need for maximal accretion is exceeded. The factorial model in Table 4 was therefore manipulated to make the lysine intake come out to 17.8 g/d, the quantity actually consumed by the gilts in the experiment. The model assumes, therefore, constant utilization (86% efficiency) of absorbed lysine between maintenance and 75% of targeted lysine accretion (i.e., 10 g/d), but decreasing utilization (at an increasing rate) between 75 and 100% of targeted lysine accretion. The average value for the utilization of absorbed lysine in this upper portion of the accretion curve was selected to be 45%, because this efficiency value would yield 6.6 g/d of lysine needed for protein accretion II, precisely the amount needed to make the factorial method sum to a total daily need of 17.8 g/d of lysine.

Efficiency of Utilizing Absorbed Amino Acids for Protein Accretion

The data in Table 4 illustrate how efficiency assumptions can affect a predicted lysine requirement via modeling. The gilts in our experiment had more lysine in whole-body protein at the end of the trial (7.3% of CP at 111 kg) than at the beginning of the trial (6.2% of CP at 55 kg), and this was an important factor

Table 4 Lysine Efficiency Calculations

Observed results[a]	
Avg daily weight gain (g)	895
Avg daily protein gain (g)	112
Avg daily lysine gain (g)	10
Avg daily feed intake (g)	2818
Avg daily lysine intake (g)	17.8
Efficiency of lysine utilization (%)[b]	56
Factorial approach (avg daily lysine need)[c]	
Maintenance (0.7[d] ÷ 0.84 ÷ 1.0) =	0.8 g
Protein accretion I (7.5 ÷ 0.84 ÷ 0.86) =	10.4 g
Protein accretion II (2.5 ÷ 0.84 ÷ 0.45) =	6.6 g
Total[e]	17.8 g/d

[a] From Hahn and Baker (1995). The corn-soybean meal diets fed contained 0.65% lysine (0.55% apparently digestible lysine) during early finishing (55 to 90 kg) and 0.595% lysine (0.50% apparently digestible lysine) during late finishing (90 to 111 kg).

[b] Average daily whole-body lysine gain divided by average daily lysine intake.

[c] Assumptions were (1) apparent digestibility of lysine in the corn-soybean meal diet is 84%; and (2) efficiencies of utilizing absorbed lysine is 100% for maintenance, 86% for the first 75% of lysine accreted, and 45% for the last 25% of lysine accreted.

[d] The maintenance requirement was extrapolated from Baker et al. (1966) assuming an average weight of 84 kg for the gilts in question.

[e] This total intake of lysine is projected to meet the need for 10 g/d of lysine accretion. If one assumes that an additional 1.9 g/d of lysine intake (total intake = 19.7 g/d) would result in all gilts in the population reaching maximal lysine accretion (average assumed to be 10.5 g/d), the efficiency of achieving this is 0.5 ÷ 1.9 = 26%. The 19.7 g/d intake (0.70% of diet) was obtained from empirical research with PIC finishing pigs (Hahn et al., 1995) and represents an average of the determined requirements for early and late finishing pigs.

in their accretion of 10 g/d of lysine. In most modeling scenarios, the lysine accretion value is not known, but is instead calculated. For example, one might assume that lysine in whole-body pig protein is 6.8%. This assumed concentration is multiplied by *estimated* whole-body protein gain to arrive at a lysine accretion value. Several lysine concentration values (percent of whole-body pig protein) have been determined for pigs in the weight range of 45 to 110 kg (Batterham et al., 1990; Krick et al., 1993; Hahn and Baker, 1995), and they range from 6.2 to 7.4%. The value selected to calculate whole-body lysine accretion has a significant effect on the predicted lysine requirement, just as the assumption regarding the efficiency of depositing absorbed lysine has a major effect on the predicted lysine requirement.

Generally speaking, the efficiencies of utilizing (absorbed) amino acids other than lysine are lower than that determined for lysine. Batterham (1994) and Batterham and Anderson (1994) reviewed the efficiencies of utilizing ileal digestible lysine, threonine, methionine, tryptophan, leucine, and isoleucine in pigs fed soybean meal. Their calculation (amino acid retained ÷ ileal digestible amino acid intake) included the maintenance component. The determined values were 75%

for lysine, 73% for isoleucine, 67% for leucine, 64% for threonine, 45% for methionine, and 38% for tryptophan.

While the previous efficiencies of amino acid use for maintenance plus accretion need verification, at the very least they point strongly to the conclusion that efficiencies are not the same for all amino acids. Thus, because amino acids like tryptophan, SAA, and threonine are used less efficiently than lysine from body amino acid pools (derived from protein catabolism plus dietary consumption), the ratios of these amino acids to lysine in whole-body pig protein are lower than those actually needed by the pig.

Because of the pitfalls and complexities inherent in determining an accurate lysine requirement for pigs of a given genetic potential that are consuming a given diet in a given environment, even though data are plentiful for lysine in contrast to the paucity of data for other amino acids, it would seem patently unwise to rely upon either empirical data *or* modeling approaches to estimate requirements for amino acids other than lysine. Instead, it would seem logical that emphasis should be placed on the lysine need as influenced by genetics, sex, environment, disease, feed intake, and other factors and then use ideal ratios to predict the needs for the remaining amino acids.

Feed Intake Influences on Amino Acid Requirements

Feed intake increases or decreases that are caused by caloric density changes or by the inherent capacity of a given genetic line for lean growth potential will change amino acid requirements. But how does one translate a daily requirement (e.g., for lysine) determined at one feed intake to another feed intake. The evidence is mounting that stress-related decreases in feed intake, within a genetic line, will lower the lean growth potential of the pigs in question. Stresses such as crowding (Hahn et al., 1995; Kornegay et al., 1993), increased numbers of pigs per pen (Chappel, 1993; Hahn et al., 1995), heat stress (Han and Baker, 1993; Meyer and Bucklin, 1994), and disease (Stahly, 1994) have been shown to depress feed intake and to also lower lean growth potential.

To meet the lysine need for virtually all of the gilts in the population sampled required 21.1 g/d lysine during early finishing (55 to 90 kg) and 17.2 g/d during late finishing (90 to 110 kg) in the example pigs described in Table 4 (Hahn and Baker, 1995; Hahn et al., 1995). During both early and late finishing, the gilts consumed an average of 2820 g of feed per day. This calculates to be a dietary concentration of 0.75% for early finishing and 0.61% for late finishing. If these pigs are now stressed such that their daily feed intake falls to 2500 g/d, do they still require 21.1 g/d (0.85% of diet) during early finishing and 17.2 g/d (0.69%) during late finishing? Alternately, do they now require 18.75 g/d (0.75%) during early finishing and 15.25 g/d (0.61%) during late finishing? The evidence would suggest that the stressed gilts in question no longer require 21.2 and 17.2 g/d during early and late finishing, respectively. Thus their lean gain potential has been reduced, so their daily lysine needs are reduced as well. In fact, the stressed pigs consuming 2500 g of feed per day may still require close to 0.75 and 0.61% lysine during early and late finishing, respectively.

IDEAL PROTEIN FOR BROILER CHICKS

Large broiler producers are making several diet changes during the 6- to 8-week grower period. Also, some have recognized that female broilers have substantially lower requirements for amino acids than male chicks. Clearly, solid requirement data are not available for most amino acids, particularly during the growth period beyond 21 d of age. Therefore, it makes sense to use the large body of literature to set ideal ratios of each essential amino acid to lysine rather than to worry about specific requirements. Thus, if the ideal ratio of digestible threonine to digestible lysine is 67% for birds from hatching to 21 d posthatching, this *ratio* would hold (1) for both males and females, (2) for diets that are either high or low in metabolized energy (ME) or crude protein, (3) for strains with both high or low lean growth potential, and (4) for all environmental conditions. The same cannot be said for the threonine *requirement.* Indeed, the digestible threonine requirement (percent of ME) would be higher for males than for females and higher at high levels of CP than at lower levels of CP.

Why Use Lysine as a Reference Amino Acid

In the practical setting, SAA, lysine, threonine, valine, and arginine are the only amino acids that are of potential practical importance for broiler production (Han et al., 1992). In a corn-soybean meal scenario, soybean meal contributes at least 80% of the total dietary amino acids, while corn contributes less than 20%. Our recent work with corn and soybean meal (Fernandez et al., 1994) has shown that the order of amino acid limitation in soybean meal for chick growth is SAA (1), threonine (2), lysine and valine (3), and amino nitrogen (4). In corn, the order of limitation is lysine (1); threonine (2); tryptophan (3); and valine, isoleucine, and arginine (4).

Among the essential amino acids, lysine is a clear choice to be the reference amino acid for the following reasons:

1. Following SAA, lysine is the second limiting amino acid in broiler diets. It is also economically feasible as a dietary supplement.
2. Unlike SAA, analysis of lysine in feedstuffs is straightforward.
3. Unlike SAA, lysine has only one function in the body, i.e., protein accretion.
4. There is a good body of information on the lysine requirement of birds under a variety of dietary, environmental, and body compositional circumstances.

Illinois Ideal Chick Protein

For over 30 years, work has been ongoing at the University of Illinois on proper amino acid ratios for broiler chicks. Thus both requirement ratios and slope ratios (amino acid intake regressed on body weight gain) have provided a good basis for setting ideal ratios (to lysine) of essential amino acids. Most of this work, however, has been done with chicks between hatching and 21 d posthatching. Clearly, ideal ratios of some amino acids (cystine, threonine, tryptophan, valine, arginine, isoleucine) will have to be somewhat higher for late growth (21 to 42 d) than for early growth, because recent research in our laboratory has shown that ideal ratios of these amino acids to lysine are higher for maintenance than for the total requirement (protein accretion + maintenance). Maintenance as a percentage of the total requirement is very small (probably 3 to 8%) for young birds, but it increases as birds advance in age and weight.

Calculated estimates of ideal amino acid ratios for both early and late growth of broilers are shown in Table 5. These ratios have been used to calculate amino acid requirements for each growth phase. Obviously, the digestible lysine requirement is very important because it is now the basis for setting requirements for all the other essential amino acids. The digestible lysine requirement for early growth is based upon the work of Han and Baker (1991, 1993). For the 3- to 6-week growth period, the lysine requirement of Ross × Ross broiler chicks was found to be 0.89% of the diet (ME = 3200 kcal/kg) for males and 0.84% of the diet for females (Han and Baker, 1994). Assessment of breast yield indicated that the lysine level required to maximize this parameter was similar to that required to maximize feed efficiency.

Illinois ideal chick protein (IICP) for the early growth period (0 to 3 weeks) was recently compared with the profiles listed in the NRC (1984) and NRC (1994) requirement publications (Baker and Han, 1994). This work verified that IICP is a superior profile in that gain per unit of indispensable amino acid nitrogen was higher than that observed for the NRC (1984) or NRC (1994) amino acid profiles. Based upon these comparisons, however, we have made slight modifications in IICP (histidine and leucine) compared with that presented in our 1992 review paper on ideal protein for broilers (Baker et al., 1992).

The ideal ratios shown for birds between 21 and 42 d of age also differ somewhat from those presented in a recent review (Baker, 1994). Changes were made in SAA, arginine, valine, and isoleucine based upon recent unpublished research from our laboratory dealing with maintenance requirements of these amino acids relative to the maintenance requirement for lysine.

Does voluntary feed intake and weight-gain potential affect the lysine requirement? Not necessarily. In our 1991 study (Han and Baker, 1991), a slow-growing broiler strain (New Hampshire × Columbian) was compared with a fast-growing broiler strain (Hubbard × Hubbard). The fast-growing strain ate twice as much feed as the slow-growing strain during the second and third week of life, and they also gained twice as fast. The lysine requirement, however, expressed as a percent of the diet was identical. Hence, the fast-growing strain required twice as much lysine in absolute amounts, but they met this larger need by consuming more feed. The key to predicting whether different broiler strains have different amino acid requirements lies in body composition, and, in the previous comparison, protein gain as a percent of total weight gain was 16.9% for both strains. If, however, a given broiler strain has more protein and less fat in its weight gain than some other strain, then the leaner bird will have higher amino acid requirements. Thus the lysine requirement of very lean male chicks might be higher than 0.89% during the 3- to 6-week period, and all the other amino acid requirements would be higher

Table 5 Digestible Amino Acid Requirements of Broiler Chickens

Amino acid	Ideal ratio[a] (0-21 d)	Requirement (0–21 d) Male	Requirement (0–21 d) Female	Ideal ratio[a] (21–42 d)	Requirement (21–42 d) Male	Requirement (21–42 d) Female
		(% of diet)			(% of diet)	
Lysine[b]	100	1.12	1.02	100	0.89	0.84
Methionine + cystine	72	0.81	0.74	75	0.67	0.63
Methionine	36	0.405	0.37	36	0.32	0.30
Cystine	36	0.405	0.37	39	0.35	0.33
Arginine	105	1.18	1.07	108	0.96	0.91
Valine	77	0.86	0.79	80	0.71	0.67
Threonine	67	0.75	0.68	70	0.62	0.59
Tryptophan	16	0.18	0.16	17	0.15	0.14
Isoleucine	67	0.75	0.68	69	0.61	0.58
Histidine	32	0.36	0.33	32	0.28	0.27
Phenylalanine + Tyrosine	105	1.18	1.07	105	0.93	0.88
Leucine	109	1.22	1.11	109	0.97	0.92

[a] Ideal ratios are expressed as percentages of true digestible lysine.

[b] The lysine requirements of male and female broilers were taken from the data of Han and Baker (1993) for the period 0 to 21 d posthatching; the lysine requirement for the 21- to 42-d period was obtained from Han and Baker (1994).

as well. Also, male chicks have higher amino acid requirements than female chicks because they have more protein and less fat in their whole-body weight gain (Han and Baker, 1991, 1993). The advantage of using the concept of ideal protein is that ideal ratios would remain the same for birds of all genetic potential, although requirements would differ.

Heat stress is another condition that can affect voluntary feed intake. In our 1993 study (Han and Baker, 1993), heat-stressed birds (37°C) ate 22% less feed than birds housed in a comfortable environment (24°C), but there was little evidence that heat stress impacted the lysine requirement expressed as a percent of diet.

REFERENCES

ARC (Agricultural Research Council), *The Nutrient Requirements of Pigs,* Commonwealth Agricultural Bureaux, Farnham Royal, Slough, U.K., 1981.

Baker, D. H., Efficiency of amino acid utilization in the pig, in *Manipulating Pig Production IV,* Batterham, E. S., Ed., Australasian Pig Science Association, Werribee, 1993, 191.

Baker, D. H., Ideal amino acid profile for maximal protein accretion and minimal nitrogen excretion in swine and poultry, *Proc. Cornell Nutr. Conf.,* 56, 134, 1994.

Baker, D. H., Becker, D. E., Nortin, H. W., Jensen, A. H., and Harmon, B. G., Quantitative evaluation of the tryptophan, methionine, and lysine needs of adult swine for maintenance, *J. Nutr.,* 89, 441, 1966.

Baker, D. H. and Chung, T. K., Ideal protein for swine and poultry, *Biokyowa Technical Review #4,* Biokyowa Inc., Chesterfield, MO, 1992.

Baker, D. H. and Han, Y., Bioavailable level (and source) of cysteine determines protein quality of a commercial enteral product: adequacy of tryptophan but deficiency of cysteine for rats fed an enteral product prepared fresh or stored beyond shelflife, *J. Nutr.,* 123, 541, 1993.

Baker, D. H. and Han, Y., Ideal amino acid profile for broiler chicks during the first three weeks posthatching, *Poultry Sci.,* 73, 1441, 1994.

Baker, D. H., Parsons, C. M., Fernandez, S., Aoyagi, S., and Han, Y., Digestible amino acid requirements of broiler chickens based upon ideal protein considerations, in *Proc. Ark. Nutr. Conf.,* Vol., 22, 1992.

Batterham, E. S., Ileal digestibility of amino acids in feedstuffs for pigs, in *Amino Acids in Farm Animal Nutrition,* D'Mello, J. P. F., Ed., CAB International, Wallingford, U.K., 1994, 113.

Batterham, E. S. and Anderson, L. M., Utilization of ileal digestible amino acids by growing pigs: isoleucine, *Br. J. Nutr.,* 71, 531, 1994.

Batterham, E. S., Anderson, L. M., Baigent, D. R., and White, E., Utilization of ileal digestible amino acids by growing pigs: effect of dietary lysine concentration on efficiency of lysine retention, *Br. J. Nutr.,* 64, 81, 1990.

Bicker, P., Verstegen, W. A., and Campbell, R. G., Digestible lysine requirement of gilts with high genetic potential for lean gain, in relation to the level of energy intake, *J. Anim. Sci.,* 72, 1744, 1994.

Black, J. L. and Davies, G. T., Ideal protein: its variable composition, in *Manipulating Pig Production III,* Batterham, E. S., Ed., Australasian Pig Science Association, Werribee, 1991, 111.

Burgoon, K. G., Knabe, D. A., and Gregg, E. J., Digestible tryptophan requirements of starting, growing and finishing pigs, *J. Anim. Sci.,* 70, 2493, 1992.

Chappel, R. P., Effect of stocking arrangement on pig performance, in *Manipulating Pig Production IV,* Batterham, E. S., Ed., Australasian Pig Science Assoc., Werribee, 1993, 87.

Chung, T. K. and Baker, D. H., Ideal amino acid pattern for 10-kilogram pigs, *J. Anim. Sci.,* 70, 3102, 1992a.

Chung, T. K. and Baker, D. H., Efficiency of dietary methionine utilization in young pigs, *J. Nutr.,* 122, 1862, 1992b.

Cole, D. J. A., Amino acid nutrition of the pig, in *Recent Advances in Animal Production,* Harresign, W. and Lewis, D., Eds., Butterworths, London, 1978, 59.

Fernandez, S. R., Aoyagi, S., Han, Y., Parsons, C. M., and Baker, D. H., Limiting order of amino acids in corn and soybean meal for growth of the chick, *Poultry Sci.,* 73, 1887, 1994.

Fuller, M. F., Amino acid requirements for maintenance, body protein accretion and reproduction in pigs, in *Amino Acids in Farm Animal Nutrition,* D'Mello, J. P. F., Ed., CAB International, Wallingford, U.K., 1994, 155.

Fuller, M. F., Present knowledge of amino acid requirements for maintenance and production, in *Protein Metabolism and Nutrition,* EAAP Publisher, Herning, Denmark, 1991, 116.

Fuller, M. F., McWilliam, R., Wang, T. C., and Giles, L. R., The optimum dietary amino acid pattern for growing pigs. 2. Requirements for maintenance and for tissue protein accretion, *Br. J. Nutr.,* 62, 255, 1989.

Hahn, J. D. and Baker, D. H., Optimum ratio to lysine of threonine, tryptophan, and sulfur amino acids for finishing swine, *J. Anim. Sci.,* 73, 482, 1995.

Hahn, J. D., Biehl, R. R., and Baker, D. H., Ideal digestible lysine level for early and late-finishing pigs, *J. Anim. Sci.,* 73, 773, 1995.

Han, Y. and Baker, D. H., Effects of sex, heat stress, body weight and genetic strain on the lysine requirement of broiler chicks, *Poultry Sci.,* 72, 701, 1993.

Han, Y. and Baker, D. H., Lysine requirement of fast and slow growing broiler chicks, *Poultry Sci.,* 70, 2108, 1991.

Han, Y. and Baker, D. H., Lysine requirement of male and female broiler chicks during the period three to six weeks posthatching, *Poultry Sci.,* 73, 1739, 1994.

Han, Y., Chung, T. K., and Baker, D. H., Tryptophan requirement of pigs in the weight category 10 to 20 kilograms, *J. Anim. Sci.,* 71, 139, 1993.

Han, Y., Suzuki, H., Parsons, C. M., and Baker, D. H., Amino acid fortification of a low protein corn-soybean meal diet for maximal weight gain and feed efficiency of the chick, *Poultry Sci.,* 71, 1168, 1992.

Kornegay, E. T., Lindemann, M. D., and Ravindran, V., Effects of dietary lysine levels on performance and immune response of weanling pigs housed at two floor space allowances, *J. Anim. Sci.,* 71, 522, 1993.

Krick, B. J., Boyd, R. D., Roneker, K. R., Beerman, D. H., Bauman, D. E., Ross, D. A., and Meisinger, D. J., Porcine somatotropin affects the dietary lysine requirement and net lysine utilization for growing pigs, *J. Nutr.,* 123, 1913, 1993.

Meyer, R. O. and Bucklin, R. A., Effect of Increased Dietary Lysine on Performance and Carcass Characteristics of Growing-Finishing Swine Reared in a Hot, Humid Environment, University of Florida Swine Field Day, Gainesville, FL, 1994, 52.

Mitchell, H. H., *Comparative Nutrition of Man and Domestic Animals,* Academic Press, New York, 1964.

Mitchell, H. H. and Block, R. J., Some relations between the amino acid contents of proteins and their nutritive values for the rat, *J. Biol. Chem.,* 163, 599, 1946.

Moughan, P. J., Simulation of the daily partitioning of lysine in the 50 kg liveweight pig — a factorial approach to estimating amino acid requirements for growth and maintenance, *Res. Dev. Agric.,* 6, 7, 1989.

National Research Council (NRC), *Nutrient Requirements of Poultry,* 8th edition, National Academy of Science, Washington, D.C., 1984.

National Research Council (NRC), *Nutrient Requirements of Poultry,* 9th edition, National Academy of Science, Washington, D.C., 1994.

Stahly, T., Impact of immune system activation on growth and nutrient needs of pigs, *Proc. Minn. Nutr. Conf.,* 1994, 243.

Wang, T. C. and Fuller, M. F., The optimum dietary amino acid pattern for growing pigs. I. Experiments by amino acid deletion, *Br. J. Nutr.,* 62, 77, 1989.

Wang, T. C. and Fuller, M. F., The effect of the plane of nutrition on the optimum dietary amino acid pattern for growing pigs, *Anim. Prod.,* 50, 155, 1990.

CHAPTER 6

Characterization of Gastrointestinal Amino Acid and Peptide Transport Proteins and the Utilization of Peptides as Amino Acid Substrates by Cultured Cells (Myogenic and Mammary) and Mammary Tissue Explants

J. C. Matthews, Y. L. Pan, S. Wang, M. Q. McCollum, and K. E. Webb, Jr.

INTRODUCTION

Increasing demands for our dwindling natural resources are causing agriculturalists to consider the environmental consequences of our current production practices. The addition of nitrogen (N) to surface water and groundwater degrades the quality of water available for human and animal consumption. Because the N contained in the feces and urine produced by livestock can contribute to the addition of N to our water resources, it is imperative that the industries involved with animal agriculture focus their attention on ways to maximize N utilization and minimize N excretion through improved feeding practices.

When one contemplates the subject of protein, amino acid, and N utilization by ruminants, it is soon realized that an extraordinary amount of effort has been expended to research this phenomenon. The current feeding standards have been developed through these investigations of N metabolism. Even when the best possible formulations are fed to animals, N retention is still only a fraction of N intake. Obviously, there must be some obligatory loss of N, but the magnitude of loss occurring is strong evidence that there is still considerable room for improving our understanding of how best to supply the amino acid needs of ruminants.

Whereas much is known about the digestibility of feedstuffs in ruminants, relatively few of the biological mechanisms involved in the absorption and utilization of protein N as free and peptide-bound amino acids have been identified in digestive tract and peripheral tissues. Even less is known about the whole animal and cellular events that regulate these processes and, therefore, which ultimately control the extent of N retention.

1-56670-199-6/96/$0.00+$.50

CHARACTERIZATION OF FORESTOMACH ABSORPTION OF PEPTIDE-BOUND AND FREE AMINO ACIDS

Historically, the potential for absorption of significant quantities of amino acids from the ruminant forestomach has been regarded as minor. It is generally thought that there is rapid metabolism of amino acids by rumen microbes and that forestomach epithelia have little capacity for amino acid absorption. Consequently, the fate of dietary amino acids has been discussed in terms of their conversions to ammonia N, incorporation into microbial protein, absorption by the intestine, and excretion.

Expected Amino Acid and Peptide Substrate Concentrations in Forestomach Liquor

A review of the literature indicates that prefeeding levels of free amino acid concentrations in strained ruminal fluid of 0.12 to 1.5 mg/dl and postfeeding concentrations of 0.72 to 6 mg/dl would be available to drive nonmediated absorption of free amino acids in sheep (Table 1). Interestingly, much higher concentrations of peptide N have been reported in prefeeding (1.5 to 6 mg/dl) and postfeeding (10 to 27 mg/dl) ruminal fluid (Table 2). From this information it appears that more peptide-bound than free amino acids exist in ruminal liquor before and after the feeding of common diets. If the data in Tables 1 and 2 are representative of *in vivo* substrate concentrations, then the *in vivo* osmotic driving force for peptide amino acid absorption would exceed that for free amino acids across the ruminal epithelium.

Because omasal liquor amino acid and peptide N concentration values are unknown, it is not possible to predict what would be the potential relative driving forces for amino acid or peptide absorption across the omasal epithelium. However, the absorption of water by omasal tissues would presumably result in the concentration of rumen liquor solutes, thereby potentially reestablishing (or generating greater) omasal liquor-to-blood concentration gradients and solvent-drag forces that were present across ruminal epithelia, depending on the relative water and substrate absorption rates (Smith, 1984).

Forestomach Absorption of Peptide-Bound Amino Acids

The absorption of L-carnosine and L-methionine (Met) and L-methionylglycine (Met-Gly; using [^{35}S]L-methionine and [^{35}S]L-methionylglycine as representative markers) across ruminal and omasal epithelia collected from sheep was studied using parabiotic chambers that were repeatedly sampled over a 240-min incubation. Absorption of carnosine by ruminal and omasal epithelia was measured as the mean appearance of carnosine in the serosal chamber buffers as a function of time (Figure 1) and mucosal substrate concentration (Figure 2). For both tissues, the increased appearance of carnosine had both linear ($p < 0.01$, $p < 0.01$) and quadratic ($p < 0.01$, $p < 0.06$) components as time and substrate concentrations

Table 1 Amino Acid N Concentrations in Strained Ruminal Fluid

Nitrogen supplement	Species	Crude protein (%)	Amino acid N (mg/dl) Prefeeding	Postfeeding	Ref.
None	Sheep	8.3[a]	0.12	0.72 (3)[b]	Lewis, 1955
None	Sheep	13.8[c]	0.28	0.93 (3)	Lewis, 1955
Casein	Sheep	21.3[d]	0.34	1.4 (3)	Lewis, 1955
Casein	Sheep	15.0[e]	1.5	6.0 (2)	Annison, 1956
None	Sheep	4.4[f]	0.22	1.0 (1)	Leibholz, 1969
Meat meal	Sheep	16.1[g]	0.28	1.8 (1)	Leibholz, 1969
Skim milk	Sheep	16.1[h]	0.32	1.7 (1)	Leibholz, 1969
None	Sheep	16.1[i]	0.50	2.5 (1)	Leibholz, 1969
None	Sheep	18.2[j]	0.48	5.1 (1)	Leibholz, 1969
Urea	Sheep	15.6[k]	0.23	0.73 (1)	Broderick and Wallace, 1988
Casein	Sheep	15.6[l]	0.25	0.98 (1)	Broderick and Wallace, 1988

[a] Diet consisted of 0.9 kg of hay. Animal had access to feed for 2 h.

[b] The time (h) of ruminal fluid collection after feeding. The time listed corresponds to the only (Lewis, 1955; Leibholz, 1969) or maximal (Annison, 1956; Broderick and Wallace, 1988) amino acid N concentrations reported.

[c] Diet consisted of 0.4 kg of hay and 0.5 kg of dried grass. Animal had access to feed for 2 h.

[d] Diet consisted of 0.7 kg of hay and 0.15 kg of casein. Animal had access to feed for 2 h.

[e] Diet consisted of 0.60 kg of alkali-washed straw (total N .2 to .3%), 0.15 kg of starch, 0.10 kg of sucrose, 0.20 kg of molasses, 0.50 kg of water, and 0.10 kg of casein.

[f] Diet consisted of 1 kg of wheat chaff consumed within 1 h.

[g] Diet consisted of 0.72 kg of wheat chaff and 0.28 kg of meat meal consumed within 1 h.

[h] Diet consisted of 0.64 kg of wheat chaff and 0.36 kg of dried skim milk consumed within 1 h.

[i] Diet consisted of 0.6 kg of alfalfa hay and 0.4 kg of barley consumed within 1 h.

[j] Diet consisted of 1 kg of alfalfa hay consumed within 1 h.

[k] Diet consisted of 0.67 kg of ryegrass and 0.33 kg of concentrate divided over two feedings (32% of crude protein provided by urea, dry matter basis).

[l] Diet consisted of 0.67 kg of ryegrass and 0.33 kg of concentrate divided over two feedings (34% of crude protein provided by casein, dry matter basis).

increased, respectively. From 15 through 240 min, the rate of carnosine absorption by omasal tissues was on average 7.4 times greater ($p < 0.01$, time × tissue interaction) than that for ruminal tissues (Figure 1). Similarly, approximately 7.2 times more carnosine ($p < 0.01$, concentration × tissue interaction) appeared in omasal tissue serosal chambers than in ruminal (Figure 2). After 240 min, histidine was not detectable in serosal buffers above background levels, which suggests that carnosine passed through both tissues without hydrolysis. The appearance of carnosine in the serosal buffers indicates that both tissues possess the ability to absorb carnosine. Clearly, the omasal tissue demonstrated a greater ability to absorb carnosine than did ruminal epithelia. Despite the relatively high (physiologically) concentrations used (6 to 96 m*M*), the amount of carnosine that passed through ruminal and omasal tissues increased in proportion as the initial

Table 2 Peptide-Bound Amino Acid N Concentrations in Strained Ruminal Fluid

Nitrogen supplement	Species	Crude protein (%)	Peptide N (mg/dl) Prefeeding	Peptide N (mg/dl) Postfeeding	Ref.
Casein	Sheep	15.0[a]	1.5	10 (2)[b]	Annison, 1956
Casein	Sheep	15.6[c]	1.5	27 (1)	Broderick and Wallace, 1988
SBM[d]-FM[e]	Lactating cows	17.8[f]	5.4[g]	15 (2)	Chen et al., 1987
SBM-ESB[h]	Lactating cows	17.8[f]	6.0[g]	16 (2)	Chen et al., 1987
SBM	Lactating cows	17.8[f]	5.4[g]	22 (2)	Chen et al., 1987
None	Dry cows	17.0[i]	—	18[j](0-6)[k]	Broderick et al., 1990
DDGS[l]-CGM[m]	Dry cows	18.1[i]	—	22[j](0-6)[k]	Broderick et al., 1990
ESBM[n]	Dry cows	18.1[i]	—	23[j](0-6)[k]	Broderick et al., 1990
SBM	Dry cows	18.9[i]	—	24[j](0-6)[k]	Broderick et al., 1990

[a] Diet consisted of 0.60 kg of alkali-washed straw (0.2 to 0.3% total N), 0.15 kg of starch, 0.10 kg of sucrose, 0.20 kg of molasses, 0.50 kg of water, and 0.10 kg of casein.

[b] The time (h) of ruminal fluid collection after feeding. The time listed corresponds to the only (Chen et al., 1987; Broderick et al., 1990) or maximal (Annison, 1956; Broderick and Wallace, 1988) peptide-bound amino acid N concentrations reported by the researchers.

[c] Diet (as fed) consisted of 0.67 kg of ryegrass and 0.33 kg of concentrate (34% casein, dry matter basis) divided over two feedings.

[d] Soybean meal.

[e] Fishmeal.

[f] Diet dry matter (DM) contained 41.8% corn silage, 8.8% haylage, barley (at least 20%), and mineral mix. The portion of barley added depended on the CP percentage supplied by the protein supplements: 17.3% barley with 18% SBM; 17.9% barley with 9.5 and 12%, respectively, SBM-ESB; 18.1% barley with 7.6 and 6.3%, respectively, SBM-FM.

[g] Samples taken 14 h after the daily feeding.

[h] Extruded soybean meal.

[i] The TMR DM contained 58% alfalfa silage, 34% high moisture corn, and 1% mineral mix. The portion of corn grain added depended on the CP percentage supplied by the protein supplements: 7% (no supplement), 1.6% (4.2% DDGS plus 1.5% CGM), 2.6% (4.6% ESBM), and 0% (7.2% SBM).

[j] Values were calculated from reported millimoles peptide-bound amino acids per liter and presented assumptions that one peptide contained three amino acid residues.

[k] Values were reported as the mean of five sampling periods (0, 1, 2, 4, and 6 h postfeeding).

[l] Distillers dried grains with solubles.

[m] Corn gluten meal.

[n] Expeller soybean meal.

amount of carnosine in the mucosal buffers was increased. This observation indicates that carrier-mediated transport of carnosine, if present, was not detractable at these concentrations.

In order to determine and to compare the total accumulation and translocation capacities of ruminal and omasal epithelia, the mean quantities of Met and Met-Gly that appeared in serosal buffers after 4 h were added to the mean quantities that had accumulated in tissues. The resulting data (Figure 3) represent the total

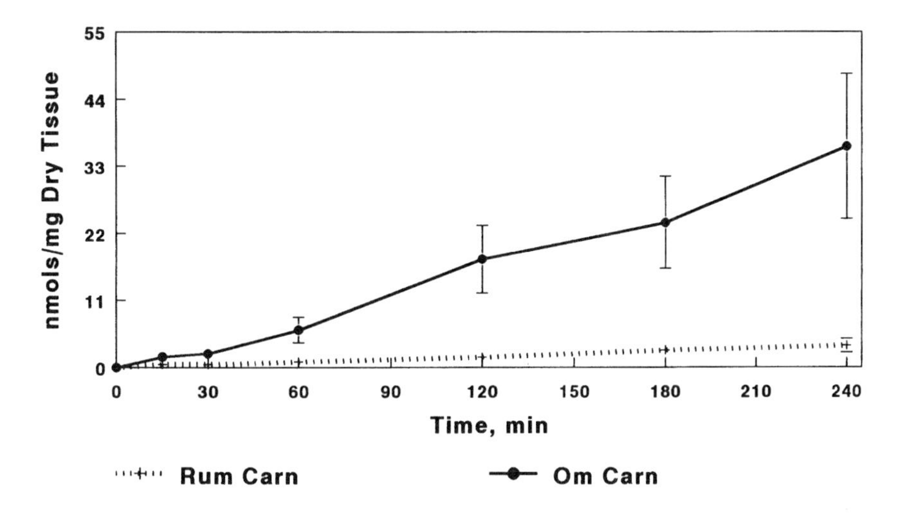

Figure 1 Effect of time on carnosine appearance in serosal buffers. Data were transformed using the natural log function before analysis. Nontransformed means and SE are presented in the figure and are the means of four animals across concentrations and represent the quantity of carnosine (Carn) that appeared in the serosal chamber of the parabiotic units. Rum = ruminal, Om = omasal. Linear $(p < 0.01)$ and quadratic $(p < 0.01)$ time effect. Tissues differed $(p < 0.01)$. Time × tissue interaction $(p < 0.01)$.

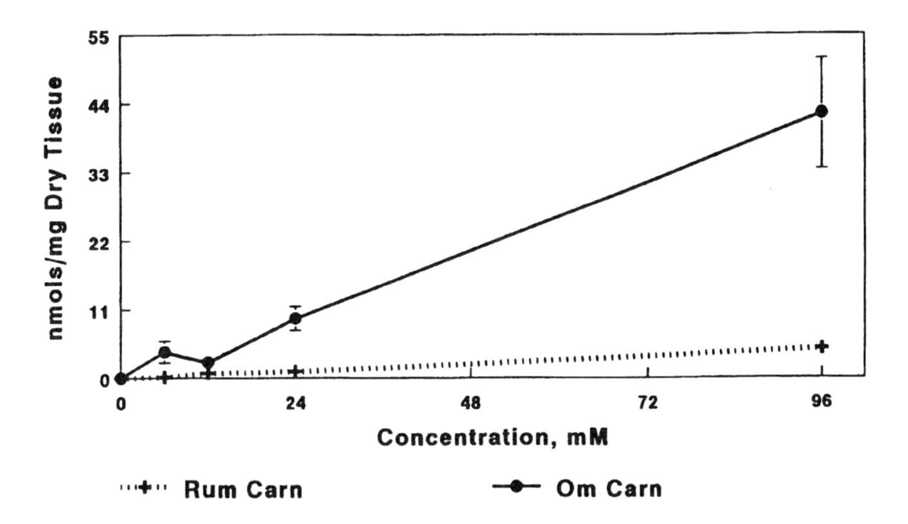

Figure 2 Effect of increasing mucosal buffer substrate concentration on carnosine (Carn) appearance in serosal buffers. Data were transformed using the natural log function before analysis. Nontransformed means and SE are presented in the figure and are the means of four animals across time and represent the quantity of carnosine that appeared in the serosal chamber of the parabiotic units. Rum = ruminal, Om = omasal. Linear $(p < 0.01)$ and quadratic $(p < 0.06)$ concentration effect. Tissues differed $(p < 0.01)$. Concentration × tissue interaction $(p < 0.01)$.

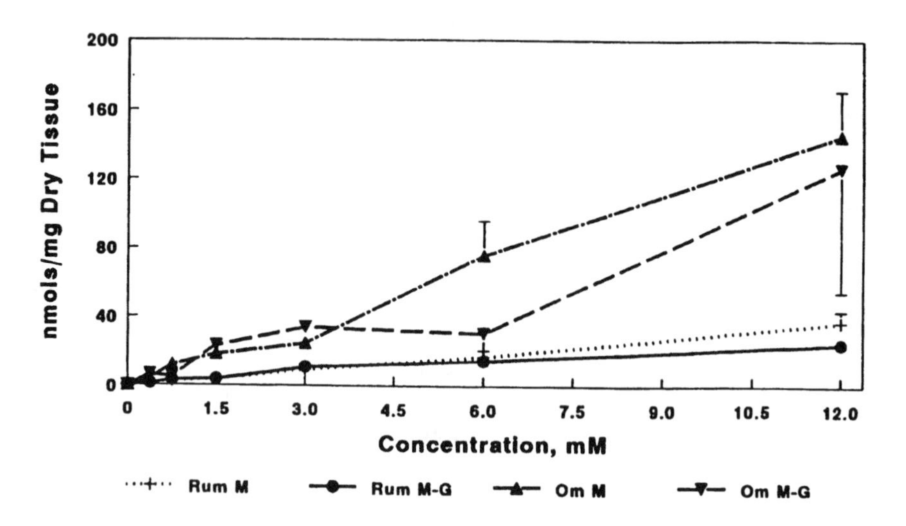

Figure 3 Total absorption of methionine (Met) and methionylglycine (Met-Gly). Data were transformed using the natural log function before analysis. Nontransformed means and SE are presented in the figure and are the means of seven animals for each Met and Met-Gly and represent the quantity of Met and Met-Gly that accumulated in tissues and that appeared in the serosal buffers of the parabiotic units after 240 min incubation, assuming all ^{35}S was in the form of ^{35}S-Met and ^{35}S-Met-Gly acting as representative markers for Met and Met-Gly, respectively. Rum = ruminal, Om = omasal, M = methionine, M-G = methionylglycine. Linear $(p < 0.01)$ and quadratic $(p < 0.01)$ concentration effect. Tissues differed $(p < 0.01)$. Tissue × concentration interaction $(p < 0.01)$.

absorption of Met and Met-Gly. For both tissues, the increase in mean total absorption of both Met and Met-Gly had linear $(p < 0.01)$ and quadratic $(p < 0.01)$ components as their initial concentrations in mucosal chambers were increased. Between tissues, the total absorption of both Met and Met-Gly by omasal tissues was approximately 3.9 times more $(p < 0.01)$ than by ruminal tissues. For omasal tissue, the vast majority of absorbed Met (91.4%) and Met-Gly (91.3%) passed through the epithelium into the serosal buffers. In contrast, approximately half of the Met (54.6%) and Met-Gly (46.1%) absorbed was measured in the serosal buffers and half in the epithelium of ruminal tissues. Quantitatively, both ruminal and omasal epithelia tended to absorb more (23%; $p < 0.12$) Met than Met-Gly. The data indicate that omasal tissue has a greater overall capacity for Met and Met-Gly absorption than does ruminal tissue, that both substrates pass through omasal tissue in greater quantities than through ruminal, that both tissues displayed a slightly greater tendency to absorb Met than Met-Gly, and that the flux of Met and Met-Gly across tissues was not saturable (0.375 to 12.0 m*M*).

In summary, the results from these studies suggest that the absorption of peptide-bound and free amino acid N is independent of mediated transport and is greater by omasal than by ruminal epithelial tissues. Uptake was studied at postfeeding substrate concentrations reported to exist in the ruminal liquor of sheep and cows (Tables 1 and 2). If the previous characterization of *in vitro* forestomach epithelial absorption capacity accurately represents *in vivo* absorption processes, then proportionately more amino acids would be expected to be absorbed across the forestomach in the form of peptides than as free amino acids because more peptide-bound than free amino acids are expected to be present in forestomach liquor (Tables 1 and 2).

Expression of mRNA for a Peptide Transport Protein(s) from Omasal Epithelium

From a teleological perspective, the lack of demonstrated mediated absorption of peptide-bound amino acids across forestomach tissues was surprising. Ions reported to be involved in active, secondary transport of small peptides (H^+, Na^+) in eukaryotes (Webb and Matthews, 1994) are constituents of forestomach liquor. Omasal liquor is essentially acidic (Prins et al., 1972), and ruminal liquor can develop pH levels of 5.5 or less (Whitelaw et al., 1970). Proton gradients of this magnitude have been used to demonstrate the presence of carrier-mediated dipeptide transport by intestinal and renal epithelial brush border membrane vesicles (BBMV) (Ganapathy and Leibach 1983; Daniel et al., 1991) and in cultured colon cells (Saito and Inui, 1993). Additionally, the Na^+/H^+ exchanger and Na^+/K^+ ATPase proteins, considered to be essential in maintaining proton gradients in other epithelial cells, are reported to function in both ruminal and omasal epithelia (Martens and Gabel, 1988). Perhaps the multicell layered morphological structure of the forestomach epithelium prevented the detection and characterization of peptide transport proteins by whole-tissue flux measurements (Figures 1, 2, and 3).

In order to test this hypothesis and to potentially express proteins that may be responsible for the mediated absorption of small peptides, size-fractionated poly(A)$^+$ RNA (RNA) isolated from omasal epithelial tissue of sheep (average body weight (BW) = 67.5 kg) were injected into defolliculated *Xenopus laevis* oocytes. RNA were extracted from omasal rather than ruminal epithelium because of the greater *in vitro* ability of omasal epithelium to absorb carnosine and Met-Gly (Figures 1, 2, and 4). The ability of oocytes injected with RNA or water to absorb [^{14}C]glycyl-L-sarcosine (Gly-Sar) from media (usually pH 5.5) was compared. After 4 d ($p < 0.02$) of culture, specific RNA fractions induced an increased ($p < 0.02$) rate of Gly-Sar absorption, as compared to water-injected oocytes. The dependency of Gly-Sar uptake on the presence of a pH gradient was evaluated at pH 5.0, 5.5, 6.0, 6.5, and 7.5. Inducible Gly-Sar uptake increased ($p < 0.001$) as the concentration of protons was increased, whereas endogenous uptake of Gly-Sar decreased ($p < 0.001$). At pH 5.5, induced Gly-Sar uptake was saturable (K_t = 0.4 m*M*), but endogenous uptake was not. The specificity of Gly-Sar absorption was studied by the coincubation of 0.1 m*M* Gly-Sar with 5 m*M* levels of competing substrates (pH 5.5). Induced uptake was inhibited ($p < 0.05$) 44% by carnosine, 94% by methionylglycine, 91% by glycylleucine, but not by glycine. Incubation of RNA with DNA oligomers that were complementary to the cloned rabbit intestinal transporter completely inhibited ($p < 0.05$) induced Gly-Sar uptake. These results indicate

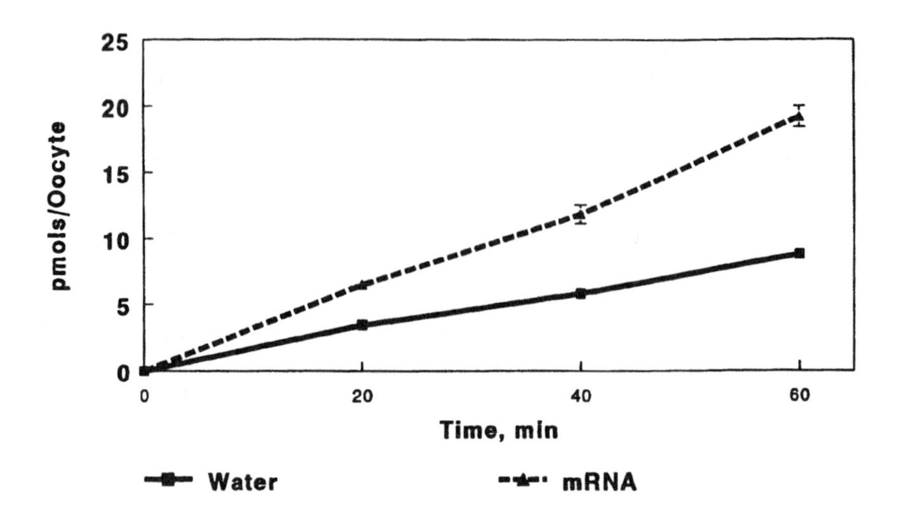

Figure 4 Time course for lysine absorption by defolliculated *Xenopus* oocytes injected with water or size-fractionated sheep omasal epithelium RNA (RNA). Oocytes were assayed in a media containing 0.05 m*M* L-lysine. Each time period measurement represents group (n = 1) absorption means ± SE of oocytes injected with water (n = 4 to 5) or RNA (n = 4 to 5). Linear time effect $(p < 0.001)$. Time × treatment effect $(p < 0.001)$.

that sheep omasal epithelial cells express mRNA that encode for proteins that are capable of H^+-dependent dipeptide transport activity.

Forestomach Absorption of Free Amino Acids

The ability of the ruminant forestomach to absorb amino acids has not been well characterized. In parabiotic chambers, the absorption of histidine across rumen epithelial sheets was not saturable from 0.66 to 20 m*M* (Leibholz, 1971). However, histidine transfer across this tissue may have been at least partially mediated because methionine (50%), arginine (50%), and glycine (40%) inhibited histidine passage when coincubated at equal concentrations (0.66 m*M*). In hindsight, that arginine inhibited the absorption of histidine indicates that these substrates may have been competing for recognition by the Na^+-independent (y^+) or Na^+-dependent (Y^+) cationic transporter. Alternatively, and/or concomitantly, the demonstration that arginine, methionine, and glycine inhibited histidine absorption parallels the competitive profile displayed by the Na^+-independent ($b^{o,+}$) or Na^+-dependent ($B^{o,+}$) cationic and neutral transporter system (Van Winkle et al., 1988). Additional evidence that ruminal epithelia of sheep may possess proteins capable of facilitative transport of cationic amino acids is supported by the observation that the flux of lysine and arginine across ruminal tissue sheets was saturable from 0.3 to 30 m*M* (Fejes et al., 1991).

In order to study the potential for lysine absorption across ruminal and omasal epithelia, tissue sheets were collected from six wethers and mounted in parabiotic chambers whose Na^+-containing buffers were repeatedly sampled throughout a

60-min incubation. The potential for L-lysine absorption was investigated because of its importance to the nutrition of ruminants and because it is reported to be present in relatively high concentrations in ruminal liquor (Wright and Moir, 1967; Leibholz, 1969). The appearance of lysine (using ^{3}H-L-lysine as a representative marker) in serosal chamber buffers increased linearly *($p < 0.0001$)* with time for both tissues. More *($p < 0.0001$)* lysine appeared in omasal serosal chamber buffers than in ruminal serosal chamber buffers. The tissues responded differently *($p < 0.0001$)* to increases in the mucosal chamber buffer concentrations (0.094, 0.118, 0.375, 0.75, 1.5, and 3.0 m*M)* of lysine. The appearance of lysine in the serosal chamber buffers of ruminal tissue increased proportionally as the concentration of lysine increased in the mucosal chamber buffers. However, the appearance of lysine in the serosal chamber buffers of omasal tissue increased proportionally up to a substrate concentration of 1.5 m*M* and then plateaued. After 60 min of incubation, the accumulation of lysine in the ruminal epithelium was greater *($p < 0.0001$)* than the accumulation in the omasal epithelium. For both tissues, the accumulation of lysine increased linearly *($p < 0.011$)* as mucosal buffer substrate concentrations were increased. Total absorption (tissue accumulation plus serosal buffer appearance) was greater *($p < 0.0001$)* for omasal than for ruminal epithelia. For ruminal tissue, total absorption increased linearly; however, for omasal tissue, absorption increased linearly up to 1.5 m*M* *($p < 0.0009$)* and then plateaued. The linear increase in total absorption of lysine throughout the range of substrate concentrations by the ruminal tissue suggests that lysine absorption was nonmediated. In contrast, the fact that lysine uptake by the omasal tissue was saturable suggests that the omasal epithelial cells possess at least one transport system capable of mediated absorption of lysine.

The potential for mediated absorption of lysine in omasal forestomach epithelial tissue was further evaluated by characterizing the functional expression of omasal mRNA in *X. laevis* oocytes. Compared to water-injected oocytes, oocytes injected with omasal epithelial RNA displayed a greater ability to absorb (picomoles per oocyte·35/min^{-1}) lysine from both Na^+-free and Na^+-containing buffers. Within RNA fractions, the amount of induced lysine absorbed did not differ between buffers. Therefore, it was concluded that induced uptake was by Na^+-independent processes. In spite of this observation, lysine uptake was further characterized in Na^+-containing media because forestomach liquor contains Na^+. The linear *($p < 0.001$)* rate of lysine (0.05 m*M)* uptake by RNA-injected oocytes was approximately two times greater *($p < 0.001$)* than that demonstrated by oocytes injected with water (Figure 4). In order to determine whether this RNA-induced uptake was the result of an increase in mediated absorption ability, the uptake of 0.05 m*M* lysine by water- and RNA-injected oocytes was evaluated in the presence of media that also contained 5 m*M* leucine, 5 m*M* glutamate, or 0.2 m*M* cysteine (Figure 5). The absorption of lysine (picomoles · oocyte^{-1} · 35 min^{-1}). by oocytes injected with water or RNA was, essentially, completely inhibited *($p < 0.05$)* by leucine and not affected by glutamate, thus displaying characteristic $b^{o,+}$-like transport activity. However, that cysteine stimulated *($p < 0.05$)* the uptake of lysine in oocytes injected with omasal RNA suggests that the omasal epithelium may possess an unusual isoform of this transport protein.

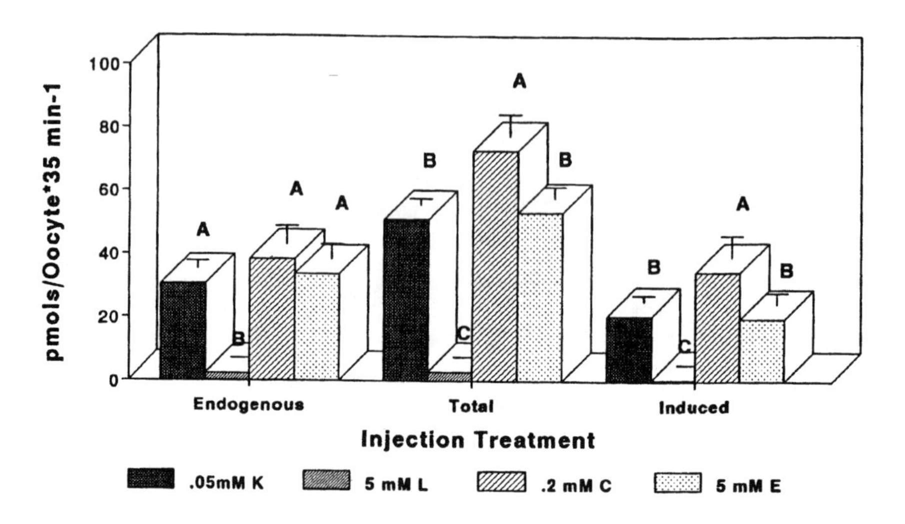

Figure 5 Competitive lysine absorption by *Xenopus* oocytes injected with water (endogenous) or size-fractionated sheep omasal epithelium RNA (Total). Oocytes were assayed in a media containing 0.05 m*M* L-lysine (K), L-leucine (L), L-cysteine, (C), or L-glutamate (E) at indicated concentrations. Individual histogram values represent across group (n = 2) absorption means ± SE of oocytes injected with water (n = 5 to 7) or RNA (n = 11 to 19). At 0.05 m*M*, uptake of lysine by oocytes injected with RNA (n = 15) was greater ($p < 0.001$) than for water-injected oocytes (n = 5). Within treatments, bars lacking a common letter differ ($p < 0.05$).

From these results, it is preliminarily concluded that the omasal epithelial tissue possesses mRNA that encode for proteins that are capable of mediating the Na^+-independent absorption of lysine.

TISSUE UTILIZATION OF PEPTIDE-BOUND AMINO ACIDS

Support of Protein Accretion in Myogenic and Mammary Epithelial Cells

L-Methionine-containing peptides were evaluated for their ability to be a source of methionine to support protein accretion in C_2C_{12}, mouse muscle myogenic, and MAC-T, bovine mammary epithelial, cells. The cell cultures were incubated for 72 h at 37°C in a humidified environment of 90% air: 10% CO_2 for C_2C_{12} cells or 95% air: 5% CO_2 for MAC-T cells. The basal medium contained methionine-free Dulbecco's Modified Eagle's Medium (DMEM) and 6% desalted fetal bovine serum. Treatments included the basal medium, the basal medium supplemented with an L-methionine-containing peptide, or the basal medium supplemented with free L-methionine (methionine).

Expressed as a percentage of the response to free methionine (Figure 6), growth of C_2C_{12} cells differed due to the type of dipeptide (11 to 108%). Methionylmethionine, methionylvaline, and leucylmethionine were utilized as

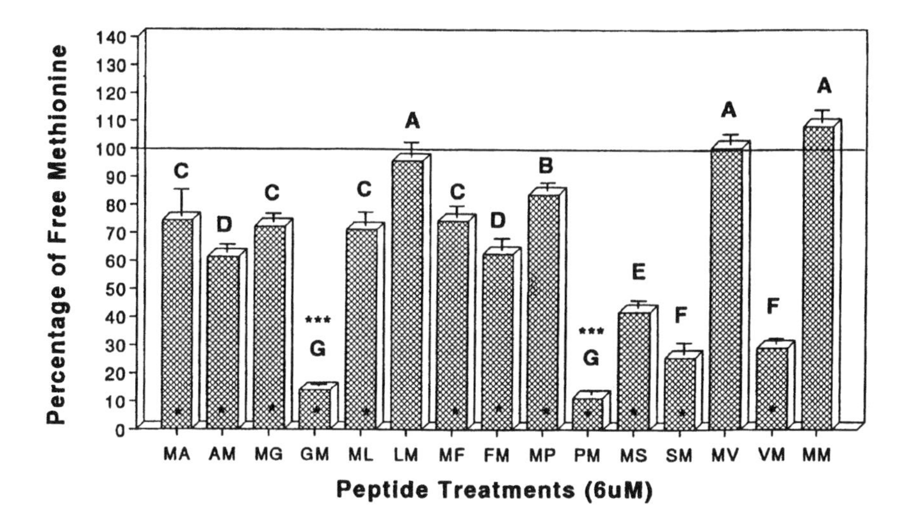

Figure 6 Effect of L-methionine-containing dipeptides on protein accretion in C_2C_{12} cells. Values represent means ± SE (n = 9) which are expressed as the percentage of the response promoted by L-methionine. The horizontal line indicates the response of L-methionine. Bars lacking a common letter differ ($p < 0.05$). Means (*) differ ($p < 0.05$) from the L-methionine treatment. Means (***) did not differ from the initial level. All peptides were constituted with L-α-amino acids. One-letter abbreviations for amino acids: A, alanine; F, phenylalanine; G, glycine; L, leucine; M, methionine; P, proline; S, serine; V, valine.

efficiently as free methionine. Prolylmethionine and glycylmethionine were poorly utilized by C_2C_{12} cells. The remaining peptides were utilized at a rate of about 62 to 86% of the rate of free methionine, except for methionylserine, serylmethionine, and valylmethionine, which were utilized at 26 to 43% of the rate of free methionine.

The results presented in Figure 7 indicate that all the methionine-containing dipeptides were able to support protein accretion of cultured MAC-T cells with the response ranging from 35 to 122% of the free methionine growth response. Methionylvaline, methionylleucine, methionylmethionine, and leucylmethionine supported greater ($p < 0.05$) protein accretion than did free methionine. Phenylalanylmethionine, methionylphenylalanine, alanylmethionine, methionylalanine, methionylserine, and methionylglycine were utilized as effectively as free methionine. Glycylmethionine, prolylmethionine, and serylmethionine were the least ($p < 0.05$) utilized peptides in MAC-T cells.

These studies indicate that C_2C_{12} myogenic and MAC-T mammary epithelial cells are able to utilize small methionine-containing peptides as sources of methionine to support protein accretion. Further examination of the data collected in this study indicated that dipeptides with methionine at the N-terminus appear to be preferred substrates and that hydrophobicity or some related characteristic of dipeptides is also moderately related to the efficiency with which these dipeptides are used to promote protein accretion.

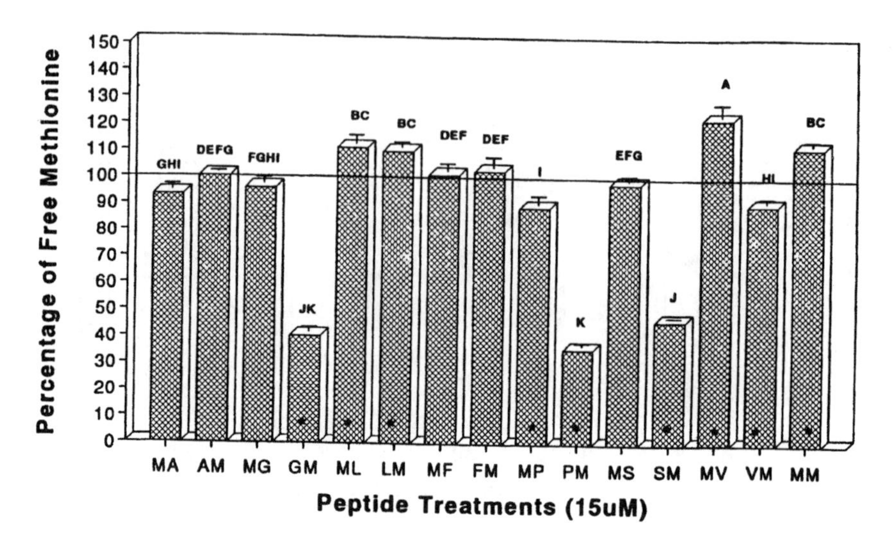

Figure 7 Effect of L-methionine-containing dipeptides on protein accretion in MAC-T cells. Values represent means ± SE (n = 4) which are expressed as the percentage of the response promoted by L-methionine. The horizontal line indicates the response of L-methionine. Bars lacking a common letter differ ($p < 0.05$). Means (*) differ ($p < 0.05$) from the L-methionine treatment. All peptides were constituted with L-α-amino acids. One-letter abbreviations for amino acids: A, alanine; F, phenylalanine; G, glycine; L, leucine; M, methionine; P, proline; S, serine; V, valine.

To determine whether serumal factors may be involved in regulating the use of peptides as amino acid substrates for protein accretion and cell proliferation, C_2C_{12} and MAC-T cells were incubated in a basal medium containing methionine-free DMEM supplemented with 0.4% bovine serum lipids, 1% chemically defined lipid concentrate, bovine insulin (1 µg/ml), 3% low protein serum replacement (LPSR-1), or 6% desalted animal serum (human, horse, chicken, pig, or rabbit). Treatment media included basal media supplemented with no L-methionine, L-methionine, or one of the L-methionine-containing peptides. Results from this study demonstrated that adult animal sera from humans, horses, chickens, pigs, and rabbits can promote the utilization of most methionine-containing peptides. There were some differences in the growth responses of the MAC-T and C_2C_{12} cells on the same peptides in the presence of different animal sera, suggesting that some species differences may exist. By themselves, neither insulin nor serum lipids were able to facilitate peptide utilization.

Support of Protein Accretion in Ovine Myogenic Satellite Cells

Primary cultures of ovine myogenic satellite cells were evaluated for their ability to use peptide-bound methionine as a source of methionine for protein accretion and cell proliferation after isolation from skeletal muscle and culture at

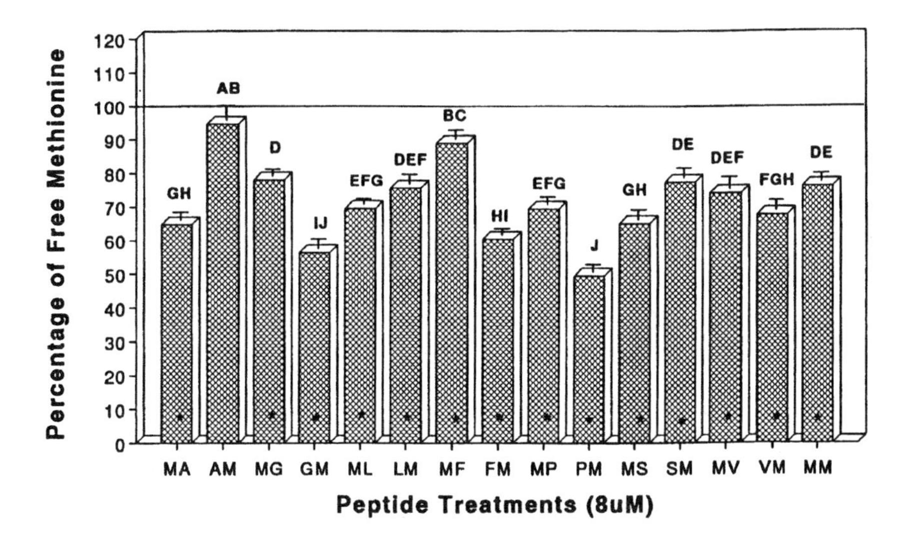

Figure 8 Effect of L-methionine-containing dipeptides on protein accretion in primary cultures of ovine skeletal muscle. Values represent means ± SE (n = 16) which are expressed as the percentage of the response promoted by L-methionine. The horizontal line indicates the response of L-methionine. Bars lacking a common letter differ *($p < 0.05$)*. Means (*) differ *($p < 0.05$)* from the L-methionine treatment. All peptides were constituted with L-α-amino acids. One-letter abbreviations for amino acids: A, alanine; F, phenylalanine; G, glycine; L, leucine; M, methionine; P, proline; S, serine; V, valine.

37°C for 5 to 6 d in a humidified environment of 90% air: 10% CO_2. The basal medium contained methionine-free DMEM supplemented with 6% desalted fetal bovine serum. Treatment media included the basal medium supplemented with no L-methionine, L-methionine, or one of several L-methionine-containing peptides. The cultured myogenic cells were able to utilize all the methionine-containing dipeptides tested for protein accretion with responses ranging from about 49 to 95% of the response for free methionine (Figure 8). This is consistent with the concept that peptide-bound amino acids can serve as amino acid sources for protein accretion in sheep skeletal muscle.

In some cases, the molecular arrangement of the dipeptides with the same amino acid composition influenced the relative ability of the dipeptides to serve as methionine sources. For all peptides studied, however, only alanylmethionine was utilized to support protein accretion as well as free methionine.

Support of Protein Secretion by Lactating Mammary Gland

Mammary tissue explants from lactating (10 to 11 d) CD-1 mice were used to study the ability of methionine- and lysine-containing peptides to substitute for free methionine and lysine, respectively, for the synthesis of secreted proteins.

Explants were incubated in medium containing [^{3}H]L-leucine and either L-methionine or an L-methionine-containing peptide or L-lysine or an L-lysine-containing peptide. The ability of methionine or lysine substrates to promote incorporation of [^{3}H]leucine into secreted proteins was quantified.

Mammary tissue explants utilized methionine from all peptides studied (Figure 9). However, differences ($p < 0.0001$) were observed among these peptides in their ability to promote incorporation of [^{3}H]leucine into secreted proteins. Eleven of these peptides promoted 15 to 76% greater ($p < 0.05$) incorporation of [^{3}H]leucine into secreted proteins than did free methionine, with dipeptides containing either valine or serine showing the highest [^{3}H]leucine incorporation. The remaining six peptides were not different ($p > 0.05$) from free methionine in promoting [^{3}H]leucine incorporation. Tissue explants incubated in methionine-free medium (negative control) also showed some incorporation of [^{3}H]leucine into secreted proteins. However, the incorporation rate was lower ($p < 0.05$) than that promoted by methionine and all the peptides.

Mammary explants were able to utilize lysine from all of the peptides studied for the synthesis of secreted proteins (Figure 10). These lysyl peptides generally were similar to free lysine in promoting [^{3}H]leucine incorporation into secreted proteins. The incorporation of [^{3}H]leucine promoted by glycylhistidyllysine was about 17% greater ($p < 0.05$) than that promoted by lysine. The other peptides were not different ($p > 0.05$) from lysine in promoting [^{3}H]leucine incorporation, which ranged from 91 to 108% of the incorporation promoted by lysine. Within each of the three peptide pairs, aspartyllysine and lysylaspartate, glycyllysine and lysylglycine, valyllysine and lysylvaline, the location of the lysyl residue at either the N- or C-terminal position did not affect ($p > 0.05$) [^{3}H]leucine incorporation (Figure 10).

These studies demonstrate that protein secretion by mammary explants from the lactating mouse can be supported by lysine and methionine derived from small peptides. Compared to free methionine, many of the methionyl peptides studied promoted a greater secretion of proteins. In contrast to methionine, the form of lysine supplementation did not greatly affect the amount of protein secretion by mouse mammary tissue.

Support of Protein Secretion in Cultured Mammary Epithelial Cells

Cultured MAC-T bovine mammary epithelial cells were used to study the ability of methionine-containing peptides to substitute for free methionine in the synthesis of secreted proteins. The cells were allowed to grow for 3 or 8 d and then were incubated in medium containing [^{3}H]L-leucine and L-methionine or one of the L-methionyl peptides for 3, 6, or 24 h. The ability of methionine substrates to promote incorporation of [^{3}H]leucine into secreted and cell proteins was quantified.

All of the methionyl peptides were utilized by the MAC-T cells as sources of methionine for the synthesis of both secreted and cellular proteins. The pattern of promotion of the incorporation of [^{3}H]leucine into cellular proteins and secreted proteins by the various peptide substrates was similar. It also was observed that,

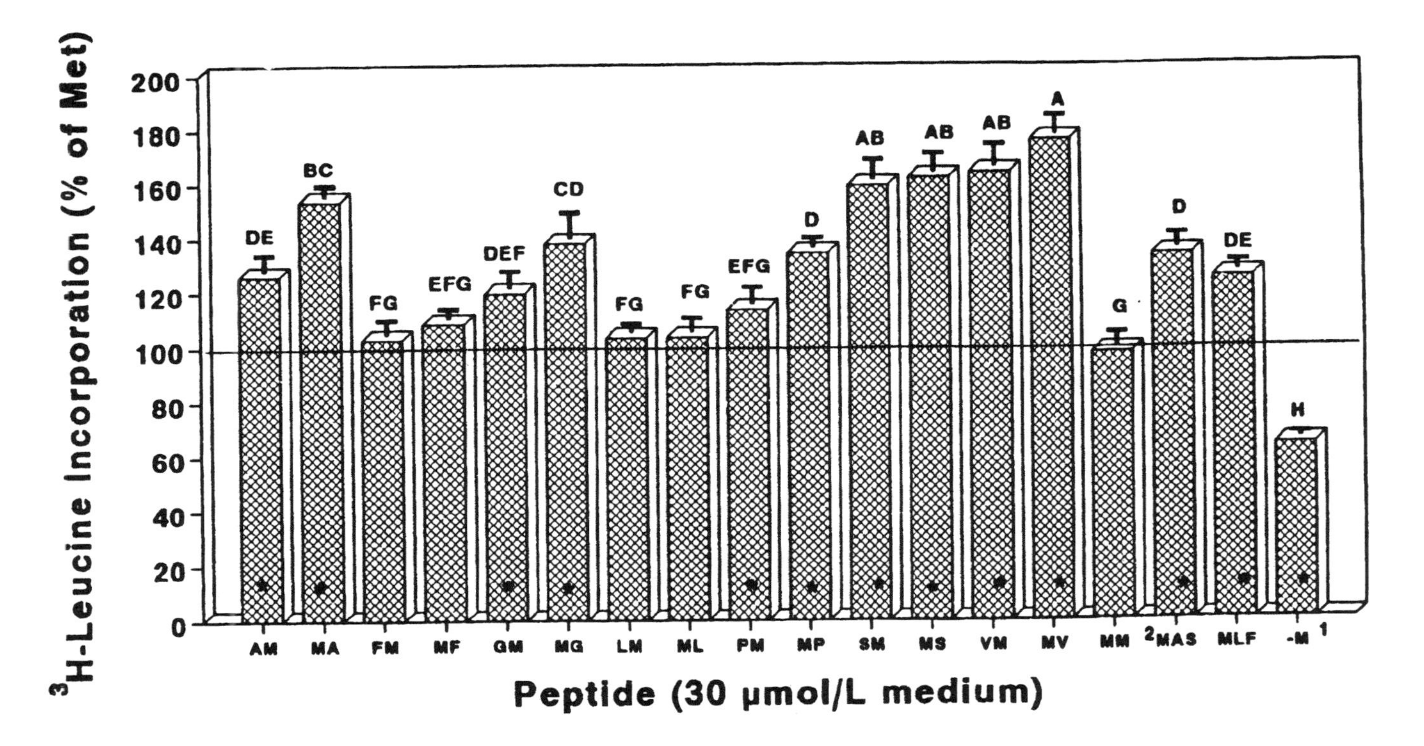

Figure 9 Incorporation of [^{3}H]L-leucine into secreted proteins promoted by L-methionine-containing di- and tripeptides by mammary explants from lactating mice. Values represent means ± SE (n = 12) which are expressed as the percentage of [^{3}H]leucine incorporation promoted by L-methionine. The horizontal line indicates the amount of [^{3}H]leucine incorporation promoted by L-methionine. Bars lacking a common letter differ ($p < 0.05$). Means (*) differ from L-methionine treatment ($p < 0.05$). [1]M = no methionine substrate in the incubation medium (negative control). [2]MM = 15 µmol/l medium. All peptides were constituted with L-α-amino acids. One-letter abbreviations for amino acids: A, alanine; F, phenylalanine; G, glycine; L, leucine; M, methionine; P, proline; S, serine; V, valine.

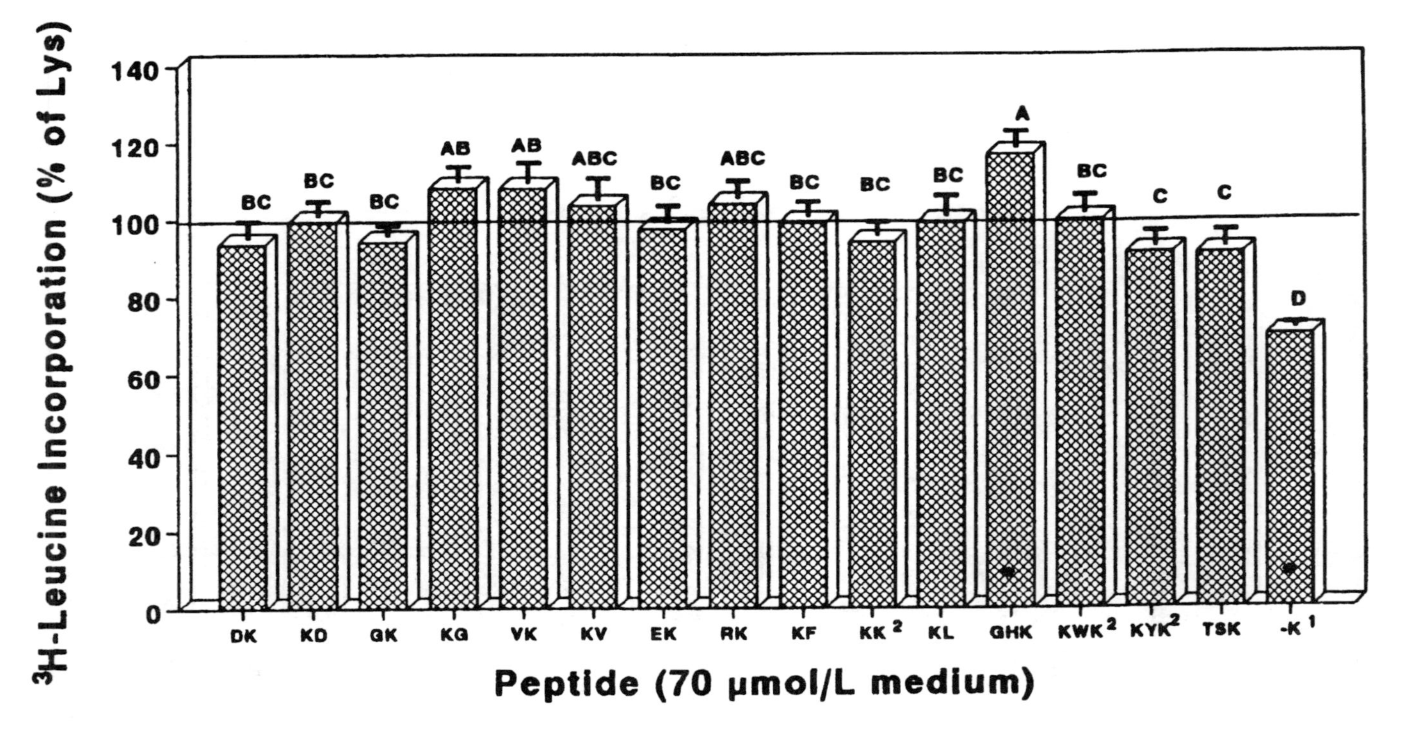

Figure 10 Incorporation of [³H]L-leucine into secreted proteins promoted by L-lysine-containing di- and tripeptides by mammary explants from lactating mice. Values represent means ± SE (n = 12, two repeated experiments) which are expressed as the percentage of [³H]leucine incorporation promoted by L-lysine. The horizontal line indicates the amount of [³H]leucine incorporation promoted by L-lysine. Bars lacking a common letter differ ($p < 0.05$). Means (*) differ ($p < 0.05$) from L-lysine treatment. [1]K, no lysine substrate in the incubation medium (negative control). [2]KK, KWK, KYK, 35 µmol/l medium. All peptides were constituted with L-α-amino acids. One-letter abbreviations for amino acids: D, aspartic acid; E, glutamic acid; F, phenylalanine; G, glycine; H, histidine; K, lysine; L, leucine; R, arginine; S, serine; T, threonine; V, valine; W, tryptophan; Y, tyrosine.

when incubation time increased from 3 to 24 h, the ratio of [^{3}H]leucine-labeled secreted proteins to cell proteins decreased from 1:11 to 1:5, indicating that a large portion of the cell proteins were proteins destined for secretion. Most of the methionine-containing peptides were as efficient as free methionine in promoting protein synthesis.

These results are consistent with those previously discussed for lactating mouse mammary tissue. Together, these studies indicate that a wide range of peptide-bound methionine and lysine substrates can support the synthesis of milk proteins by mammalian epithelial cells, at least as well as free methionine and lysine.

IMPLICATIONS

Evidence continues to accumulate that indicates the ruminant gastrointestinal tract absorbs small peptides that peripheral tissues can utilize. Research from this laboratory has demonstrated that the ruminant forestomach appears to possess the mechanisms and the ability to absorb several different dipeptides without hydrolysis and that many methionine- and lysine-containing peptides are utilized by ruminant myogenic and mammary epithelial cells and mouse mammary tissue for the synthesis of cellular and secreted proteins.

Therefore, a more thorough understanding of the mechanisms involved with peptide absorption and utilization, and the magnitude of these processes, may reveal new opportunities for increasing N absorption and utilization in ruminants by designing new nutritional management regimens. One consequence of increased N retention will be a decrease in the amount of N from livestock waste to pollute our natural ecosystems.

REFERENCES

Annison, E. F., Nitrogen metabolism in the sheep. Protein digestion in the rumen, *Biochemistry,* 64, 705, 1956.

Broderick, G., Ricker, D. B., and Craig, W. M., Expeller soybean meal and corn byproducts versus solvent soybean meal for lactating dairy cows fed alfalfa silage as a sole forage, *J. Dairy Sci.,* 73, 453, 1990.

Broderick, G. A. and Wallace, R. J., Effects of dietary nitrogen source on concentrations of ammonia, free amino acids and fluorescamine-reactive peptides in the sheep rumen, *J. Anim. Sci.,* 66, 2233, 1988.

Chen, G. H., Sniffen, C. J., and Russell, J. B., Concentration and estimated flow of peptides from the rumen of dairy cattle: effects of protein quantity, protein solubility, and feeding frequency, *J. Dairy Sci.,* 70, 983, 1987.

Daniel, H., Morse, E. L., and Adibi, S. A., The high and low affinity transport systems for dipeptides in kidney brush border membrane respond differently to alterations in pH gradient and membrane potential, *J. Biol. Chem.,* 266, 19917, 1991.

Fejes, J., Faixova, Z., Varady, J., and Cibula, M., In vitro transport of amino acids across the rumen mucosa in sheep, *Vet. Med. Praha.,* 36, 551, 1991.

Ganapathy, V. and Leibach, F. H., Role of pH gradient and membrane potential in dipeptide transport in intestinal and renal brush-border membrane vesicles from the rabbit, *J. Biol. Chem.*, 258, 14189, 1983.

Leibholz, J., Effect of diet on the concentration of free amino acids, ammonia and urea in the rumen liquor and blood plasma of sheep, *J. Anim. Sci.,* 29, 628, 1969.

Leibholz, J., The absorption of amino acids from the rumen of the sheep. II. The transfer of histidine, glycine, and ammonia across the rumen epithelium in vitro, *Aust. J. Agric. Res.,* 22, 647, 1971.

Lewis, D., Amino-acid metabolism in the rumen of the sheep, *Br. J. Nutr.,* 9, 215, 1955.

Martens, H. and Gabel, G., Transport of Na and Cl across the epithelium of ruminant forestomachs: rumen and omasum: A review, *Comp. Biochem. Physiol.,* 90A, 569, 1988.

Prins, R. A., Hungate, R. E., and Prast, E. R., Function of the omasum in several ruminant species, *Comp. Biochem. Physiol.,* 43A, 155, 1972.

Saito, H. and Inui, K.-I., Dipeptide transporters in apical and basolateral membranes of the human intestinal cell line Caco-2, *Am. J. Physiol.*, 265, G289, 1993.

Smith, R. H., Microbial activity in the omasum, *Proc. Nutr. Soc.*, 43, 63, 1984.

Van Winkle, L. J., Campione, A. L., and Gorman, J. M., Na^+-independent transport of basic and zwitterionic amino acids in mouse blastocysts by a shared system and by processes which distinguish between those substrates, *J. Biol. Chem.,* 263, 3150, 1988.

Webb, K. E., Jr. and Matthews, J. C., The absorption of amino acids and peptides, in *Principles of Protein Nutrition of Ruminants,* Asplund, M., Ed., CRC Press, Boca Raton, FL, 1994, chapter 7.

Whitelaw, F. G., Hyldgaard-Jensen, J., Reid, R. S., and Kay, M. G., Volatile fatty acid production in the rumen of cattle given an all-concentrate diet, *Br. J. Nutr.,* 24, 179, 1970.

Wright, D. E. and Moir, R. E., Amino acid concentrations in rumen fluid, *Appl. Microbiol.,* 15, 148, 1967.

CHAPTER 7

Advancing Our Understanding of Amino Acid Utilization and Metabolism in Ruminant Tissues

J. C. MacRae

INTRODUCTION

Twenty years ago, the drive in European agriculture was to produce more food. Today, the goals have changed as society has become increasingly aware of social, environmental, and particularly health issues relating to the food that they consume (NACNE, 1983; James, 1993), and producers strive to accommodate many of these considerations within the agricultural enterprises that generate that food.

In the U.K., ruminant production (approximately £7 billion in 1993) represents over 40% of gross agricultural income, and meat and dairy products comprise approximately 25% of the family food budget. Yet producers face challenges from a number of quarters: Animal welfare lobbyists are concerned about issues such as the castration of beef animals (mainly carried out as a management practice) and the transport of animals to and the methods used during slaughter. Economists have been arguing for some time that overproduction and the consequent stockpiling of surpluses is highly uneconomical, and indeed quotas were introduced for milk production in 1984 to curb this overproduction. Environmentalists are concerned about the pollution of water courses by grassland fertilizer runoff, animal excreta, and silage effluent, and, perhaps most important of all, the medical profession have labeled animal products in general, and ruminant (saturated) fat in particular, as detrimental to health and implicated them in cardiovascular and other health problems (James, 1993). As a result, the majority of ruminant products presented for sale have had the fat trimmed (meat) or skimmed off (milk) in order to make them more consumer acceptable, a process that, although helping to safeguard consumer acceptability, is inherently uneconomic as the deposition/secretion of this fat has cost the farmer dearly during the production process.

Against this background, agricultural scientists in Europe have focused on ways of improving the efficiency of resource management, including adoption of the popular concept of "sustainable agriculture" involving maximum use of forage energy and nitrogen (N) produced from minimal inputs of fertilizer N. Unfortunately, such systems impose constraints on animal performance because, where

1-56670-199-6/96/$0.00+$.50

meat and dairy products are derived from predominantly forage-based systems (where grazed herbage or grass silage provides a substantial proportion of both the energy and N intake of the animal), ruminant productivity is inefficient, and even in the high-yielding dairy cow less than 20% of ingested protein is converted into saleable products. Furthermore, while the level of production can be boosted by addition of supplementary protein (e.g. fishmeal), efficiency of utilization is not enhanced (Beever et al., 1992), with the result that excretion of N waste products is increased, representing an additional burden on environmental pollution.

To reduce this pollution and so enhance the environment, we will need a better understanding of why inefficiencies in dietary protein utilization occur, and so this goal is likely to be achieved in conjunction with increased rates of protein deposition (growth) and secretion (lactation) in animal products.

This chapter will attempt to assess the present understanding of the regulation of efficiency of utilization of dietary protein in forage-fed ruminants and will suggest possible ways toward producing higher protein, lower fat animal products.

RUMEN FERMENTATION OF DIETARY PROTEIN

Amino acid (AA) nutrition of ruminants is complicated by the need to consider the requirements of the microbial populations in the rumen as well as the host animal itself. Indeed, the efficiency of conversion of dietary protein into microbial products has considerable bearing on the net supply of AA for subsequent host-animal digestion and absorption. Originally, the ruminant's pregastric microbial fermentation allowed it to forage on poorer quality indigenous herbages containing large amounts of complex polysaccharides, upon which the mammalian digestive enzyme systems have little effect; however, on higher-quality forages this fermentation, while still essential to the utilization of the energy-generating nutrients, can contribute to inefficiencies in the utilization of the protein components.

It has been many years since the proteolytic properties of rumen microbes was first identified (Annison and Lewis, 1959; Barnet and Reid, 1961) and since it was realized that rumen fermentation could have a major influence on the amount and nature of AA presented to the main site of host animal absorption (Clarke et al., 1966). MacRae and Reeds (1980), in reviewing the influence of the chemical and physicochemical composition of foodstuffs on the transfer of AA to the small intestine, identified the importance of "synchrony of release" of protein degradation products and energy to the conversion of dietary into microbial protein. Earlier, MacRae (1976) had argued that the poor transfer of AA to the small intestine of sheep given fresh forage (MacRae and Ulyatt, 1974) probably resulted from the rapid release of ammonia from the herbage exceeding the rate of release of energy (mainly from structural carbohydrates). This asynchrony seems to be particularly problematic in better quality (high nitrogen) fresh herbages because of the highly soluble nature of the plant proteins. Indeed, comparison across

Table 1 Comparisons of Digestion Data from Sheep Given Fresh, Frozen, and Dried Clover

			Dried[a]	
	Fresh[b]	Frozen[a]	Wafers	Pellets
Amino acid intake (g/d)	127	127	124	123
AA entering the small intestine (g/d)	80	133	147	175
AA apparently absorbed (g/d)	50	95	78	96

[a] From Beever et al., 1971.

[b] From MacRae and Ulyatt, 1974.

separate studies seems to suggest that where solubility of the forage is altered, for example, as occurs on frozen storage (MacRae et al., 1975) or during heat denaturation in the production of dried grass, the asynchrony appears to be alleviated with a resultant increase in passage of AA to the small intestine. Thus Table 1 presents data from studies where similar forage (white clover) was fed either fresh (MacRae and Ulyatt, 1974) or frozen and dried (Beever et al., 1971) on an isonitrogenous basis to sheep. The greater transfer of amino N to the small intestine in the sheep given the processed forages probably indicated a greater capture of dietary N by microbes. Indeed, when Beever et al. (1974) compared the rates of microbial protein synthesis and the amounts of AA entering the small intestine on fresh, frozen, and dried forages, they reported 26 and 57% higher rates of microbial protein synthesis and 15 and 51% higher amounts of AA entering the small intestine on the frozen and dried grass compared with the same herbage fed freshly cut.

Where substantial proportions of the dietary protein are lost as ammonia absorption from the rumen, the availability of AA for tissue metabolism is reduced, and the amount of N excreted in the urine is increased. In theory, there are two obvious ways of alleviating this asynchrony: addition of dietary components containing rapidly fermented carbohydrate to alter the temporal pattern of available energy for microbial metabolism or, in some way reducing the solubility of the forage proteins to slow the release of ammonia and therefore help the microbes capture a greater proportion for microbial protein synthesis. However, in practice, things are not quite so simple, and consideration needs to be given as to when soluble carbohydrates should be fed and whether lower solubility of forage protein can be achieved while maintaining sward ecology and sustainability.

In the dairy industry, cereals containing higher levels of soluble carbohydrate are used routinely to increase the energy density of the ration and therefore boost performance. Here it is interesting to note that, whereas in the past these were routinely fed twice daily at milking, farmers with higher-producing herds now find benefit from feeding concentrates more frequently, using out of parlor feeders, or even from the provision of total mixed rations where the concentrate and forage are mixed together prior to these being offered to the animal in a zero-grazing situation (Gill, 1979).

There seem to have been few attempts to alter protein solubility in forages, perhaps because of difficulties associated with plant viability. But some years ago,

Barry and colleagues in New Zealand tried the alternative approach of protecting the soluble proteins in their high-quality grasses and clovers from rumen degradation (or at least slow down the rate of degradation). First, they sprayed formaldehyde-type materials onto the pastures (Barry, 1973) and later introduced tannin-containing plants (Barry and Reid, 1986; Barry et al., 1986) that perform virtually the same function of complexing with the soluble plant proteins and therefore protect them from rapid microbial degradation. Both these approaches have resulted in increases in AA supply to the small intestine, and production studies with both sainfoin and lotus species have indicated increased rates of protein deposition (Egan and Ulyatt, 1980; John and Lancashire, 1981; Barry et al., 1986) and milk protein secretion (G. C. Waghorn, personal communication). Unfortunately, subsequent attempts to develop agronomic mixtures of plant species that could allow the tannin from one plant (e.g., lotus species) to crossreact with the soluble protein of other species (e.g., ryegrass or clover) have met with difficulties; the former are not particularly sward hardy and are not able to compete, at least in the grazing situation, with the other species.

Utilization of Absorbed AA

Asynchrony in the rumen, leading to a poor supply of AA to the small intestine, is only part of the reason for the low productivity of forage-fed ruminants. Despite the fact that current ration formulation schemes indicate that 0.4 to 0.8 of the AA absorbed from the small intestines are utilized for productive purposes (ARC, 1980; AFRC, 1984, 1992; Jarrige, 1989; NRC, 1989), efficiencies found in practice have been more in the range 0.25 to 0.6 for sheep given forage rations (Coehlo da Silva et al., 1972; MacRae and Ulyatt, 1974; MacRae et al., 1985, 1995a). Indeed, even where dairy cows and goats have been given abomasal infusions of casein, the response in milk protein output has fallen well below that predicted using these schemes (Clarke, 1975; MacRae et al., 1988a).

These efficiencies are of course considerably lower than corresponding values for nonruminants. Table 2 gives data on the marginal efficiency of protein retention per unit extra protein absorbed by ruminants and pigs. While reasons for these low efficiencies in ruminants are not yet fully understood, following the development of tracer kinetic procedures for the measurement of rates of protein synthesis in different tissues of the body, it has been possible to distinguish differences in the way ruminants and nonruminants utilize their absorbed AA. Figure 1 illustrates the turnover of protein in the splanchnic tissues, muscle, and skin of young lambs derived from the protein synthesis measurements of Attaix et al. (1987) and the accretion data of MacRae et al. (1993). While 0.62 of the total net accretion of protein occurs in the main saleable tissues (skeletal muscle and wool), the partition of whole-body protein synthesis (WBPS) to these tissues, including skin metabolism to produce the wool, represents only 0.42 of WBPS. The gastrointestinal tract (GIT) represents 0.27 of WBPS, even though accretion of protein in GIT represents less than 0.03 of total body accretion. The liver also accounts for another 0.1 of WBPS, twice

Table 2 Efficiency of Utilization of Absorbed Amino Acids in Ruminants and Nonruminants

Species	Ratio $\frac{\text{Extra CP retained}}{\text{Extra CP absorbed from SI}}$	Ref.
Sheep		
Fresh grass	0.4	MacRae and Ulyatt (1974)
Dried grass	0.5	MacRae et al. (1985; 1995a)
Dried grass plus conc.	0.6	MacRae et al. (1995a)
Cattle		
Forage plus conc.	0.47	Lobley et al. (1980)
Pigs	0.75	Batterham et al. (1990)
	0.7	Bikker et al. (1994)

its proportional rate of protein deposition. These splanchnic tissues (GIT and liver) are known to be wasteful in terms of the N economy of the animal, both in terms of AA oxidation and, in the case of the GIT, in terms of many secretory proteins (digestive enzymes and glycoproteins of mucins) that enter the lumen of the tract and are not fully redigested and resorbed. Another feature of Figure 1 is that although muscle protein deposition represents 0.36 of total protein retention, this is only approximately one third (0.29) of the protein that is synthesized daily by muscle.

These two features, a large proportional distribution of AA to splanchnic tissues and a relatively poor retention of synthesized protein by muscle, may both contribute to the inefficiency of utilization of AA by ruminants. Comparison of these aspects of the data from lambs with corresponding data from 250-kg heifers

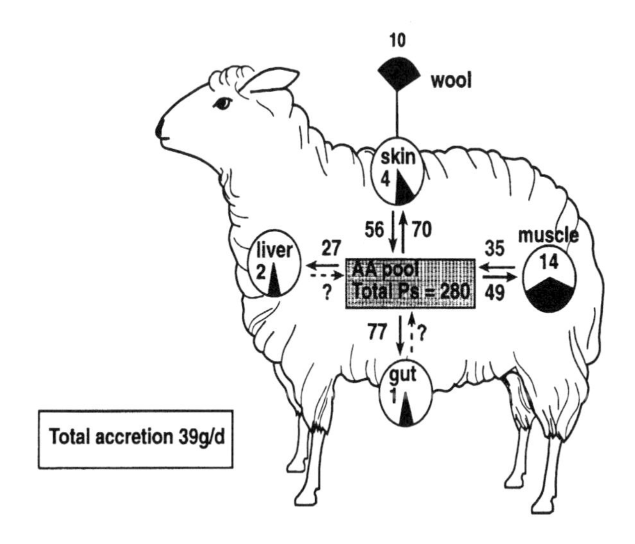

Figure 1 Schematic diagram representing rates (g/d) of protein synthesized (arrow to tissue), accreted (segment in tissue), and degraded (arrow from tissue; calculated by difference) by various tissues of a 20-kg weaned lamb. (Adapted from data of Attaix et al., 1987 and MacRae et al., 1993.)

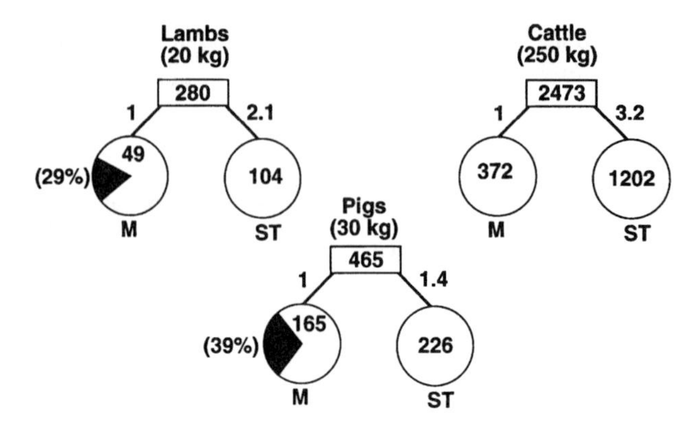

Figure 2 Comparative estimates of partition of amino acid to protein synthesis in muscle (M) and splanchnic tissues (ST) of lambs, cattle, and pigs. Also, proportion of protein synthesized in muscle which is retained (segment) in lambs and pigs. (Adapted from data of Edmunds et al., 1980; Lobley et al., 1980; Attaix et al., 1987; and MacRae et al., 1993.)

(Lobley et al., 1980) and 30-kg pigs (Edmunds et al., 1980; Figure 2) indicates a greater partition of AA for protein synthesis in the splanchnic tissues vs. muscle in the ruminants (2:1 in the lambs and 3:1 in the heifers) than in the pigs (1.4:1). At the same time, the proportional retention of the synthesized protein in skeletal muscle is greater in the pigs than in the lambs. The pigs synthesized 165 g of protein per day (0.35 of WBPS) in muscle and retained 65 g/d (0.39 of synthesis or 0.71 of total body accretion). In comparison, lambs synthesized 49 g of protein per day (0.17 of WBPS) in muscle, but retained only 14 g/d (0.28 of synthesis or 0.36 of total body accretion). Unfortunately, rates of muscle accretion were not determined in the heifers.

Comparisons given in Figure 2 seem to suggest the bases for a research strategy aimed at altering the rates of protein deposition in ruminants. When MacRae and Lobley (1991) reviewed ways of manipulating growth in ruminants, they concluded that it was possible to alter both the partition of AA to muscle vs. other organs and the rates of protein synthesis and degradation in skeletal muscle using hormone interventions. For example, growth hormone seems to increase WBPS (Eisemann et al., 1989), anabolic steroids reduce protein degradation without altering synthesis (Lobley et al., 1985), and β-agonists appear to increase synthesis and reduce degradation simultaneously (MacRae et al., 1988b). The β-agonists also appear to increase rates of protein deposition in skeletal muscle at the expense of deposition in other tissues (Williams et al., 1987). Unfortunately, in the present climate of consumer-driven agriculture where political considerations can often outweigh scientific judgment, there seems little possibility of using direct hormone manipulation of growth, at least in Europe. Thus the need is to try to understand how to manipulate these processes in more consumer-acceptable ways.

The Protein Metabolism Group at the Rowett Institute in Aberdeen, Scotland, is currently adopting an approach focused primarily on the control of splanchnic tissue metabolism and the intrinsic regulation of muscle protein turnover. Splanchnic tissues are being studied using trans-organ *(in vivo)* preparations, which, by a measurement of arteriovenous (AV) differences and the use of mass isotope tracers, provide information on the net fluxes of AA across the GIT and liver and their use for protein synthesis in these organs. The muscle metabolism studies utilize a combination of laboratory animal *(in vivo)*, tissue *(in vitro)* and cell culture techniques to try to identify the possible roles of anabolic hormones (e.g., IGF-I) and neuropeptides in the regulation of muscle protein turnover, as well as the function of various second messenger pathways that link membrane receptor signals to the synthetic machinery of the cell. So far, the programs have produced some interesting preliminary data.

Gastrointestinal Tract Metabolism

One potential problem in ruminants is the imbalance in the AA composition of certain biological tissue proteins. MacRae and Lobley (1991) identified wool and certain GIT secretions as having the potential to create limiting AA situations for other tissues under certain conditions. For example, in undernourished sheep, wool protein synthesis still continues even though the animals may be mobilizing muscle protein to service the myriad of functional processes. Here, presumably, synthesis of 1 g of wool protein would require mobilization of 4 g of muscle protein to provide the very high levels of sulfur AA (10% of total protein in wool) required (MacRae, 1989).

Tagari and Bergman (1978) seemed to indicate a similar problem associated with absorptive metabolism when they compared the rates of absorption of essential amino acids (EAA) from the small intestine with their subsequent appearance in the portal drained viscera (PDV; the venous drainage from the whole GIT). They found their recoveries ranged from 0.6 for lysine and phenylalanine to less than 0.25 for leucine and histidine (Figure 3a), and they went on to suggest this apparently preferential and selected use of certain EAA during absorptive metabolism was a possible contributor to inefficiency in terms of its potential to create a limiting AA situation at the peripheral tissue level, i.e., alter the biological value of the absorbed AA mixture. This selective loss of EAA during absorptive metabolism was not substantiated, however, in recent experiments where sheep were given a pelleted lucerne ration. Recoveries of EAA across the PDV were much more consistent, averaging 0.70 ± 0.044 of the EAA disappearance from the small intestine (MacRae et al., 1996; Figure 3b). Indeed, net fluxes across the mesenteric drained viscera (MDV, which drains the majority of the small intestines) accounted for 1.04 ± 0.022 of apparent absorption, indicating that the tissues of the small intestine are in approximate net balance in terms of protein turnover (i.e., the rates of EAA used for protein synthesis, from both luminal and arterial sources, equate with the rates of EAA released from protein degradation into the venous drainage). This being the case, then, as all the MDV flux enters the PDV, the difference between the MDV and PDV fluxes must indicate that the tissues anterior and posterior to those drained by the

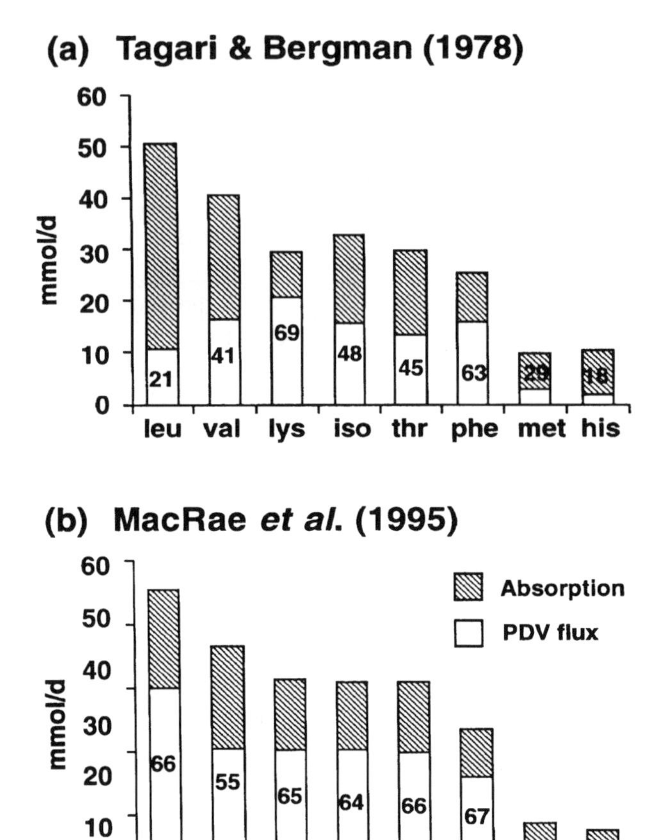

Figure 3 Comparative recoveries of essential amino acids disappearing from the small intestine as the net flux of amino acid at the portal vein in sheep given forage diets as reported by (a) Tagari and Bergman, 1978 and (b) MacRae et al., 1996.

MDV (i.e., the forestomachs and hindgut) are net users of EAA for protein synthesis in terms of the AV difference of blood-free AA.

One explanation for such a phenomenon could be the release of GIT protein degradation products that are in peptide or protein form. Indeed, absorption of peptide AA from the forestomachs has been reported (Webb et al., 1992, 1993; Koeln et al., 1993). However, when AV differences of peptide AA of <1500 Da were determined in this experiment, no positive absorption could be detected across the PDV (Hipolita-Reis et al., 1994). An alternative explanation for the lower PDV fluxes would involve a proportion of the degradation components of GIT tissue protein turnover in these regions entering the lumen of the tract rather than the venous drainage. If this occurs in the forestomachs and abomasum, these proteins/peptides/AA could be subsequently redigested and resorbed from the small intestine, but the subsequent fate of the EAA from hindgut tissue degradation would probably be excretion in feces. Based on the relative rates of protein synthesis in the forestomachs and hindgut (3:1; Attaix et al., 1987, 1992), perhaps 0.15 of duodenal EAA flow could be of endogenous origin (MacRae et al., 1996), a factor that is largely ignored in the current metabolizable protein schemes (NRC, 1989; Jarrige, 1989; AFRC, 1992).

One interesting feature of the tracer studies carried out as part of this study was that, with the exception of phenylalanine, the proportional use of arterially and luminally derived EAA for protein synthesis by GIT tissues and secretions was heavily toward the former (ratio of 4 or 5:1, MacRae et al., 1996), and so the GIT perhaps needs to be considered more as a competitive peripheral organ, extracting its EAA from arterial sources (i.e., in competition with skeletal muscle, skin, and other peripheral tissues), rather than as a tissue which enjoys preferential advantage from using these EAA during absorptive metabolism.

Metabolic Consequences of Altering GIT Protein Metabolism

Although the worry that the GIT might be causing imbalances in the EAA mixtures available for other tissues seems to be ill founded, the fact that the rate of protein turnover by GIT tissues is so high (approximately 0.4 of WBPF) and competitive perhaps makes this phenomenon a rather sensitive aspect of interorgan coordination and hence a potential regulator of animal productivity. The ability of the GIT to compete for circulating AA and the metabolic consequences of this is perhaps most apparent where GIT environmental conditions are changed in ways that perturb its requirements. For example, where animals are subjected to subclinical intestinal parasitism (e.g., *T. colubriformis),* GIT protein synthesis increases (measurements have only, to date, been made in laboratory species; Symons et al., 1981), and one consequence of this hyperactivity is an increased loss of protein from the small intestine into the hindgut (Poppi et al., 1986; Kimambo et al., 1988). Under these conditions, animals exhibit reduced rates of protein synthesis in muscle, wool, and kidneys and, as a consequence, growth retardation. Only in the liver is protein synthesis maintained or increased, probably in order to service the immunological response to the infection (see review by MacRae, 1993).

If hyperactivity of GIT metabolism in parasitism penalizes the other organs, probably in terms of reducing the availability of EAA for their protein synthesis, this raises the possibility that agents that reduce normal GIT protein turnover might afford a benefit to these other organs. Antimicrobial agents such as Flavomycin and Avoparcin are known to reduce the mitotic index of intestinal mucosal cells (Parker, 1990). Some years ago, when Flavomycin was given to lambs with and without the β-agonist Clenbuterol (MacRae, unpublished data), it significantly increased the rate of protein deposition, with the two agents together giving an additive effect (Table 3). As yet, it is too early to say unequivocally that this increased N retention and hence reduced excretion of N waste products is derived exclusively from reducing GIT protein turnover with a consequent increase in the flux of EAA to the PDV and onward for the metabolism of other

Table 3 Protein Retention of Lambs Fed Dried Grass and Administered Clenbuterol[a] and/or Flavomycin[b]

	Control	Clenbuterol	Difference
Control	20.6	35.0	14.4 ($p < 0.05$)
Flavomycin	25.0	39.4	14.4 ($p < 0.001$)

[a] 3 mg/d.

[b] 20 mg/d.

peripheral tissues (e.g., skeletal muscle), but this type of manipulation might be worthy of further investigation.

Liver Metabolism

One of the characteristics of forage-fed ruminants is the substantial absorption of ammonia from the rumen, particularly where asynchrony of dietary protein degradation and energy release occurs (see pages 74–76). This ammonia enters the PDV and must be converted (detoxified) to urea in the liver because ammonia is toxic to the central nervous system. The biochemical pathways involved in ureagenesis determine that only one of the two N atoms of urea can come directly from the condensation of ammonia with bicarbonate, via carbamoyl phosphate. The other N has to come from aspartate. The question of whether the aspartate N can itself be derived from ammonia via glutamate through the glutamate dehydrogenase (GLDH) reaction, or whether the incorporation of the aspartate N into urea represents net loss of AA N through amino transferase reactions is a topic of considerable current interest.

In cattle given forage-based rations, only a proportion (0.3 to 0.9) of the urea N produced could be attributed to ammonia detoxification (Wilton and Gill, 1988; Fitch et al., 1989; Huntington, 1989; Maltby et al., 1991; Reynolds et al., 1991), suggesting that other N sources, perhaps AAs, are required for urea synthesis. Indeed, where Maltby et al. (1991) supplemented corn silage with urea, they observed that the 0.6 increase in ammonia absorption was accompanied by an increased rate of EAA extraction by the liver. The work of Reynolds and Tyrrell (1989) seemed to indicate that this phenomenon could be diet specific; on a high concentrate ration (25% lucerne, 75% maize and soybean meal), lower ammonia absorption was accompanied by lower hepatic AA extraction. This could relate to the capacity of GLDH to transfer ammonia N to aspartate at a rate that can equal the carbamoyl phosphate reaction. There is evidence that the rate of the latter reaction is reduced in the presence of higher levels of propionate, as might be expected to occur on concentrate rations. However, the relative rates of the two reactions can also relate to the acid-base status of the animal (Lobley, personal communication). Recent studies have provided direct evidence that ureagenesis related to ammonia removal can lead to increased AA catabolism (Lobley et al., 1995), and therefore where animals experience high ammonia loads, such as in the asynchrony situations of fresh pasture or silage, this detoxification may influence the AA availability for anabolic purposes.

Peripheral Tissue Utilization of Amino Acids

The anabolic potential of an animal is not solely related to AA supply to peripheral tissues, but factors that alter GIT and liver metabolism and hence splanchnic release of EAA do have a major influence on production. There are many reports of milk production responses to supplementary infusions of casein into the abomasum (Clarke, 1975), and recent conjoint work from the Rowett Institute and Reading University industrially funded consortium has demonstrated the ability to increase milk protein concentration as well as milk yield and milk protein output with jugular and mesenteric vein infusions of mixtures of either total or just essential AA (Metcalf et al., 1996; Reynolds et al., 1995). However, there are also situations where increased supply of AA to the mammary gland, resulting from dietary supplementation, has not resulted in major increases in milk protein output (Metcalf et al., 1994). Therefore, some of the mechanistic aspects of peripheral tissue protein metabolism, including cell transporter function, the responses of cells (muscle or mammary gland) to changes in the amount and composition of nutrients presented to them, and indeed the control of vasodilation and hence capillary blood flow within the tissues undergoing net anabolic drive, are extremely complicated and as yet ill defined. Furthermore, there are some situations where nutrient status per se has a major influence on the tissue responses to metabolic regulators and where this response can be compromised if nutritional considerations are ignored. A prime example of this is the anabolic response of muscle tissue to bovine somatotrophin (bST) in the growing animal.

There would seem to be little doubt about the galactopoietic effect of bST in the dairy cow (see reviews by Bauman and McCutcheon, 1986; Hart, 1986), which, at least in the long term, seems to be driven through increasing nutrient intake rather than altering the biological efficiency with which these nutrients are utilized by the animal, i.e., the economic benefits come from producing more milk from less cows, thus diluting the "maintenance costs" of this production.

In contrast, where bST has been applied to growing animals, results have been much more equivocal, and, until it was realized that both in pigs (Etherton, 1989) and ruminants (MacRae et al., 1991) the protein status of the animal is important in terms of its anabolic response to bST, many trials failed to demonstrate substantial responses to treatment. The reasons for this phenomenon could be multifactorial. Breier et al. (1988) reported the GH receptor affinity in the liver (leading to IGF-I production) was influenced by nutrition with high-affinity receptors being activated by higher protein status, but being absent in animals fed diets leading to lower protein status, such as fresh pasture. However, it is likely that postreceptor events are also responsive to AA status. Eisemann et al. (1989) found that the GH-IGF-I anabolic response in cattle was achieved by increasing protein synthesis. It would seem reasonable to argue therefore that the responsive tissues would require extra AA to synthesize into the protein. When MacRae et al. (1991) administered bST to lambs sustained by intragastric infusion, where the intake of protein as casein infused into the abomasum could be rigorously controlled, they were unable to obtain sustained responses to bST in animals given

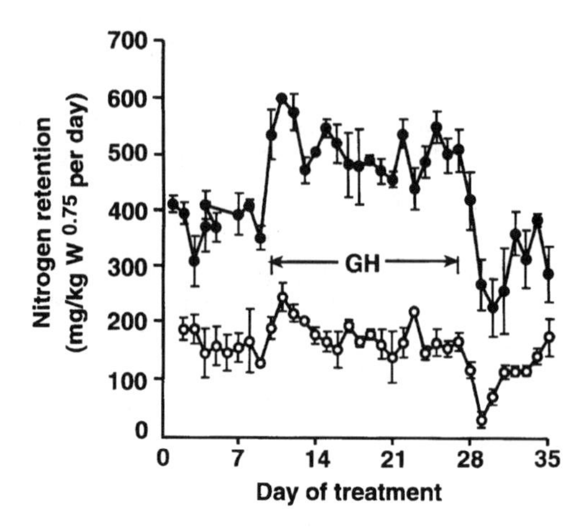

Figure 4 Responses in nitrogen retention to infusion of bovine somatotropin (4 mg/d) in lambs receiving either 1200 (~) or 600 mg (O) casein N/kg W 0.75/d. (Data of MacRae et al., 1991.)

casein at the level of 1.2 × N equilibrium, but obtained sustained (30%) anabolic response in animals infused with twice that amount of casein (Figure 4). In both infusions, the maximum response was obtained at days 2 to 3, and, following withdrawal of bST, N retention was reduced to values considerably below control levels for the initial 3 to 4 d prior to reestablishment at the control level. This might be consistent with a very rapid response in protein synthesis followed by a slower, coordinated change in protein degradation as the animal's metabolism signals the need for the extra AA, which are either available from dietary sources, thus allowing the sustained response on the higher protein intake, or have got to be derived from increased turnover of body tissues, as probably occurred on the lower protein intake.

Responses to anabolic steroids and β-agonists seem to be more independent of protein status, perhaps because the steroids seem to target protein degradation and reduce AA oxidation (Lobley et al., 1985), while the β-agonists seem to repartition AA toward skeletal muscle anabolism at the expense of GIT and skin metabolism (Williams et al., 1987; Nash et al., 1994). In this respect, part of the response could be similar to that suggested for the antimicrobial agents.

CONCLUSIONS

Ruminant protein metabolism and the efficiency of conversion of dietary protein into animal products is an inefficient process, leading to uneconomic use of the expensive part of rations and a substantial contribution to environmental pollution. Current research is attempting to gain a better understanding of the regulation of tissue protein metabolism in ruminants, particularly those given forage-based rations, in the hope that this knowledge will identify ways of altering the nutrient partition in growing animals toward the production of carcasses with greater lean and less fat. Similarly, the dairy industry would benefit from increased milk protein concentration, particularly in those areas where large volumes have to be processed into cheeses and yogurts, with the ability to process milk into cheese and yogurts being problematical if protein concentration is less than 30 g/kg.

Although the manipulations of growth can be achieved by endocrinological means, the exogenous administration of "hormones" or hormone-based additives is unacceptable to consumers in many parts of the world. In the longer term, the alternative would seem to be through a knowledge of gene expression and perhaps accelerated breeding, but in the short term it might be possible to accomplish some manipulations by alterations of the immune system. For example, manipulation of the GH axis via immunological suppression of somatostatin (Spencer, 1983) and enhancement of GH activity by a combination of monoclonal antibodies (Pell et al., 1989) are interesting and potentially valuable ways of manipulating protein metabolism (see review by Flint, 1992). However, there is still a need for more understanding of the nutritional and other requirements for response to such manipulations before we can develop practical schemes.

Although the way ahead will not be easy and will need clear thinking and judgment both by those researching the techniques and by those developing policies relating to the animal industry, the difficulties involved should not deter the work, because there is little doubt that if the goals can be achieved they would make a substantial contribution to enhancing the environment in areas of the world where ruminant production from forages is a major sector of the agricultural industry.

REFERENCES

Agricultural and Food Research Council (AFRC), in *The Nutrient Requirements of Ruminant Livestock, Supplement No. 1, Report of the Protein Group of the ARC Working Party,* Commonwealth Agricultural Bureau, Farnham Royal, Slough, U.K., 1984.

Agricultural and Food Research Council (AFRC), Technical Committee on Responses to Nutrients, Report No. 9, Nutritive requirements of ruminant animals: protein, *Nutr. Abst. Rev.*, 62B, 787, 1992.

Agricultural Research Council (ARC), in *The Nutrient Requirements of Ruminant Livestock, Technical Reviews,* Commonwealth Agricultural Bureau, Farnham Royal, Slough, U.K., 1980.

Annison, E. F. and Lewis, D., *Metabolism in the Rumen,* Methuen, London, 1959, 92.

Attaix, D., Aurousseau, E., Bayle, G., Maughebati, A., and Arnal, M., Protein synthesis and degradation in growing lambs, in *Protein Metabolism and Nutrition,* EAAP Publication No. 35, Rostock, Wiss, Z.WPU, 1987, 24.

Attaix, D., Aurousseau, E., Rosolowska-Huszez, D., Bayle, G., and Arnal, M., In vivo longitudinal variations in protein synthesis in developing ovine intestine, *Am. J. Phys.*, 263, R1318, 1992.

Barnett, A. J. G. and Reid, D. L., *Reactions in the Rumen,* Arnold, London, 1961, 107.

Barry, T. N., Effect of treatment with formaldehyde and intraperitoneal supplementation with D-L-methionine on the digestion and utilization of a hay diet by sheep. 1. The digestion of energy and nitrogen, *N. Z. J. Agric. Res.*, 16, 185, 1973.

Barry, T. N., Manly, T. R., and Duncan, S. J., The role of condensed tannins in the nutritional value of Lotus pedunculatus for sheep. 4. Sites of carbohydrate and protein digestion as influenced by dietary reactive tannin concentration, *Br. J. Nutr.*, 55, 123, 1986.

Barry, T. N. and Reid, C. S. W., Nutritional effects attributable to condensed tannins cyanogenic glycosides and oestrogenic compounds in New Zeland forages, in *Forage Legumes for Energy Efficient Animal Production,* Barnes, R. F., Ball, P. R., Brougham, R. W., Marten, G. L., and Mension, D. J., Eds., USDA Technical Information Service, Springfield, VA, 1986, 251.

Batterham, E. S., Andersen, L. M., Baignent, D. R., and White, E., Utilization of ileal digestible amino acid by growing pigs: effect of dietary lysine concentration on efficiency of lysine retention, *Br. J. Nutr.*, 64, 81, 1990.

Bauman, E. E. and McCutcheon, S. M., The effect of growth hormone and prolactin on metabolism, in *Control of Digestion and Metabolism in Ruminants,* Milligan, L. P., Grovum, W. J., and Dobson, A., Eds., Prentice-Hall, Englewood Cliffs, NJ, 1986, 436.

Beever, D. E., Camel, S. B., and Wallace, A., The digestion of fresh, frozen and dried perennial ryegrass, *Proc. Nutr. Soc.,* 33, 73A, 1974.

Beever, D. E., Dawson, J. M., and Buttery, P. J., Control of fat and lean deposition in forage fed cattle, in *The Control of Fat and Lean Deposition*, Boorman, K. N., Buttery, P. J., and Lindsay, D. B., Eds., Butterworth-Heinemann, Oxford, 1992.

Beever, D. E., Thomson, D. J., and Harrison, D. G., The effects of drying and the comminution of red clover on its subsequent digestion by sheep, *Proc. Nutr. Soc.,* 30, 86A, 1971.

Bikker, P., Verstegen, M. W. A., Campbell, R. G., and Kemp, B., Digestible lysine requirements of gilts with high genetic potential for lean gain in relation to energy intake, *J. Anim. Sci.,* 72, 1744, 1994.

Breier, B. H., Gluckman, P. D., and Bass, J. J., The somatotrophic axis in young steers: influence of nutritional status and estradiol-17α on hepatic high- and low-affinity somatotrophic binding sites, *J. Endocrinology,* 116, 169, 1988.

Clarke, J. H., Lactational responses to postruminal administration of proteins and amino acids, *J. Dairy Sci.,* 50, 1621, 1975.

Clarke, E. M. W., Ellinger, G. M., and Phillipson, A. T., The influence of diet on the nitrogenous components passing to the duodenum and through the lower ileum of sheep, *Proc. Royal Soc.,* Ser. B, 166, 63, 1966.

Coelho da Silva, J. F., Seeley, R. C., Beever, D. E., Prescott, J. A. H., and Armstrong, D. J., The effect in sheep of physical form and stage of growth on the sites of digestion of the dried grass, *Br. J. Nutr.,* 28, 357, 1972.

Edmunds, B. K., Buttery, P. J., and Fisher, C., Protein and energy metabolism of the growing pig, in *Energy Metabolism,* Mount, L. E., Ed., Butterworths, London, 1980, 129.

Egan, A. R. and Ulyatt, M. J., Quantitative digestion of fresh herbage by sheep. VI. Utilization of nitrogen in five herbages, *J. Agric. Sci. Camb.,* 94, 47, 1980.

Eisemann, J. H., Hammond, A. C., Rumsey, T. S., and Bauman, D. E., Nitrogen and protein metabolism and metabolites in plasma and urine of beef steers treated with somatotropin, *J. Anim. Sci.,* 67, 105, 1989.

Etherton, T. D., The mechanisms by which porcine growth hormone improves pig growth performance, in *Biotechnology in Growth Regulation,* Heap, R. P., Prosser, C. G., and Lamming, G. E., Eds., Butterworths, London, 1989, 97.

Fitch, N. A., Gill, M., Lomax, M. A., and Beever, D. E., Nitrogen and glucose metabolism by the liver of forage- and forage-concentrate-fed cattle, *Proc. Nutr. Soc.,* 48, 76A, 1989.

Flint, D. J., The control of fat and lean deposition by the immune system, in *The Control of Fat and Lean Deposition,* Boorman, K. N., Buttery, P. J., and Lindsay, D. B., Eds., Butterworths, Heinemann, Oxford, 1992, 229.

Gill, M., The principles and practice of feeding ruminants on complete diets, *Grass Forage Sci.,* 34, 155, 1979.

Hart, I. C., Altering the efficiency of milk production of dairy cows with somatotropin, in *Nutrition and Lactation in the Dairy Cow,* Garnsworthy, P. C., Ed., Butterworths, London, 1986, 232.

Hipolita-Reis, M., MacRae, J. C., and Backwell, F. R. C., Peptide uptake from the gastrointestinal tract of sheep, *Anim. Prod.,* 58, 451, 1994.

Huntington, G. B., Hepatic urea synthesis and site and rate of urea removal from blood of beef steers fed alfalfa hay or a high concentrate diet, *Can. J. Anim. Sci.,* 69, 215, 1989.

James, W. P. T., *The Scottish Diet,* Scotland's Health, Scottish Office, 1993.

Jarrige, R., Protein: the PDI system, in *Ruminal Nutrition: Recommended Allowances and Feed Tables,* INRA and John Libbey, 1989, 33.

John, A. and Lancashire, J. A., Aspects of the feeding and nutritive value of Lotus species, *Proc. N. Z. Grass. Assoc.,* 42, 152, 1981.

Kimambo, A. E., MacRae, J. C., Walker, A., Watt, C. F., and Coop, R. L., Effect of prolonged subclinical infection with Trichostronglyus colubriformis on the performance and nitrogen metabolism of growing lambs, *Vet. Parasitol.,* 28, 191, 1988.

Koeln, L. L., Schlagheck, T. G., and Webb, K. E., Jr., Amino acid flux across the gastrointestinal tract and liver of calves, *J. Dairy Sci.,* 76, 2275, 1993.

Lobley, G. E., Connell, A., Mollison, G. S., Brewer, A., Harris, C. I., and Buchan, V., The effect of a combined implant of trenbolone acetate and estradiol 17α on protein and energy metabolism of growing beef steers, *Br. J. Nutr.,* 54, 681, 1985.

Lobley, G. E., Milne, V., Lovie, J. M., Reeds, P. J., and Pennie, K., Whole body and tissue protein synthesis in cattle, *Br. J. Nutr.,* 43, 491, 1980.

Lobley, G. E., Weijs, P. J. M., Connell, A., Calder, A. G., Brown, D. S., and Milne, E., Influence of diet quality on the fate of absorbed and exogenous ammonia in young growing sheep, *Br. J. Nutr.,* in press, 1995.

MacRae, J. C., Utilization of the protein of green forage by ruminants at pasture, in *From Plants to Animal Protein — Reviews in Rural Science 2,* Sutherland, T. M., McWilliams, J. R., and Leng, R. A., Eds., University of New England Publishing Unit, Armidale, Australia, 1976, 93.

MacRae, J. C., Protein metabolism relationships with body reserves, *Cornell Nutr. Conf. Feed Manuf.,* 1989, 52.

MacRae, J. C., Metabolic consequences of intestinal parasitism, *Proc. Nutr. Soc.,* 52, 121, 1993.

MacRae, J. C., Bruce, L. A., and Brown, D. S., Efficiency of utilization of absorbed amino acids in growing lambs given forage and forage:barley rations, *Anim. Sci.,* 61, 277, 1995.

MacRae, J. C., Bruce, L. A., Brown, D. S., and Farningham, D. A. H., Comparison of apparent absorption of essential amino acids from the small intestine and their net flux across the mesenteric and portal drained viscera of lambs, *Proc. Nutr. Soc.,* 55, 63A, 1996.

MacRae, J. C., Bruce, L. A., Hovell, F. D. deB., Hart, I. C., Inkster, J., Walker, A., and Atkinson, T., Influence of protein nutrition on the response of growing lambs to exogenous bovine growth hormone, *J. Endocrinol.,* 130, 53, 1991.

MacRae, J. C., Buttery, P. J., and Beever, D. E., Nutrient interactions in the dairy cow, in *Nutrition and Lactation in the Dairy Cow,* Garnsworthy, P. C., Ed., Butterworths, London, 1988a, 55.

MacRae, J. C., Campbell, D. R., and Eadie, J., Changes in the biochemical composition of herbage upon freezing and thawing, *J. Agric. Sci. Camb.,* 84, 125, 1975.

MacRae, J. C. and Lobley, G. E., Physiological and metabolic implications of conventional and novel methods for the manipulation of growth and production, *Livestock Prod. Sci.,* 27, 43, 1991.

MacRae, J. C. and Reeds, P. J., Prediction of protein deposition in ruminants, in *Protein Deposition in Animals,* Buttery, P. J. and Lindsay, D. B., Eds., Butterworths, London, 1980, 225.

MacRae, J. C., Skene, P. A., Connell, A., Buchan, V., and Lobley, G. E., The action of the β agonist clenbuterol on protein and energy metabolism in fattening wether lambs, *Br. J. Nutr.,* 59, 457–465, 1988b.

MacRae, J. C., Smith, J. S., Dewey, P. J. S., Brewer, A. C., Brown, D. S., and Walker, A., The efficiency of utilization of metabolizable energy and apparent absorption of amino acids in sheep given spring- and autumn-harvested dried grass, *Br. J. Nutr.,* 54, 197, 1985.

MacRae, J. C. and Ulyatt, M. J., Quantitative digestion of fresh herbage by sheep 1. The site of digestion of some nitrogenous constituents, *J. Agric. Sci. Camb.,* 82, 309, 1974.

MacRae, J. C., Walker, A., Brown, D. S., and Lobley, G. E., Accretion of total protein and individual amino acid by organs and tissues of growing lambs and the ability of nitrogen balance techniques to quantitate protein retention, *Anim. Prod.,* 57, 237, 1993.

Maltby, S. A., Lomax, M. A., Beever, D. E., and Pippard, C. J., The effect of increased ammonia and amino acid supply on post-prandial portal drained viscera and hepatic metabolism in growing steers fed maize silages, in *Energy Metabolism of Farm Animals,* EAAP Publication No. 58, Wenk, C. and Boessinger, M., Eds., ETH-Zurich, 1991, 20.

Metcalf, J. A., Beever, D. E., Sutton, J. D., Wray-Cahen, D., Evans, R. T., Humphries, D. J., Backwell, F. R. C., Bequette, B. J., and MacRae, J. C., The effect of supplementary protein on in vivo metabolism of the mammary gland in lactating dairy cows, *J. Dairy Sci.,* 77, 1816, 1994.

Metcalf, J. A., Crompton, L. A., Wray-Cahen, D., Lomax, M. A., Bequette, B. J., MacRae, J. C., Backwell, F. R. C., Lobley, G. E., Sutton, J. D., and Beever, D. E., Responses in milk constituent secretion to intravenous administration of two mixtures of amino acids in dairy cows, *J. Dairy Sci.,* 79, 1425, 1996.

Nash, J. E., Rocha, H. J. G., Buchan, V., Calder, A. G., Milne, E., Quirke, J. F., and Lobley, G. E., Effects of hormones on protein synthesis, *Br. J. Nutr.,* 71, 501, 1994.

National Advisory Committee on Nutrition Education, Discussion Paper on Proposals for Nutritional Guidelines for Health Education in Britain, James, W. P. T., Ed., Health Education Council, London, 1983.

National Research Council (NRC), *Nutrient Requirements of Dairy Cows,* 6th edition, National Academy Press, Washington, D.C., 1989.

Parker, D. S., Manipulation of the functional activity of the gut by dietary and other means (antibiotic/probiotics) in ruminants, *J. Nutr.,* 120, 639, 1990.

Pell, J. M., Elcock, C., Walsh, A., Trig, T., and Aston, R., Potentiation of growth hormone activity using a polyclonal antibody of restricted specificity, in *Biotechnology and Growth Regulation,* Heap, R. B., Prosser, C. G., and Lamming, G. E., Eds., Butterworths, London, 1989, 259.

Poppi, D. P., MacRae, J. C., Brewer, A., and Coop, R. L., Nitrogen transactions in the digestive tract of lambs exposed to the intestinal parasite Trichostronglyus colubriformis, *Br. J. Nutr.,* 55, 593, 1986.

Reynolds, C., Crompton, L., Firth, K., Beever, D., Sutton, J., Lomax, M., Wray-Cahen, D., Metcalf, J., Chettle, E., Bequette, B., Backwell, C., Lobley, G., and MacRae, J., Splanchnic and milk protein responses to mesenteric vein infusions of three mixtures of amino acids in lactating dairy cows, *J. Anim. Sci.,* 73 (Suppl. 1), 274, 1995.

Reynolds, C. K. and Tyrrell, H. F., The effect of forage to concentrate ratio and intake on visceral tissue and whole body energy metabolism of growing beef heifers, in *Energy Metabolism of Farm Animals,* EAAP Publication No. 43, van der Horning, Y. and Close, W. H., Eds., Publ. Pudoc, Wageningen, The Netherlands, 1989, 151.

Reynolds, C. K., Tyrrell, H. F., and Reynolds, P. J., The effects of dietary forage to concentrate ratio and intake on energy metabolism in growing beef heifers: net nutrient metabolism in visceral tissues, *J. Nutr.,* 121, 1004, 1991.

Spencer, G. S. G., Garssen, G. J., and Hart, I. C., A novel approach to growth promotion using auto-immunization against somatostatin. 1. Effects of growth hormone levels in lambs, *Livestock Prod. Sci.,* 10, 25, 1983.

Symons, L. E. A., Steele, J. W., and Jones, W. O., Tissue protein metabolism in parasitized animals, in *Isotopes and Radiation in Parasitology,* Vienna, International Atomic Energy Authority, 4, 171, 1981.

Tagari, H. and Bergman, E. N., Intestinal disappearance and portal blood appearance of amino acids in sheep, *J. Nutr.,* 108, 790, 1978.

Webb, K. E., Jr., DiRienzo, D. B., and Matthews, J. C., Recent developments in gastrointestinal absorption and tissue utilization of peptides: a review, *J. Dairy Sci.,* 76, 351, 1993.

Webb, K. E., Jr., Matthews, J. C., and DiRienzo, D. B., Peptide absorption: A review of current concepts and future perspectives, *J. Anim. Sci.,* 70, 3248, 1992.

Williams, P. E. V., Pagliani, L., Innes, G. M., Pennie, K., Harris, C. I., and Garthwaite, P., Effect of a beta-agonist (Clenbuterol) on growth, carcass composition, protein and energy metabolism of veal calves, *Br. J. Nutr.,* 57, 417, 1987.

Wilton, J. C. and Gill, M., Uptake of ammonia across the liver of forage fed cattle, *Proc. Nutr. Soc.,* 47, 153A, 1988.

CHAPTER 8

Nutrient Extraction by the Ruminant Mammary Gland

Joseph H. Herbein

INTRODUCTION

Current knowledge of amino acid (AA) utilization by the mammary gland for milk protein synthesis in the dairy cow is based on approximately 25 years of research. Recently practical application has been derived from portions of the research dealing with transfer efficiencies for each of the essential AA (EAA). The efficiencies, which range from 1.00 to 0.42 when expressed as the ratio of output (grams per day, g/d) in milk protein to uptake (g/d) from blood by the mammary gland, are currently used in an empirical computer model (O'Connor et al., 1993) to evaluate dietary nutrient availability vs. requirement. The difference between uptake and output of an EAA represents catabolism within the gland to form nonessential AA or other metabolic intermediates used to support milk synthesis. Metabolic fate of AA at the cellular level is described in considerable detail by a mechanistic computer model (Baldwin et al., 1994) that was designed primarily for planning and evaluating basic research. Although these models presently may be a simplification of complex interactions within the lactating cow, they have the potential to consolidate and improve our understanding of AA utilization for the purpose of reducing nitrogen (N) loss in urine. A primary objective of current research is to reduce excretion of absorbed AA N by reducing catabolism of AA in the mammary gland and other tissues.

After considerations of N partitioning in lactating cows and interrelationships between the primary nutrients that serve as substrates for milk synthesis, the focus of this chapter will be on methods used to evaluate utilization of EAA or other nutrients by the mammary gland. The methods include determination of arteriovenous (AV) concentration differences, extraction ratios, mammary blood or plasma flow, supply, and ratio of uptake from blood to output in milk protein. Simultaneous sampling of blood from an artery and the subcutaneous mammary vein in conjunction with AA analysis of whole blood or plasma will provide an AV difference. Sources of arterial blood have included the carotid, internal iliac, and coccygeal. Extraction ratio is calculated by dividing AV difference (micromolar, μm) by arterial concentration (μm). Supply (micromoles per minute, μmol/min) of an EAA is estimated by multiplying arterial concentration by blood flow (liters per minute, L/min); whereas uptake of an EAA is estimated by multiplying AV difference by blood flow.

1-56670-199-6/96/$0.00+$.50

Determination of mammary blood flow obviously is a critical factor for calculation of supply and uptake of EAA by the gland. Accurate estimates of flow are necessary if we expect computer models to predict required amounts of EAA that must be available for absorption in the gastrointestinal tract. Currently, average milk production/cow/lactation in many herds is 12,000 kg or more, but available estimates of EAA supply and uptake are based on research conducted with cows producing much less. Therefore, current estimates of mammary blood flow and EAA uptake must be extrapolated to account for anticipated rates of milk protein synthesis in high-producing cows during early and peak lactation

NITROGEN PARTITIONING DURING LACTATION

Improvements in models intended to predict relative quantities of each of the EAA needed to optimize efficiency of milk protein synthesis depend on refinement of estimates presently used to describe AA flux within and between major tissues, because AA supply to the mammary gland is the culmination of AA transport and metabolism throughout the body. Deamination and catabolism of nonessential and EAA provides metabolic intermediates for functions other than protein synthesis, such as hepatic gluconeogenesis or energy via complete oxidation. Estimates of EAA catabolism in mammary and other tissues vs. incorporation into milk protein were reported by Armentano (1994), who expressed quantities of each EAA secreted in milk protein relative to its flow into the duodenum. Approximately 30 to 50% of all EAA, except Arginine, flowing to the intestine are absorbed and conserved for milk protein synthesis. Arginine (Arg) may be considered highly conserved also because the mammary gland converts nearly half of the Arg to proline (Pro) for secretion in milk protein (Clark et al., 1975). Assuming protein digestibility is 65%, 46 to 77% of absorbed EAA apparently are conserved. The remainder will be catabolized, primarily in liver, and the N will eventually be excreted in urine.

In terms of environmental considerations, improvements in efficiency of AA utilization for lactation could reduce N loss in urine. Approximately 65% of intake N is absorbed by the lactating cow during early lactation (Figure 1) and is used to meet requirements for maintenance, tissue growth or replenishment, and milk synthesis. Our results indicated 46 and 48% of absorbed N available for productive functions (absorbed N minus maintenance N) were partitioned to milk N in first-lactation and older cows, respectively. On the basis of total absorbed N, 40 and 43% were recovered in milk N. Spires et al. (1975) reported recovery of 49% of absorbed N when five cows received abomasal infusions of casein compared with 44% recovery when infused with a glucose/glutamate/urea mixture. Also, urine N output as a percentage of total N intake was reduced during casein infusion. Thus Spires et al. (1975) introduced two important concepts that prompted other investigators to conduct similar studies. First, there apparently is potential for improving recovery of absorbed N as milk N when an ideal balance of EAA

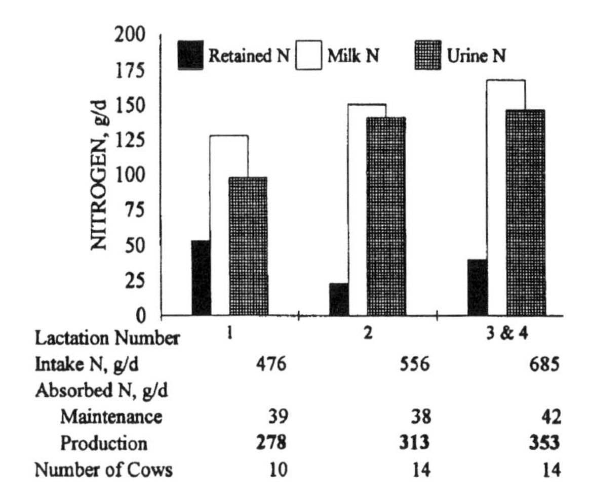

Figure 1 Partitioning of absorbed N available for productive functions in lactating Holstein cows between 30 and 45 d in lactation. Diets contained 16.2% crude protein and 1.61 Mcal NE_L per kilogram of dry matter.

is contained in protein available for intestinal digestion and absorption. Second, improved uptake of EAA by the mammary gland can be achieved through an elevated arterial concentration, which appears to be associated with a greater AV difference.

SUPPLY OF EAA TO THE MAMMARY GLAND

In regard to total AA utilization by the mammary gland, Cant et al. (1993a) and Wonsil (1993) determined that 70 to 79% of N in AA uptake is recovered as milk protein N. When their data were expressed on the basis of total AA mass, recovery was approximately 90%, suggesting minimal opportunity for improving efficiency of AA utilization by the mammary gland. However, many investigators postulated that milk protein synthesis could be increased by supplementing all EAA or selected EAA thought to be limiting the rate of synthesis. Supplemental EAA were provided via the diet or by infusion into the duodenum or blood. Guinard et al. (1994) infused 0 to 762 g of casein per day into the duodenum, which resulted in N intake from 91 to 125% of the requirement. Although increasing amounts of casein resulted in linear increases in milk and milk protein yield, the percentage of total EAA uptake transferred to milk protein decreased linearly. Guinard and Ruilquin (1994a) confirmed that the ratios of uptake to output for individual EAA also increased linearly, suggesting inefficient utilization of excess EAA. Metcalf et al. (1994) also reported increases in the ratios of uptake to output for EAA, but without significantly increasing milk protein yield, when they raised the crude protein content of a basal diet from 14 to 17% by addition of fishmeal. Elevated plasma and milk urea concentrations when cows were fed the fishmeal indicated inefficient utilization of AA in the mammary gland and perhaps increased catabolism to support hepatic gluconeogenesis.

In order to avoid inefficiencies associated with an excess AA supply that resulted from duodenal casein infusion in other studies, Metcalf et al. (1991) used direct infusion of a mixture of selected EAA (threonine, Thr; methionine, Met; leucine, Leu; phenylalanine, Phe; and lysine, Lys) into the arterial blood supply of the mammary gland. Uptake of each of the infused EAA was increased, but milk and milk protein yield were not improved. However, the authors noted tendencies for higher milk and milk protein yield and speculated that a concurrent increase in energy supply is necessary for the mammary gland to respond to additional EAA. Guinard and Ruilquin (1994b) infused 0 to 63 g of Lys per day into the duodenum. Infused Lys supplemented the dietary Lys content, and the sum of the two sources provided 80, 90, 100, or 120% of the requirement. Milk protein percentage increased quadratically, peaking when Lys supply was at 100% of the requirement, but protein yield was not significantly influenced. Arterial Lys concentration, AV difference, and uptake increased linearly as Lys supply increased from 80 to 100% of the requirement. However, Lys uptake at 120% compared with 100% of the requirement was similar due to a reduced extraction ratio.

Overall, it appears that uptake of EAA by the mammary gland was a function of supply, which was proportional to arterial concentration in the studies cited earlier. Unfortunately, further assessment of the potential of supplemental EAA to favorably influence N partitioning to milk is limited by the low number of cows (4 or 5), variability, different methods for determining blood flow, and lack of information concerning N retained by other tissues or N flow to urine in these studies. As suggested by Spires et al. (1975), it may be inappropriate to attribute any favorable response, whether statistically significant or not, to the supply of potentially limiting EAA without considering their utilization in other parts of the body, interactions with other major nutrients, and potential effects on the endocrine system.

EAA INTERRELATIONSHIPS WITH OTHER NUTRIENTS

Much of our present understanding of nutrient uptake was derived from a comprehensive study by Bickerstaffe et al. (1974). Blood flow, AV differences, mammary uptakes and outputs, and whole-body turnover of precursors for milk synthesis were determined for two Jersey and two Fresian cows producing 12 to 26 kg of milk per day. Concepts presented by the authors have been verified by other investigators over the past 20 years. Their results indicated that glucose uptake by the mammary gland accounted for over 60% of whole-body entry rate, with 10 to 20% oxidized by the gland and the remainder converted to lactose. Approximately 30% of acetate uptake was oxidized, with the remainder used primarily for *de novo* synthesis of all fatty acids with chain lengths up to 14 carbons and approximately 63% of the palmitic acid. The remainder of the palmitate and all of the stearate and oleate, with 57% of the oleate derived from desaturation of stearate, were extracted from plasma triacylglycerides.

With respect to EAA, Bickerstaffe et al. (1974) simply noted that uptake seemed to be sufficient to account for their output in milk protein. They also noted a large uptake of Arg relative to output in milk and speculated that the excess was deaminated to provide N for synthesis of nonessential AA. Mepham (1982) reviewed his earlier work with lactating goats indicating that Arg was hydrolyzed to Orn and urea, with some of the Arg carbon used to synthesize Pro. In addition, he suggested grouping EAA according to their ratio of uptake to output. Group1 included Met, Phe, Trp, and Tyr, which are taken up in amounts essentially equal to output in milk. Group 2 included the remaining EAA (Arg; histidine, His; isoleucine, Ile; Leu; Lys; and valine, Val), which are taken in excess of output. Ratios reported in the literature vary, depending on the method used to determine blood flow, but a typical range for group 2 is from 1.1 to 1.9:1, whereas the ratio for Arg is usually 2.5 to 3.0:1.

Lipid Precursors

Thompson and Christie (1991) reported that plasma triacylglycerols (TG) are a primary source of long-chain fatty acids for the mammary gland, and extraction ratios for stearate and palmitate from plasma TG are 74 and 48%, respectively. In general, AV differences for most mammary lipid precursors increase linearly with increasing arterial concentration. Miller et al. (1991a) reported linear relationships for acetate, TG, and nonesterified fatty acids (NEFA) when using combined data from cows with and without bovine somatotropin (bST) treatment. Their data also indicated a negative AV difference for NEFA at lower concentrations, suggesting hydrolysis of TG by lipoprotein lipase released excess fatty acids that elevated NEFA concentration in venous blood. The relationship between AV difference and arterial concentration of β-hydroxybutyrate (BHB) also was described by a linear equation, but a Michaelis-Menten equation provided a better correlation. Cant et al. (1993b) found elevated concentrations of TG, NEFA, and BHB in blood plasma due to feeding supplemental fat and reported that all relationships between AV difference and arterial concentration were linear. Gagliostro et al. (1991) reported that postprandial concentrations of NEFA and TG in plasma were elevated in response to duodenal rapeseed oil infusion in midlactation, but only TG was elevated during early lactation. Using jugular vein samples to estimate arterial concentrations of the lipid precursors, they found linear relationships for AV differences and concentrations of TG, NEFA, and BHB. The AV difference for TG was consistently greater when oil was infused, regardless of stage of lactation.

Current feeding recommendations usually include supplemental fat, which increases dietary energy density and generally increases milk yield. Unfortunately, supplemental fat also has the potential to reduce milk protein concentration. The response is often described as protein dilution because enhanced milk production is not accompanied by a proportional increase in protein yield. Whether the lack of a concurrent increase in protein synthesis is directly related to increased extraction of lipid precursors has not been determined. Wu and Huber (1994) recently reviewed literature dealing with changes in milk protein percentages associated with dietary fat supplementation. They briefly discussed hypotheses presented by others as explanations for the response and then concluded that AA supply to the mammary gland was the limiting factor. Casper and Shingoethe (1989) found lower milk protein percentages and lower concentrations of 11 of 20 AA assayed in arterial and venous blood when cows (n = 47) fed supplemental fat in the form of whole oilseeds were compared with control cows (n = 50). Lower AA concentrations in most cases were accompanied by lower AV differences. Uptake, however, apparently was not reduced, possibly because they estimated blood flow based on 400 L of plasma or serum flow per kilogram of milk. Erickson et al. (1992) and Cant et al. (1993a) also estimated AA availability at the mammary gland in relation to decreased milk protein percentage when cows were fed calcium salts of long-chain fatty acids or yellow grease. Blood flow in these studies, however, was estimated on the basis of AV differences for Phe plus Tyr, which are group 1 AA extracted from blood in amounts equal to output in milk, as suggested by Mepham (1982). Erickson et al. (1992) found no major changes in EAA concentration, AV difference, or uptake when cows were fed supplemental fat. Cant et al. (1993a) reported that concentrations of AA in arterial plasma tended to be lower and plasma urea N higher, suggesting increased utilization of AA for hepatic gluconeogenesis. Uptake of AA by the mammary gland, however, was similar for control and fat-supplemented cows. Apparently, the mammary gland's capacity to increase daily milk volume in response to availability of lipid precursors is independent of factors that regulate protein synthesis. Thus attempts to maintain or increase protein concentration as milk yield increases must focus on methods to enhance overall AA supply or EAA supply to the gland by increasing arterial concentrations of AA or increasing the rate of blood flow.

Glucose Availability

Sutton and Morant (1989) reviewed factors that influence milk protein yield and concluded that ratio of forage to concentrate was the primary factor. At a constant digestible energy intake, replacement of forage with readily fermentable starch sources increased milk and milk protein yield. Increasing availability of propionate for gluconeogenesis apparently spared AA from hepatic catabolism, thus making AA available for mammary protein synthesis. Including concentrates at greater than 50% of diet dry matter, which is current practice for high-producing cows, does not appear to provide additional benefit in protein percentage. Coomer et al. (1993) confirmed the lack of a milk protein response when they fed high concentrate diets containing 17.2% crude protein and 28, 31, or 37% nonstructural carbohydrate with corresponding acid detergent fiber concentrations at 27, 24, or 21%. Although milk production (40 kg/d) and dry matter intake (25 kg/d) were not significantly different, there was a significant linear increase in milk production per kilogram of dry matter intake with increasing nonstructural carbohydrate content of the diet. The increase in efficiency was at the expense of milk fat content, which decreased from 3.5 to 3.1%, because milk protein concentration and yield did not change. Individual arterial AA concentrations, AV differences, and ratios of uptake to output also were not affected by dietary treatments. Thus factors regulating milk protein synthesis apparently are independent of glucose availability in the mammary gland when cows are fed adequate amounts of concentrate.

Peptides

A recent study indicated that 52 to 78% of the AA available for tissue uptake from arterial blood of rats, steers, and sheep are in the form of low molecular weight peptides (Seal and Parker, 1991). Backwell et al. (1994) demonstrated the ability of mammary glands of goats to utilize exogenous dipeptides as a source of AA for milk protein synthesis. Our work (Wonsil, 1993) with lactating cows indicated that approximately 10% of the available AA in whole blood were in the form of low molecular weight peptides, but there was no net uptake of peptide AA by the mammary gland. Within the peptide pool, glycine (Gly) accounted for approximately 60% of the AA. Additional studies are needed to evaluate the potential of peptide AA for enhancing the efficiency of AA utilization by the mammary gland.

MAMMARY BLOOD FLOW

Davis and Collier (1985) reviewed methods for determining mammary blood flow and emphasized the importance of flow in combination with arterial concentration of a substrate in regulating the supply of that substrate for milk synthesis. Thermodilution, electromagnetic flow meters, ultrasound flow meters, the Fick principle, and dye dilution procedures were in use at that time. Since then, tritiated water was used as a flow indicator by McDowell et al. (1987). Thermodilution (Bickerstaffe et al., 1974) and the antipyrine method (Kronfeld and Hartmann, 1973) were used to obtain some of the first estimates of mammary blood flow in lactating cows. Electromagnetic and ultrasonic flow meters can provide continuous measurement of flow, but they require extensive surgical preparations and limit the number of cows per experiment. Ultrasonic flow meters have been used in many recent studies (Prosser and Davis, 1992; Guinard et al., 1994; Guinard and Ruilquin, 1994a, 1994b; Stelwagen et al. 1994; Boirie et al., 1995). However, when Metcalf et al. (1992) compared the ultrasonic meters with the dye (*p*-amino hippuric acid) dilution method, the ultrasound estimates were 51 to 54% of the estimates from dye dilution. Wonsil (1993) compared ultrasound with the Fick principle (using Phe + Tyr) and found ultrasound estimates to be 46 to 56% of the estimates by the Fick principle. When comparing electromagnetic meters with the Fick principle (Phe + Try), however, Davis et al. (1988) found no difference between methods.

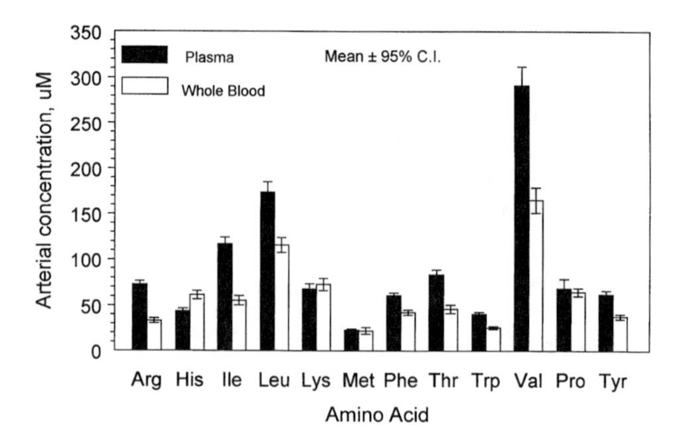

Figure 2 Concentrations of essential amino acids, Pro, and Tyr in plasma (n = 71) and whole blood (n = 24) from the coccygeal artery of Holstein cows in early through midlactation.

Presently, the Fick principle is used by many investigators, primarily because surgical procedures are not required and the number of cows per treatment is not restricted. In order to apply the Fick principle, uptake of the tracer AA by the gland must equal the amount secreted in milk. Methionine (Davis and Bickerstaffe, 1978), tryptophan (Davis and Collier, 1985), and glutamine (Meijer et al., 1993) have been suggested as tracer AA, but are rarely used. However, the sum of Phe and Try, as recommended by Spires et al. (1975), recently has been applied to the Fick principle by many investigators (Erickson et al., 1992; Cant et al., 1993a, 1993b; Wonsil, 1993; Metcalf et al., 1994). Because erythrocytes are a potential source of many AA, Hanigan et al. (1991) recommended that whole blood rather than plasma be used to determine AA supply and uptake by the mammary gland. Concentrations of EAA, Pro, and Tyr in whole blood and plasma are compared in Figure 2. Cant et al. (1993a) and Wonsil (1993) found that AA uptakes calculated using plasma or whole-blood AV differences were similar.

Milk production of cows used in studies to determine mammary supply and uptake of EAA ranged from 24 to 40 kg/d. Many commercial herds, however, have cows with peak production over 50 kg/d. Coccygeal artery and subcutaneous mammary vein samples were obtained from cows used for feeding trials over a 2-year period to estimate blood flow by applying the Fick principle (Phe + Tyr). The relationship between milk production and blood flow (Figure 3) was curvilinear, with a greater rate of increase in blood flow from 40 and 60 kg/d than from 20 to 40 kg/d. With respect to projected supply of EAA to the mammary gland, AV differences for Lys in relation to arterial concentration are presented in Figure 4 as an example. The linear relationship reflects the substantial influence of arterial concentrations of EAA on uptake. However, the decline in arterial Lys concentration at higher rates of milk production (Figure 4 insert) suggests that the curvilinear relationship between blood flow and milk production is necessary to maintain Lys supply.

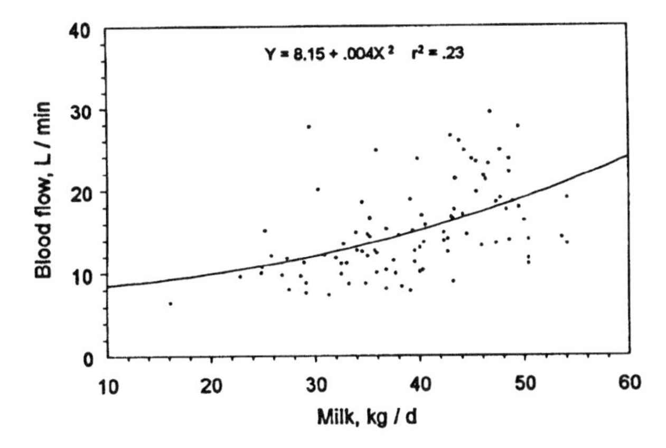

Figure 3 Predicted blood flow relative to milk production in Holstein cows (n = 95) fed diets containing 16 to 17% crude protein. Blood flow was estimated using the Fick principle, with the assumption that Phe + Tyr output in milk protein is equivalent to uptake from blood × blood flow.

Because milk production is occasionally used to estimate blood flow (Casper and Shingoethe, 1989 and Coomer et al., 1993, for example), the estimated changes in blood flow (liters per minute, L/min) in Figure 3 also were expressed as liters per kilogram (L/kg) of milk in Figure 5. A higher ratio of blood flow to milk yield by cows was noted at lower levels of production (Bickerstaffe et al., 1974) and also noted for sheep (Mepham, 1982) and goats (Fleet and Mepham, 1985). Flows from 500 to 1100 L/kg were reported for goats in late lactation. The higher ratios during late lactation probably reflect flow to maintain nonsecretory tissues, whereas higher ratios starting at 40 to 50 kg/d (Figure 5) reflect regulated flow to maintain substrate supply relative to rates of milk component synthesis.

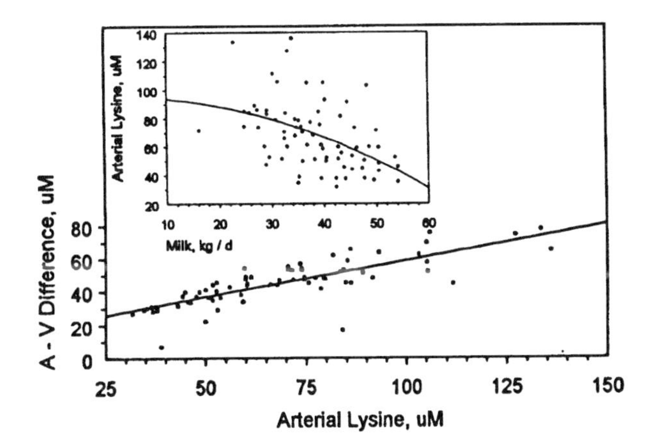

Figure 4 Changes in Lys AV difference relative to arterial concentration and changes in arterial concentration with increasing milk yield (insert).

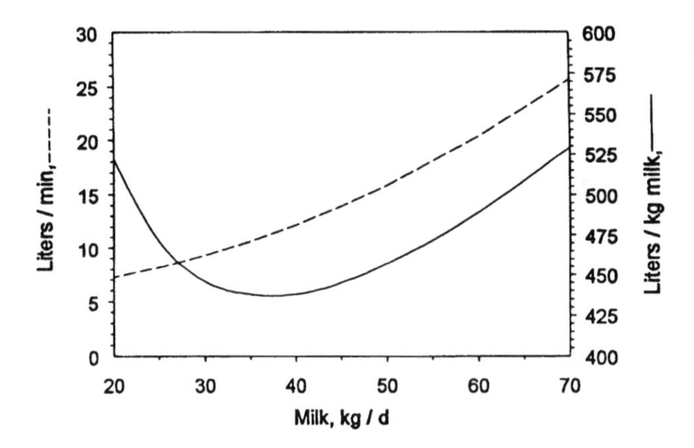

Figure 5 Changes in mammary blood flow per kilogram of milk calculated using the equation for predicted blood flow (L/min) in Figure 3.

BLOOD FLOW REGULATION

Diurnal increases in mammary blood flow have been associated with udder massage, oxytocin injection, milking (Davis and Collier, 1985; Metcalf et al., 1992), and milking frequency (Prosser and Davis, 1992). Milk accumulation in the gland decreased blood flow (Stelwagen et al., 1994). Decreases in blood flow also can occur in response to catecholamines or fasting in cows (Davis and Collier, 1985). Lower blood flow in response to fasting also was reported for rats and women (Vina et al., 1987). Information concerning local or humoral control of vascular tone and capillary recruitment in the mammary gland of cows was limited when Davis and Collier (1985) reviewed the topic. Since then, Gorewit et al. (1993) reported that angiotensins cause vasoconstriction of mammary arteries from lactating cows, and the response can be blocked by receptor agonists. Nitric oxide (NO) was shown to be an important local regulator of vascular tone and was responsible for changes in blood flow in the gastrointestinal tract and muscle of exercised rats (Tadakazu et al., 1994). The possibility that NO regulates vasodilation and/or blood flow in the mammary gland should be investigated. Arginine is the precursor for formation of NO, which has a very short half-life, but administration of Arg to dairy cows via abomasal or intravenous infusion had no effect on milk yield or composition (Vincini et al., 1988). However, intravenous Arg did induce release of somatotropin, which has been associated with stimulation of mammary blood flow and milk yield.

Davis et al. (1988) reported that mammary blood flow, determined by electromagnetic flow probe and the Fick principle (Phe + Tyr), increased 35% and milk production 25% when cows were treated with bST. Goats responded in a similar manner when treated with ovine growth hormone (Mepham, 1982) or when insulin-like growth factor-I was infused directly into the mammary arterial supply (Prosser and Davis, 1992). McDowell et al. (1987), in addition to demonstrating the blood flow response to bST in cows, determined mammary and muscle extraction of glucose, NEFA, acetate, BHB, and lactate. Arterial glucose and NEFA concentrations increased in response to bST. Glucose AV difference in muscle decreased, but the AV difference in the mammary gland remained constant. Bauman et al. (1988) did not find an elevated arterial glucose concentration, but confirmed that reduced glucose oxidation in combination with increased irreversible loss of glucose was sufficient to account for the increased rate of milk lactose synthesis. In contrast, increased oxidation of NEFA in combination with increased incorporation into milk fat accounted for the increased irreversible loss of NEFA. Miller et al. (1991b) concluded that changes in concentrations and AV differences for glucose and lipid precursors in bST-treated cows are similar to those in cows during early lactation. Regarding N partitioning, bST did not influence the proportion of intake N excreted in feces or urine (Tyrrell et al., 1988). However, the measurements were made during short-term bST administration when body tissue was mobilized as the additional source of AA for milk protein synthesis.

IMPLICATIONS

Previous work indicated little benefit from supplying the mammary gland with excess AA. Supplying combinations of EAA only in amounts required to compensate for deficiencies appears to be an efficacious approach to optimization of absorbed N partitioning to milk. Schwab et al. (1992) used this approach and demonstrated the potential for recovering a greater percentage of intake N in milk when cows in early lactation received duodenal infusions of 10, 20, or 30 g of Lys in combination with 10 g of Met on a daily basis. Essential AA protected from degradation in the rumen may provide a practical method to compensate for deficiencies, as indicated by greater yield and concentration of milk protein when diets of cows were supplemented with ruminally protected Lys and Met during the first 100 d of lactation (Armentano et al., 1993). However, the disadvantage associated with supplementing one or two EAA thought to be most limiting is that one of the remaining EAA then becomes first limiting or colimiting.

Exogenous bST effectively and efficiently enhances nutrient supply to the mammary gland by increasing blood flow. Endogenous regulation of blood flow during high rates of milk production in early lactation (Figure 5) may be mediated in a similar manner. Other factors such as NO, therefore, should be investigated for their potential to sustain higher rates of blood flow and milk yield after peak lactation. Future considerations also should include investigations of the capacity of the mammary gland to synthesize protein, which may be subject to alteration prior to parturition. For example, a β-adrenergic agonist administered to rats during mammary gland development enhanced production of cytoplasmic mRNAs for casein and whey proteins throughout lactation (Choi and Han, 1993).

REFERENCES

Armentano, L. E., Impact of metabolism by extragastrointestinal tissues on secretory rate of milk proteins, *J. Dairy Sci.,* 77, 2809, 1994.

Armentano, L. E., Swain, S. M., and Ducharme, G. A., Lactation response to ruminally protected methionine and lysine at two amounts of ruminally available nitrogen, *J. Dairy Sci.,* 76, 2963, 1993.

Backwell, F. R. C., Bequette, B. J., Wilson, D., Calder, A. G., Metcalf, J. A., Wray-Cahen, D., MacRae, J. C., Beever, D. E., and Lobley, G. E., Utilization of dipeptides by the caprine mammary gland for milk protein synthesis, *Am. Phys. Soc.,* 267, R1, 1994.

Baldwin, R. L., Emery, R. S., and McNamara, J. P., Metabolic relationships in the supply of nutrients for milk protein synthesis: integrative modeling, *J. Dairy Sci.,* 77, 2821, 1994.

Bauman, D. E., Peel, C. J., Steinhour, W. D., Reynolds, P. J., Tyrrell, H. F., Brown, A. C. G., and Haaland, G. L., Effect of bovine somatrotropin on metabolism of lactating dairy cows: influence on rates of irreversible loss and oxidation of glucose and nonesterified fatty acids, *J. Nutr.,* 118, 1031, 1988.

Bickerstaffe, R., Annison, E. F., and Linzell, J. L., The metabolism of glucose, acetate, lipids and amino acids in lactating cows, *J. Agric. Sci.,* 82, 71, 1974.

Boirie, Y., Fauquant, J., Rulquin, H., Maubois, J., and Beaufrere, B., Production of large amounts of [^{13}C]leucine-enriched milk proteins by lactating cows, *J. Nutr.,* 125, 92, 1995.

Cant, J. P., DePeters, E. J., and Baldwin, R. L., Mammary amino acid utilization in dairy cows fed fat and its relationship to milk protein depression, *J. Dairy Sci.,* 76, 762, 1993a.

Cant, J. P., DePeters, E. J., and Baldwin, R. L., Mammary uptake of energy metabolites in dairy cows fed fat and its relationship to milk protein depression, *J. Dairy Sci.,* 76, 2254, 1993b.

Casper, D. P. and Schingoethe, D. J., Model to describe and alleviate milk protein depression in early lactation dairy cows fed a high fat diet, *J. Dairy Sci.,* 72, 3327, 1989.

Choi, Y. J. and Han, I. K., The effects of the β-adrenergic agonist cimaterol (CL 263, 780) on mammary differentiation and milk protein gene expression, *J. Nutr. Biochem.,* 4, 274, 1993.

Clark, J. H., Derrig, R. G., Davis, C. L., and Spires, H. R., Metabolism of arginine and ornithine in the cow and rabbit mammary tissue, *J. Dairy Sci.,* 58, 1808, 1975.

Coomer, J. C., Amos, H. E., Williams, C. C., and Wheeler, J. G., Response of early lactation cows to fat supplementation in diets with different nonstructural carbohydrate concentrations, *J. Dairy Sci.,* 76, 3747, 1993.

Davis, S. R. and Bickerstaffe, R., Mammary glucose uptake in the lactating ewe and the use of methionine arterio-venous difference for the calculation of mammary blood flow, *Aust. J. Biol. Sci.,* 31, 133, 1978.

Davis, S. R. and Collier, R. J., Mammary blood flow and regulation of substrate supply for milk synthesis, *J. Dairy Sci.,* 68, 1041, 1985.

Davis, S. R., Collier, R. J., McNamara, J. P., Head, H. H., and Sussman, W., Effects of thyroxine and growth hormone treatment of dairy cows on milk yield, cardiac output, and mammary blood flow, *J. Anim. Sci.,* 66, 70, 1988.

Erickson, P. S., Murphy, M. R., and Clark, J. H., Supplementation of dairy cow diets with calcium salts of long-chain fatty acids and nicotinic acid in early lactation, *J. Dairy Sci.,* 75, 1078, 1992.

Fleet, I. R. and Mepham, T. B., Mammary uptake of amino acids and glucose throughout lactation in Friesland sheep, *J. Dairy Res.,* 52, 229, 1985.

Gagliostro, G., Chilliard, Y., and Davicco, M., Duodenal rapeseed oil infusion in early and midlactation cows. 3. Plasma hormones and mammary apparent uptake of metabolites, *J. Dairy Sci.,* 74, 1893, 1991.

Gorewit, R. C., Jiang, J., and Aneshansley, D. J., Responses of the bovine mammary artery to angiotensins, *J. Dairy Sci.,* 76, 1278, 1993.

Guinard, J. and Rulquin, H., Effect of graded levels of duodenal infusions of casein on mammary uptake in lactating cows. 2. Individual amino acids, *J. Dairy Sci.,* 77, 3304, 1994a.

Guinard, J. and Rulquin, H., Effects of graded amounts of duodenal infusions of lysine on the mammary uptake of major milk precursors in dairy cows, *J. Dairy Sci.,* 77, 3565, 1994b.

Guinard, J., Rulquin, H., and Verite, R., Effect of graded levels of duodenal infusions of casein on mammary uptake in lactating cows. 1. Major nutrients, *J. Dairy Sci.,* 77, 2221, 1994.

Hanigan, M. D., Calvert, C. C., DePeters, E. J., Reis, B. L., and Baldwin, R. L., Whole blood and plasma amino acid uptakes by lactating bovine mammary glands, *J. Dairy Sci.,* 74, 2484, 1991.

Kronfeld, D. S. and Hartmann, P. E., Glucose redistribution in lactating cows given dexamethasone, *J. Dairy Sci.,* 56, 903, 1973.

McDowell, G. H., Gooden, J. M., Leenanuruksa, D., Jois, M., and English, A. W., Effects of exogenous growth hormone on milk production and nutrient uptake by muscle and mammary tissues of dairy cows in mid-lactation, *Aust. J. Biol. Sci.,* 40, 295, 1987.

Meijer, G. A., van der Meulen, J., and van Vuuren, A. M., Glutamine is a potentially limiting amino acid for milk production in dairy cows: a hypothesis, *Metabolism,* 42, 358, 1993.

Mepham, T. B., Amino acid utilization by lactating mammary gland, *J. Dairy Sci.,* 65, 287, 1982.

Metcalf, J. A., Sutton, J. D., Cockburn, J. E., Napper, D. J., and Beever, D. E., The influence of insulin and amino acid supply on amino acid uptake by the lactating bovine mammary gland, *J. Dairy Sci.,* 74, 3412, 1991.

Metcalf, J. A., Roberts, S. J., and Sutton, J. D., Variations in blood flow to and from the bovine mammary gland measured using transit time ultrasound and dye dilution, *Res. Vet. Sci.,* 53, 59, 1992.

Metcalf, J. A., Beever, D. E., Sutton, J. D., Wray-Cahen, D., Evans, R. T., Humpries, D. J., Backwell, F. R. C., Bequette, B. J., and MacRae, J. C., The effect of supplementary protein on in vivo metabolism of the mammary gland in lactating dairy cows, *J. Dairy Sci.,* 77, 1816, 1994.

Miller, P. S., Reis, B. L., Calvert, C. C., DePeters, E. J., and Baldwin, R. L., Patterns of nutrient uptake by the mammary glands of lactating dairy cows, *J. Dairy Sci.,* 74, 3791, 1991a.

Miller, P. S., Reis, B. L., Calvert, C. C., DePeters, E. J., and Baldwin, R. L., Relationship of early lactation and bovine somatotropin on nutrient uptake by cow mammary glands, *J. Dairy Sci.,* 74, 3800, 1991b.

O'Conner, J. D., Sniffen, C. J., Fox, D. G., and Chalupa, W., A net carbohydrate and protein system for evaluating cattle diets: IV. Predicting amino acid adequacy, *J. Anim. Sci.,* 71, 1298, 1993.

Prosser, C. G. and Davis, S. R., Milking frequency alters the milk yield and mammary blood flow response to intra-mammary infusion of insulin-like growth factor-I in the goat, *J. Endocrinol.,* 135, 311, 1992.

Schwab, C. G., Bozak, C. K., Whitehouse, N. L., and Olson, V. M., Amino acid limitation and flow to the duodenum at four stages of lactation, 2. Extent of lysine limitation, *J. Dairy Sci.,* 75, 3503, 1992.

Seal, C. J. and Parker, D. S., Isolation and characterization of circulating low molecular weight peptides in steer, sheep and rat portal and peripheral blood, *Comp. Biochem. Physiol.,* 99, 679, 1991.

Spires, H. R., Clark, J. H., Derrig, R. G., and Davis, C. L., Milk production and nitrogen utilization in response to postruminal infusion of sodium caseinate in lactating cows, *J. Nutr.,* 105, 1111, 1975.

Stelwagen, K., Davis, S. R., Farr, V. C., Prosser, C. G., and Sherlock, R. A., Mammary epithelial cell tight junction integrity and mammary blood flow during an extended milking interval in goats, *J. Dairy Sci.,* 77, 426, 1994.

Sutton, J. D. and Morant, S. V., A review of the potential of nutrition to modify milk fat and protein, *Livestock Prod. Sci.,* 23, 219, 1989.

Tadakazu, H., Visneski, M. D., Kearns, K. J., Zelis, R., and Musch, T. I., Effects of NO synthase inhibition on the muscular blood flow response to treadmill exercise in rats, *J. Appl. Physiol.,* 77, 1288, 1994.

Thompson, G. E. and Christie, W. W., Extraction of plasma triacylglycerols by the mammary gland of the lactating cow, *J. Dairy Res.,* 58, 251, 1991.

Tyrrell, H. F., Brown, A. C. G., Reynolds, P. J., Haaland, G. L., Bauman, D. E., Peel, C. J., and Steinhour, W. D., Effect of bovine somatotropin on metabolism of lactating dairy cows: energy and nitrogen utilization as determined by respiration calorimetry, *J. Nutr.,* 118, 1024, 1988.

Vicini, J. L., Clark, J. H., Hurley, W. L., and Bahr, J. M., Effects of abomasal or intravenous administration of arginine on milk production, milk composition, and concentrations of somatotropin and insulin in plasma of dairy cows, *J. Dairy Sci.,* 71, 658, 1988.

Vina, J. R., Puertes, I. R., Rodriguez, A., Saez, G. T., and Vina, J., Effect of fasting on amino acid metabolism by lactating mammary gland: studies in women and rats, *J. Nutr.,* 117, 533, 1987.

Wonsil, B. J., Influence of dietary fat and protein on nutrient supply and utilization by the lactating bovine mammary gland, Ph.D. Dissertation, Virginia Polytechnic Institute and State University, Blacksburg, 1993.

Wu, Z. and Huber, J. T., Relationship between dietary fat supplementation and milk protein concentration in lactating cows: a review, *Livestock Prod. Sci.,* 39, 141, 1994.

CHAPTER 9

Evaluation of New Technologies for the Improvement of Nitrogen Utilization in Ruminants

Martin N. Sillence

INTRODUCTION

Optimizing the efficiency of animal production has never been more important, as concern grows over diminishing natural resources and the increased output of waste products into the environment. It has been recognized for a number of years that the efficiency of nitrogen (N) utilization in ruminants can vary markedly, and the use of synthetic steroid growth promotants has illustrated the vast potential for improvements in this area. The aim of this chapter is to examine technologies that may be used in the future to reduce N excretion in ruminants. The first section will describe a range of synthetic drugs and hormones, including β-agonists and somatotropin (ST), which could become available for commercial use in the next few years. Next, a number of different vaccine technologies will be described that are aimed to enhance N utilization through immunological means. Finally, a brief review of transgenic technologies will be presented. Because of the number and diversity of approaches currently being explored, together with the fact that each of the methods described has its own advantages and disadvantages, the future of these technologies is difficult to predict and will be left to the reader. Instead, this chapter seeks only to describe the powerful effects of these technologies on N metabolism and to discuss the factors that affect their efficacy, practical constraints to their use, and their potential adverse effects on animal health and meat quality.

DRUGS AND HORMONES

β-Adrenoceptor Agonists

The first family of β-agonist drugs was developed during the early 1970s, specifically to cause the relaxation of smooth muscle such as that found in the lungs and uterus (Engelhardt, 1972). Since that time, compounds like clenbuterol have been used widely to treat asthma and premature labor, both in humans and animals. The anabolic effects of β-agonists were not discovered until the 1980s,

1-56670-199-6/96/$0.00+$.50

when it was found that large doses of clenbuterol cause hypertrophy of skeletal muscle in rats (Emery et al., 1984), lambs (Baker et al., 1984), and cattle (Ricks et al., 1984). There are several factors that make clenbuterol unsafe for use in animal production, but the potential use of suitable clenbuterol analogs has been the subject of both intense interest and concern over the past decade. While several major companies have been working toward the registration of β-agonists for use in the United States, a blanket ban on the use of all growth promoters within the European Economic Community has encouraged some European farmers to obtain clenbuterol illegally. Its abuse has resulted in the hospitalization of farm workers, as well as consumers of animal products (Anonymous, 1991), with the likely result that other β-agonists will not be approved for commercial use in Europe in the foreseeable future. Nevertheless, because of their potent effects on N metabolism, the possibility that their use will be approved in the United States, and the new technologies that their discovery has spawned, the β-agonists deserve careful attention.

Anabolic Properties

β-Agonists have little or no long-term effects on feed intake or on digestive efficiency, as evidenced by unaltered fecal N excretion (Parkins et al., 1989). However, their dietary administration causes a profound increase in N retention that is reflected by a fall in urinary N excretion (Hovell et al., 1989: Parkins et al., 1989; Williams et al., 1987). The extra N retained by β-agonist-treated animals is not distributed uniformly throughout the body, but instead is used almost exclusively in a marked drive to deposit skeletal muscle protein. This appears to be true in all species and is conveniently illustrated in a study where veal calves were treated with clenbuterol. The increase in carcass weight in treated cattle was four times greater than the increase in liveweight (Williams et al., 1987). The muscle growth response to β-agonists is not uniform, however, and individual muscles from the same animal can show markedly different responses according to their fiber type. In cattle, this effect translates into an increased proportion of muscle in the hindquarters, an effect that potentially increases the value of the carcass (Moloney et al., 1994).

It has been argued that β-agonists increase muscle mass by reducing the rate of protein degradation, a mechanism that requires no additional energy or N (Reeds et al., 1986). However, the current prevailing view is that β-agonists cause a marked, if short-term, increase in protein synthesis, as observed in one of the original studies (Emery et al., 1984). In the well-fed animal, there is no significant change in energy input or output, and therefore the additional energy required for accelerated protein synthesis is obtained partly from an increased efficiency with which dietary energy is used and partly from the mobilization of body fat (Lindsay et al., 1993). The extra N deposited in muscle can be accounted for by a marked reduction in urinary N excretion, but the mobilization of other proteinaceous tissues may provide an additional source: the liver, heart, and hide often decrease in mass following β-agonist treatment (Chikhou et al.,

1993b). This ability to cause a relative shift in nutrients from one body compartment to another has led to the β-agonists being dubbed as "repartitioning agents" rather than "growth promotants." Thus, depending on the amount of fat and viscera lost in β-agonist-treated ruminants, in proportion to the amount of muscle gained, liveweight can be either unaffected (Baker et al., 1984; Ricks et al., 1984; Williams et al., 1987) or can increase by as much 25% (Vestergaard et al., 1994). Similarly, repartitioning effects of β-agonists between fat and muscle within the carcass can sometimes result in an improvement in lean yield, with no obvious increase in carcass weight (Chikhou et al., 1993b; Miller et al., 1988; Ricks et al., 1984). In contrast to ruminants, when rats are treated β-agonists they show a marked and consistent increase in liveweight that gives a quantitative measure of the amount of muscle gained. For this reason, rats have been a popular model to screen new β-agonists and have been particularly useful for examining their temporal effects and elucidating their mechanism of action.

Practical Constraints, Tissue Residues, and Withdrawal

The fact that most β-agonists are orally active is both their greatest asset and their greatest drawback. The ability to administer β-agonists by mixing them with feed is particularly convenient for livestock producers who work in intensive systems, and in the long run this practical advantage may be crucial in allowing β-agonists to be more competitive than technologies such as ST, that have to be given by repeated injection. The negative aspect of β-agonists, however, is the increased threat to the consumer who inadvertently ingests animal products that have a high drug residue content. In this regard, protein hormones such as ST, that are relatively unstable to heat (cooking) and readily broken down in the gut, would have a theoretical advantage. In order to obtain a clear perspective on this issue, it is useful to have some knowledge of the pharmacokinetics of β-agonists, a topic that was reviewed in detail by Morgan (1990).

First, it should be understood that many β-agonists are orally active by design, being derived from synthetic compounds originally intended for clinical use. These phenylethanolamine drugs are usually well absorbed from the gut, and most are resistant to attack by the enzyme catechol-*o*-methyl transferase that normally inactivates their naturally occurring counterpart, epinephrine. With most β-agonists, however, protection from biological inactivation is not complete. After these drugs are absorbed from the gut, the hepatic portal system makes their first destination the liver, where the majority of these compounds undergo a significant degree of inactivation through sulfate conjugation (Morgan, 1990). Because of this "first pass metabolism," the bioavailability of a given amount of drug can be markedly less after oral dosing, compared with parenteral administration. Although there are no reports where different routes of administration have been compared in ruminants, the importance of this factor has been well illustrated in the rat. It has been shown that β-agonists such as salbutamol (albuterol) and salmeterol are potent growth promoters in this species when given by continuous infusion, but if presented in the feed, the daily dose of these compounds needs to be increased by up to 20-fold to elicit an equivalent growth response (Choo et al., 1992; Moore et al., 1994). Similarly, although ractopamine has been designed as a β-agonist feed additive for use in pigs, its administration in the feed of rats causes no growth response at doses that increase weight gain markedly when given by injection (Smith and Paulson, 1994). Because the pharmacokinetics of β-agonists are well understood, and if the preferred method of administering them to farm animals is as a feed additive, one might question why the pharmaceutical companies have not chosen to select compounds that are particularly resistant to sulfate conjugation and hence have extremely high oral activity. The answer lies in the fact that one such compound, clenbuterol, already exists, and important lessons have been learned from its abuse.

Clenbuterol is exceptional among β-agonists because of several unusual chemical properties. The first is betrayed by the fact that a given dose of clenbuterol is equally, if not more, effective when presented orally than by a parenteral route (Moore et al., 1994). This is possible because the chlorinated phenyl ring in clenbuterol makes the drug resistant to sulfation. Other unusual features are that clenbuterol has a marked capacity for plasma protein binding (97 vs. 8% for salbutamol) and is highly lipid soluble (Morgan, 1990). The latter property makes clenbuterol prone to become concentrated in fatty tissue and then released slowly, contributing to the long plasma half-life of the drug. Residues of clenbuterol in fatty tissue and other organs have become a danger to European consumers, and an excellent perspective on this problem can be obtained from the work of Meyer and Rinke (1991). These workers found that although clenbuterol levels in muscle could be reduced to an acceptable level following a 14-d withdrawal period, excessive amounts of the drug could still be found in the liver and abdominal fat of treated cattle. Unfortunately, more prolonged withdrawal periods are likely to result in no end benefit to the producer from using β-agonists. This was observed by Barash et al. (1994), who fed cimaterol to heifers for 4 months and then withdrew the drug. Cimaterol caused an 11-kg increase in liveweight within 9 weeks, and with continuous treatment this difference was maintained for up to 13 weeks. However, within 5 weeks of cimaterol withdrawal, not only had the liveweight advantage been lost, but the animals that had formerly received the drug were 18 to 19 kg lighter than their untreated counterparts. Furthermore, this difference was still apparent 9 weeks after withdrawal.

As a result of these factors, a conscious decision has been made to develop new β-agonists that are relatively short acting and are eliminated from the body quickly (Asato et al., 1984). This should allow brief withdrawal times, no residues, and no significant attenuation of the production gain. However, inherent in this approach is that the compounds will need to be given either continuously or at very frequent intervals. In practice, this will almost certainly confine the use of β-agonists to intensive production systems. In theory, a slow-release implant of β-agonist could be made; however, based on the potency of the drugs that are currently available, such a device would need to be large, containing between 300 and 500 mg of active material in order to provide a response in cattle over 30 to

50 d. In addition, withdrawal of the implant before slaughter would be desirable, but would need to be carefully timed.

An interesting footnote to the issue of drug residues is that clenbuterol was found to reach unusually high concentrations in the eyes of treated animals (Meyer and Rinke, 1991). Screening the eyes of beef carcasses has provided a nondestructive and highly sensitive test for illegal clenbuterol use and is reported to have had a marked impact on reducing the extent of this problem (Elliot et al., 1994).

Factors that Affect the Efficacy of β-Agonists

Choice of Agonist

There are now several different β-agonists being tested in animal experiments (BRL 47672, cimaterol, clenbuterol, L-644969, ractopamine, Ro 168714, salbutamol, and salmeterol), and it is likely that these compounds differ in potency, duration of action, and selectivity to some extent. Unfortunately, because of the many factors that influence efficacy, it is difficult to make meaningful comparisons based on different publications. The few drugs that have been compared directly include clenbuterol vs. salbutamol or salmeterol in rats (Choo et al., 1992; Moore et al., 1994), clenbuterol vs. cimaterol in sheep (Hovell et al., 1989; Warriss et al., 1989), and clenbuterol vs. bitolterol in calves (Parkins et al., 1989). Evidence from these studies shows that, in principal, the "brand" of drug may be one of the less important factors, provided that the dose rate and route of administration are appropriate.

Duration of Treatment

When β_2-adrenoceptors are stimulated constantly, they decline in number at a rate proportional to the dose and potency of the compound used. Thus, in rat muscle, the population of β_2-adrenoceptors can decrease by up to 50% with a few days of clenbuterol treatment (Rothwell et al., 1987; Sillence et al., 1991). A similar response was seen to cimaterol, and this coincided with a decline in the growth response to the drug (Kim et al., 1992). As a result of this downregulation, the optimum length of treatment with clenbuterol for rats is probably only between 4 and 10 d. A similar attenuation of effect occurs in ruminants, though apparently through mechanisms involving postreceptor desensitization rather than downregulation (Sillence, 1995). O'Connor et al. (1991) observed that a cimaterol-induced increase in muscle weight in lambs was maximal within 3 weeks of treatment, while Pringle et al. (1993) reported that the major part of the muscle growth response to L-644969 in lambs occurred within 2 weeks. In cattle, the effects of clenbuterol on urinary N excretion were plainly attenuated after 5 weeks of treatment (Sillence et al., 1993), while Barash et al. (1994) reported that the liveweight gain response of cimaterol-treated heifers plateaued within 8 weeks. There may be several ways to counteract such a marked attenuation of effect, such as increasing the dose of the drug over time or pulse feeding the compound. Nevertheless, in practice, it seems fair to label β-agonists as highly effective, but short-acting drugs and therefore most suited to use as a finishing treatment.

Gender

In principal, β-agonists should work equally well in both sexes, regardless of whether an animal is entire or gonadectomized, because sex hormones have little effect on the adrenergic system. However, while there have been no reports of direct comparisons between male and female ruminants, sex differences in response to β-agonists have been reported for other species (Dunshea, 1991). In turkeys, some variables show a greater response in hens than in toms to the drug ractopamine (Wellenreiter and Tonkinson, 1990), and the same compound is reported to be effective in gilts and barrows, but not in boars (Dunshea et al., 1993b). In contrast, rats treated with clenbuterol showed muscle growth responses in the order of intact males > castrated males > ovarectomized females > intact females (Sillence et al., 1995). However, in the latter study, the rats were of the same age, but had different starting weights. With the exception of the gastrocnemius muscle growth response, most responses were directly proportional to body weight. In summary, there is little evidence to link sex hormones or gender per se with the magnitude of the response of β-agonists, but other factors that are indirectly linked to gender, such as body size, level of fatness, and feed efficiency, may have an important influence.

Age

As with apparent gender differences, the effects of age on the response of animals to β-agonists may also be accounted for largely by differences in liveweight. For example, Vestergaard et al. (1994) treated bulls with cimaterol for 90 d and examined the response at three different liveweights. The weight gain response was 10% at 162 kg , increasing to 25 and 22% at 299 and 407 kg, respectively. In absolute terms, the carcass weight response also increased with age, with gains of 18, 28, and 37 kg, respectively, at the three liveweights. However, on a proportional basis, this represented a remarkably uniform increase in carcass weight of between 13 and 14% in all three groups. An important practical implication of this observation is that the greatest net economic returns from the use of β-agonists will be obtained through their use in larger animals. This is yet another factor that destines them to be used as a finishing technology.

Diet

Of all the factors discussed in this section, diet is probably the most significant variable that can limit the response to β-agonists, as demonstrated by the linear relationship that exists between liveweight gain and dietary protein intake in ractopamine-treated pigs (Dunshea et al., 1993a; Lindsay et al., 1993) and clenbuterol-treated rats (Perez-Llamas et al., 1992). However, an unusual feature of β-agonists that sets these compounds apart from other anabolic agents is their capacity to mobilize endogenous sources of energy and nonmuscle protein and to maintain some muscle growth even when dietary energy and N are extremely low. This has little relevance to feedlot production systems, but may be significant for animals that are forced to graze on poor-quality pastures, a common seasonal problem in tropical areas. The maintenance of muscle mass at the expense of vital organs may at first seem suicidal, but in fact a decrease in rumen and liver protein mass is a natural response to chronic underfeeding. In severe situations, this may even prolong the life of the animal, as death can result from impairment of muscle, including the ability to breathe (Anonymous, 1989). Once an adequate supply of nutrients becomes available again, the vital organs receive priority for protein deposition.

In summary, though there are many factors that influence the size of response to β-agonists, few are critical. These drugs increase muscle mass in most species to a degree that is proportional to the size of the animal, they are effective for a short period of time, and their effect is lost or even reversed following prolonged withdrawal. The desire to avoid both drug residues and prolonged withdrawal periods has led to the development of drugs that are rapidly eliminated from the body and therefore need to be given frequently. Together, these factors are overwhelming in limiting the potential use of β-agonist drugs to a finishing strategy in intensive production systems.

Adverse Effects of β-Agonists

The most commonly observed adverse effects of β-agonists are muscle tremor and increased heart rate (tachycardia). Although both these effects usually subside within a few days of continuous treatment, the tachycardia in particular can be severe in some individuals (Brockway et al., 1987; Eisemann et al., 1988; Sillence et al., 1993). As well as affecting the well being of the animal, increased heart rate may also be associated with thermogenesis and with the temporary inappetence that occurs in a proportion of β-agonist-treated cattle (Sillence et al., 1993) and sheep (Brockway et al., 1987). As β_1-adrenoceptors predominate in the heart, while β_2-adrenoceptors predominate in skeletal muscle, it seems reasonable to expect that this side effect could be eliminated through the development of drugs that have greater selectivity between these two receptor subtypes. However, Hoey et al. (1995) showed that the tachycardia was not caused by a direct stimulation of cardiac β_1-adrenoceptors, but instead was reflex response to a fall in blood pressure caused by the relaxation of arterial smooth muscle. In turn, this relaxation was caused by the activation of β_2-adrenoceptors, the same receptors that cause the muscle growth response. The implication of this finding is that this adverse effect cannot be eliminated by altering the agonist without impairing its anabolic activity in skeletal muscle. Another discovery of Hoey et al. (1995) is that if β_1-adrenoceptors are blocked in the heart, part of the blood-pressure reflex response is eliminated. Thus a combination of β_2-agonist–β_1-antagonist may be used to achieve full anabolic efficacy, while producing only 50% of the tachycardia seen with the β_2-agonist alone. In practice, such a combined treatment would be difficult to register, but drugs that are capable of activating one receptor while blocking another already exist and could point the way toward more "animal friendly" treatments of the future.

A longer-term side effect of β-agonists may be impaired renal function, as increased electrolyte and water retention and reduced creatinine clearance rates have been observed in β-agonist-treated women (Morgan 1990). β-Agonists also cause cardiac hypertrophy in rats (Reeds et al., 1986; Sillence et al.,1991) and rabbits (Forsberg et al., 1989), but have the opposite effect in cattle and sheep (Kim et al., 1987; Moloney et al., 1990). In pigs treated with salbutamol, there have been observations of impaired walking soundness associated with claw horn lesions and the suggestion that β-agonists interfere with horn production (Penny et al., 1994). Other studies in pigs treated with cimaterol (Jones et al., 1985) and L-644969 (Wallace et al., 1987) have also revealed foot lesions, suggesting that this may be a class effect of β-agonists. However, no effects on walking soundness have been reported in ruminants.

Overall, despite their unpleasant short-term effects on heart rate and muscle tremor, β-agonists are among the least toxic pharmaceutical compounds. Williams et al. (1989) observed no clinical signs of toxicity in pigs treated with up to 500 ppm ractopamine for 8 weeks, and few long-term adverse effects on animal health have been reported in ruminants.

β-Agonists and Meat Quality

Of all the factors likely to influence the future of β-agonists, their effect on meat quality will be among the most important, particularly because of their destined use in feedlots that traditionally supply the most quality-sensitive markets. The importance of this single issue has been highlighted in several symposia, the proceedings of which still provide a useful insight into the effects of these drugs (Hanrahan, 1987; Wood et al., 1991).

Briefly, the anabolic action of β-agonists in skeletal muscle results from fiber hypertrophy, in contrast to the hyperplastic growth response seen following ST treatment or in compensatory growth. In particular, β-agonists increase the mass of glycolytic fibers relative to oxidative fibers, and this effect has a number of consequences of varying significance. Several variables connected with meat color are affected. Meat from clenbuterol-treated calves has been reported to be lighter than usual (Geesink et al., 1993), but when measures of reflectance are made through fiber-optic probes, slightly darker meat is more often detected in β-agonist-treated cattle and sheep (Chikhou et al., 1993a; Moloney et al., 1994; Warriss et al., 1989). The latter effect may be influenced to some extent by reduced fat or "marbling" in the muscle (Moloney et al., 1994). A relative decrease in the mass of oxidative fibers results in less myoglobin pigment in the meat and hence less redness, and this effect seems to be consistent in cattle (Moloney et al., 1994), sheep (Warriss et al., 1989), and pigs

(Warriss et al., 1990). In general though, effects of β-agonists on meat color are of minor importance, except when animals are subjected to unusually stressful procedures prior to slaughter. In such cases, β-agonist-treated animals are significantly more prone to produce "dark-cutting meat" because of their enhanced glycolytic capacity (Hanrahan et al., 1986). Effects of β-agonists on water holding capacity have been observed, but are variable, with increases (Warris et al., 1989), no change (Moloney et al., 1994), and decreases (Fiems et al., 1990) being reported. Decreases in marbling fat and the proportion of saturated to unsaturated fats have been reported and are generally regarded as beneficial effects (Thornton et al., 1985; Moloney et al., 1994). Despite the fact that β-agonist-treated animals have leaner carcasses and can chill more quickly, cold shortening does not seem to be a significant problem (Chikhou et al., 1993a; Vestergaard et al., 1994).

The most consistent effect of β-agonists, and the most potentially serious problem, is their effect of increasing the toughness of meat. Shear-force values can be increased more than threefold as measured by mechanical methods (Vestergaard et al., 1994), and while organoleptic measurements appear to be less sensitive for this variable, there is little doubt that the increase in toughness can be detected by the consumer (Moloney et al., 1994). This effect does not appear to be strongly linked to effects on collagen, as although the proportion of heat-stable collagen increases in β-agonist-treated muscle, there is generally a reduction in total collagen content (Dawson et al., 1990; Vestergaard et al., 1994). Instead, a compelling argument has been made to link meat toughness with suppressed activity of calcium-dependent protease enzymes (calpains) through the induction of the enzyme inhibitor calpastatin. Effects of β-agonists on calpastatin concentrations within a given muscle follow the same temporal pattern as changes in both muscle mass and postmortem shear force (Pringle et al., 1993). Furthermore, the variation that is seen in the anabolic response of different muscles to β-agonists corresponds to the variation in calpastatin activity and ultimately with shear force. Finally, whereas withdrawal of β-agonist treatment can restore meat quality, an attenuation of the anabolic effects of the drug occurs over the same timeframe (Hanrahan et al., 1988).

In summary, β-agonists increase the toughness of meat, and there is strong evidence to suggest this effect is inherent in the mechanism by which they promote muscle growth. Unless further research can reveal a way to divorce these two effects, the adoption of β-agonists for use in production systems that supply quality-sensitive markets could be severely impaired.

Somatotropin

The introduction of ST for commercial use as an anabolic agent may well be imminent, as bovine ST (bST) is already being used to enhance milk production in a number of countries including the United States, and porcine ST (pST) has been approved for swine production in Australia since November 1993. Purified extracts from pituitary glands were used in early studies to demonstrate the anabolic effects of ST in sheep (Davis et al., 1970), pigs (Machlin, 1972), and cattle (Car et al., 1976). Unfortunately, the high cost and low availability of this material hampered research until the 1980s, when a method was developed to produce the hormone in large quantities through molecular biotechnology. Since that time, there has been an explosion of scientific literature describing the effects of ST on animal production.

Effects on Nitrogen Metabolism, Growth, and Feed Efficiency

Compared with the data produced in swine (Etherton et al., 1987) and dairy cattle (Burton et al., 1994), early experiments with ST in meat-producing ruminants produced variable results (Moseley et al., 1992). Nevertheless, there can be no doubt of the hormone's efficacy for improving N retention. Indeed, improvements in N balance were among the first beneficial effects to be reported in ruminants, with a doubling of N retention observed in lambs (Davis et al., 1970) and steers (Car et al., 1976) treated with ST from pituitary extracts. Later confirmation of these results came through the use of recombinant ST (Pell et al., 1990; Bosclair et al., 1993). Thus the N sparing effect of ST is consistent, even when disappointing results are obtained in terms of growth (McLaughlin et al., 1993). Reduced N excretion is principally evident in the urine, with little or no effect of ST on fecal N, reflecting unaltered feed digestibility. In part, the reduced N excretion may reflect reduced feed intake in ST-treated animals, though there is also a marked effect on N utilization (for review, see Burton et al., 1994).

Despite the environmental benefits of reducing N excretion, ST is unlikely to be adopted in commercial practice because of this virtue alone, unless other economic advantages attend its use. In this respect, ST differs from its future competitors, the β-agonists, in that the extra N retained by ST-treated animals is deposited uniformly throughout the body rather than selectivity in muscle tissue or even in individual muscles. The hide and internal organs are consistently larger in ST-treated ruminants, so that a reduced dressing percentage is often reported (Early et al., 1990; Moseley et al., 1992; McLaughlin et al., 1994). Nevertheless, despite the lack of selectivity of ST toward muscle protein accretion, the majority of studies do report some increase in lean yield, total carcass protein, or muscle mass (Maltin et al., 1990; Moseley et al., 1992). The effects of ST on lipid deposition are more consistent than those on muscle. Carcasses from ST-treated animals are consistently leaner, resulting in improved grades that in some cases account for the main economic advantage of the treatment (Dalke et al., 1992; McLaughlin et al., 1994).

Effects of ST on liveweight gain reflect the sum of its actions on skeletal muscle, noncarcass protein, and body fat. As these components follow different dose-response relationships and change in different directions, it is not surprising that reported effects on weight gain are highly variable, ranging from a 38% decrease (Moseley et al., 1992) to a 23.5% increase (Fabry et al., 1987). Moseley et al. (1992) summarized the data from a number of cattle trials and suggest that the average liveweight gain response to ST is +10%, with a 9% improvement in feed conversion efficiency. In contrast to the effects of β-agonists, the action of ST on feed efficiency is mostly accounted for by sustained growth with decreased feed intake. The suppression of feed intake is linear with increasing doses of ST and, at high doses, becomes detrimental to production efficiency (Moseley et al., 1992).

Factors Likely to Influence the Adoption of ST

An important factor in favor of the use of ST is the safety of the product (for review, see Juskevich and Guyer, 1990). Although ST is now produced by recombinant means, it is a natural hormone already present in the body and is not known to accumulate in tissues in high concentrations. Being a protein, ST is easily destroyed by heat (cooking), and, if ingested accidentally by humans, it would not be absorbed from the gut in its active form. Furthermore, with the current level of acceptance over the use of ST for milk production, it is unlikely that a strong case will be made against the use of ST to improve N utilization on the grounds of safety.

Whereas the low oral activity of ST is a great advantage in terms of safety, this feature has presented the greatest challenge to animal scientists in developing a product that is convenient to use. The physiological release of ST is episodic, with spikes of high ST concentration being particularly pronounced in males of most species (Anfinson et al., 1975). This has led to the suggestion that episodic administration of the hormone is likely to be the most effective form of treatment, though data to support this theory are equivocal. Moseley et al. (1982) examined the effects of ST on N digestion and retention in cattle when given by different patterns of administration. Although continuous ST infusion decreased the magnitude of endogenous ST pulses, effects on N metabolism did not differ between infused ST, pulse injections, or infusions plus pulse injections. In contrast, Evock-Clover et al. (1992) showed that swine showed a greater response to ST injections given four times per day than to the same total dose given once daily. Despite the theoretical advantage of frequent treatment, this imposes an extra labor cost on the producer, and therefore efforts have been made to prolong the effects of ST by up to 2 weeks through the use of slow-release preparations such as Sometribove®* (McLaughlin et al., 1994) and Somidobove®* (Toutain et al., 1993), though again success obtained with these and similar agents has been variable. In short, the development of ST implants has reduced the potential labor cost of ST treatment, but at the sacrifice of some efficacy. The study of Azain et al. (1993), however, suggests that some of this loss may be recovered if ST implants were to contain higher doses of the hormone.

Because positive effects of ST treatment have been observed in both long-term (McShane et al., 1989) and short-term (MacRae et al., 1991) experiments, it appears that ST-treated livestock are less sensitive to the downregulation/withdrawal syndrome seen in β-agonist-treated animals (Barash et al., 1994); a feature

---

*Sometribove and Somidobove are registered trademarks of Monsanto, St. Louis, MO and Eli Lilly and Co., Greenfield, IN, respectively.

that may give the former technology a distinct competitive edge. Indeed, an intriguing experimental approach has been to treat animals with ST at an early stage in their development to cause semipermanent changes in growth physiology. For example, McCutcheon et al. (1994) showed that rats treated with ST for 21 d from birth had markedly less body fat at 120 d of age. Similarly, ST treatment of lambs from birth to 11 weeks of life resulted in significantly less carcass fat at 8 months of age, with a trend toward less fat at 13 months. Unfortunately, the long-term effects of ST on N metabolism were not reported, and it would be reckless to assume that they were similarly prolonged. Nevertheless, this potential action is probably worth investigating.

Having considered the economic advantages and practical constraints of using ST, it is important from an animal welfare perspective that any potential side effects of ST treatment are given careful consideration (for review, see Klasing et al., 1991). Although a marked excess of ST secretion can result in the serious condition known as acromegaly (Daughaday, 1985), the adverse effects of ST are few when the hormone is used at anabolic doses. The heart is among the organs that enlarge in ST-treated animals, and, although the increase in mass is proportional to the increase in body mass, there is no increase in wall stress. Therefore, this change is neither detrimental nor beneficial (Sacca et al., 1994).

Immunodeficiency has occurred in transgenic animals that profoundly overexpress ST (Hoskinson, 1995), but immunodeficiency can also be a symptom of ST deficiency (Burton et al., 1994). The role of ST in the control of the immune system requires further study, as at present it is only poorly understood (for review, see Arkins et al., 1993). Impaired walking soundness, associated with a disorder of bone formation (osteochondrosis), is perhaps one of the most serious adverse effects of ST, although this is only observed in pigs treated with high doses for long periods of time (Evock et al. 1988), and its relevance to ruminants remains to be determined. Finally, ulcers have been reported in the gastrointestinal tract of ST-treated pigs, with the death of an animal in a severe case (Smith and Kasson, 1990). Moseley et al. (1992) reported that abomasal ulcers also occur in ST-treated steers, increasing in incidence from 4 to 43% with increasing doses of bST. However, none of the ulcers seen in cattle were of sufficient severity to cause clinical disease.

Insulin-Like Growth Factor-I

There is evidence that many of the anabolic effects of ST are mediated through increased production and activity of insulin-like growth factor-I (IGF-I). Growth rate and plasma concentrations of IGF-I are low in ST-deficient animals, and both can be restored by injections of IGF-I (Humbel, 1990). In normal animals, ST treatment increases plasma IGF-I levels, except in underfed ruminants where a lack of inductive effect on IGF-I is associated with a reduced anabolic response to ST (Gluckman and Breier, 1989). This important link between ST and IGF-I has led to the suggestion that IGF-I alone may be an effective treatment in animal production. Indeed, it has even been argued IGF-I may be superior to ST, on the basis that IGF-I mediates the beneficial effects on N metabolism, but lacks the sometimes undesirable lipolytic effects that can be wasteful of energy (Ballard et al., 1993). However, confirming the efficacy of IGF-I in well-fed ruminants has proven difficult, not least because of the high cost of producing the hormone, even by recombinant DNA technology, which has limited its availability.

One of the few studies in which IGF-I has been administered directly to ruminants over a long period of time was done using sheep (Cottam et al., 1992). The hormone was injected daily for 8 weeks, resulting in a doubling of plasma IGF-I concentrations. Despite a trend toward a small increase in weight gain, there was no increased growth evident in carcass muscle, bone or wool. Plasma urea concentrations were decreased by IGF-I, but creatinine clearance rates showed this to be caused through increased urea N excretion rather than protein anabolism. In addition to the detrimental effect on N metabolism, IGF-I also decreased plasma insulin concentrations and caused hyperglycemia. Thus, simple IGF-I treatment seems not to be an attractive technology to enhance N utilization.

Insulin-Like Growth Factor-II

As with IGF-I , studies with insulin-like growth factor-II (IGF-II) have been limited by unavailability of the hormone, and there are no reports of long-term production trials in ruminants. It is believed that IGF-II has a major role in promoting fetal growth, but may have a negative effect in older animals if present in high concentrations. Koea et al. (1992) observed that IGF-II impaired IGF-I-stimulated protein synthesis in sheep, presumably because of inactivation of IGF-I through competition for its binding protein, BP3. Infusions of IGF-II had no significant effect on weight gain in rats (Zapf et al., 1984). In summary, IGF-II research is not a promising area for future technologies to reduce N excretion in ruminants, unless the hormone can be used to make prenatal modifications that have lasting consequences in the growing animal.

Somatotropin-Releasing Factor

Coinciding with the production of recombinant ST came the identification of a 44-amino acid peptide that stimulates endogenous ST output, somatotropin-releasing factor (SRF). Although first purified from human pancreatic tumors (Guillemin et al., 1982; Rivier et al., 1982), the presence of hypothalmic SRF was soon confirmed in cattle (Esch et al., 1983) and sheep (Brazeau et al., 1984), and its physiological importance in ruminants was demonstrated through immunoneutralization studies. Treatment of cattle with antibodies to SRF caused reductions in plasma IGF-I concentrations and weight gain, accompanied by increases in body fat and plasma urea (Armstrong et al., 1994).

Despite evidence that optimum responses to SRF are obtained by four or eight daily injections (Kensiger et al., 1987), a number of workers have tried to elicit lactogenic responses to SRF through single or twice daily injections. Fewer studies have examined growth, but small positive effects (relative to ST) have been observed with respect to increased muscle mass, decreased plasma urea concentrations, and decreased fat deposition in pigs (Etherton et al., 1986) and sheep (Beerman et al., 1990). The anabolic effects of SRF reported in cattle have been smaller still (Lapierre et al., 1991) or absent (Enright et al., 1993). Data from studies on lactation suggest that infrequent administration may account for the disappointing results achieved so far, as SRF given through an optimum number of injections can be at least as effective at increasing milk yield in dairy cows as ST (Dahl et al., 1991). Although there is not yet any sign of a practical advantage of SRF over ST, research in this area continues. Efforts have been made to enhance SRF action through coadministration with thyrotropin-releasing hormone (a hormone that has an additive effect on ST secretion) with equivocal results (Lapierre et al., 1991; Enright et al., 1993). Finally, synthetic, nonpeptidyl ST secretagoges have been developed (L-692,249 and L-692,585) and shown to cause increased plasma concentrations of ST and IGF-I in dogs (Jacks et al., 1994).

Ovine Placental Lactogen

Ovine placental lactogen (oPL) is a 198-amino acid protein that is produced naturally by fetal cells and found in both the fetal and maternal circulation. The hormone is a member of the ST/prolactin gene family, showing more similarity to prolactin than ST and more divergence among species than either hormone (Byatt et al., 1992b). Nevertheless, oPL and ST bind with similarly high affinity to a common receptor in both fetal and adult liver, and both cause the stimulation of IGF-I output (Breier et al., 1994). Although ST lacks the potent somatogenic effects of oPL in the prenatal animal, the two hormones are thought to have similar actions postnatally, and so there has been interest in the potential use of oPL as a somatotropic agent.

When ST-deficient rats, known to be highly responsive to exogenous ST, were treated with oPL or bST at equal doses, the somatogenic effects of oPL were demonstrated convincingly (Singh et al., 1992). The two hormones caused a similar increase in growth rate and carcass N, with a concomitant reduction in carcass fat. Although oPL and bST both increased IGF-I production, only bST increased the expression of mRNA for IGF-I BP3. This observation is significant in light of the fact that BP3 is thought to be essential for IGF-I action (Cottam et al., 1992), and it suggests that in fact this is either not the case or that with oPL (and perhaps ST), growth is stimulated by an IGF-I-independent mechanism.

Following the data in ST-deficient rats, further promising results were obtained in cattle, where treatment with bovine PL (bPL) over a 7-d period increased plasma IGF-I and reduced plasma urea concentrations, suggesting improved N retention (Byatt et al., 1992a). Unfortunately, longer-term studies in lambs have produced less encouraging results. Min et al. (1994) gave oPL or bST twice daily to lambs for 3 weeks, starting at 3 d of age. Treatment with oPL did increase liveweight gain significantly, an effect that is contrary to the response to bST. The inductive effect of oPL on plasma IGF-I concentrations was less than that of bST, lending support to the suggestion of an IGF-I-independent mechanism. In fact, the authors concluded that the increased weight gain in oPL-treated lambs was simply a consequence of increased voluntary feed intake, an effect opposite to that of bST and not particularly useful in the aim of reducing N excretion. In summary, oPL does not show great promise as an N sparing agent, but, with the limited amount of data currently available, further studies would be warranted.

VACCINE TECHNOLOGIES

Somatostatin Immunization

There have been several attempts to exploit the anabolic effects of ST by direct and indirect immunological means (for reviews, see Flint, 1994 and Wynn et al., 1994). One of the original strategies was to immunoneutralize a peptide that inhibits endogenous ST release from the pituitary, known as somatotropin-releasing inhibitory factor (SRIF) or somatostatin. Spencer and Williamson (1981) and Spencer et al. (1983) reported that active immunization of 3-week-old lambs against somatostatin increased plasma concentrations of ST and IGF-I and caused an increase in weight gain and carcass weight. However, the mechanism by which these anabolic effects were achieved may not have been what was originally intended. In contrast to the effects of exogenous ST, somatostatin immunization caused no reduction in carcass fat and stimulated feed intake rather than decreasing it. Positive effects on weight gain were confirmed in immunized cattle (Lawrence et al., 1986) and in further sheep studies (Bass et al., 1987; Westbrook and McDowell, 1994), but again this occurred without changes in the muscle to fat ratio and, in the study of Bass et al. (1987), with an increase in plasma concentrations of IGF-I, but not ST.

It is now known that somatostatin influences the release of a number of hormones other than ST, including thyrotropin, adrenocorticotropin, glucagon, and insulin (Flint, 1994). Perhaps more importantly though is that somatostatin may modulate the release of peptides that influence nutrient absorption in the gastrointestinal tract (Wynn et al., 1994). A theory that antisomatostatin treatment acts through a gut mechanism has been given by Wynn et al. (1994) to explain why positive responses have been obtained in more primitive breeds of sheep, while more highly selected ruminants have often failed to show a production response (Vicini et al., 1988; Trout and Schanbacher, 1990; Zainur et al., 1991). Overall, the inconsistency observed in the response of ruminants to somatostatin immunization may have discouraged the pursuit of this strategy for commercial purposes, but it has stimulated interest in gut peptide research, opening yet another door for future technology.

ST Enhancement Through Monoclonal Antibodies

Interest in this technology peaked during the late 1980s when it was demonstrated that the anabolic effect of exogenous ST was enhanced markedly in mice and marmosets by the coadministration of monoclonal antibodies (MABs) that bind to the hormone (Holder et al., 1985). This phenomenon illustrates a certain parallel between ST and IGF-I in that both hormones have endogenous binding proteins, and neither may work optimally when administered on their own. Early research on MAB-ST enhancement in animal models was soon followed by a description of an improved galacatopoietic response in sheep (Pell et al., 1989). However, this technology seems to have endured the same fate as somatostatin immunization, in that a failure to consistently demonstrate production responses in livestock has left its early promise unfulfilled. Another striking similarity to the somatostatin story is that the mechanism of action of the MABs is poorly understood and may ultimately prove to be ST independent or at least lead to new technologies that differ markedly from ST treatment.

Other ST Vaccine Technologies

Rather than being discouraged by the fact that neither somatostatin immunization nor MAB enhancement of ST have resulted in commercial vaccines, the promise and disappointments of this research appear to have spurred the imagination of investigators. As a result, several variations on the theme of a somatotropic vaccine are now being explored. Stewart et al. (1993) have adapted the antibody-enhancement approach to IGF-I, demonstrating that polyclonal antibodies to IGF-I enhance the anabolic effects of the exogenous hormone when coadministered to rats. Pell and Aston (1991) have also tried a more direct approach to ST enhancement by actively immunizing animals with peptide fragments of ST as antigens, rather than passively transferring antibodies raised against the whole ST molecule. In lambs, this treatment resulted in an increase in carcass weight and protein content, with no reduction in fat.

In a more complicated approach, Gardner et al. (1990) produced MABs to ST and then used these as antigens to produce ST-antiidiotypic antibodies. The theory that such antibodies would share enough features in common with ST to stimulate ST receptors seems to have been correct insofar as the antiidiotypic antibodies increased body weight gain in hypophysectomized rats (Gardner et al., 1990). Furthermore, the fact that no increase in plasma IGF-I concentrations occurred lends support to a popular current theory that it is possible to selectively exploit some of the effects of ST and not others. Unfortunately, Schalla et al. (1994) had less success with the ST antiidiotype approach in attempting to increase lactation in dairy cows. Finally, as the amino acid sequence of the ST receptor has now been published for many species, it should not be long before the effect of receptor peptide antigens are reported.

β-Agonist Vaccines

In theory, two significant limitations of β-agonist drugs could be overcome through the development of a β-agonist vaccine. Rather than having to be administered daily in the feed, a vaccine could be given by one or two injections, allowing this technology to be used in grazing animals. Second, β-adrenoceptor antibodies present less risk to the consumer, as they are unlikely to accumulate in animal tissues, would be unstable to heat (cooking), and are incapable of crossing the gut if ingested accidentally. In Australia, where the vast majority of cattle and sheep are raised in extensive systems, the two different approaches to the development of a β-agonist vaccine have been taken.

Both approaches rest on a relatively new concept in immunology. Traditionally, antibodies have been used simply to bind to target hormones, either prolonging their action (Holder et al., 1985) or inactivating them (Spencer and Williamson, 1981). However, through recent research into a number of immunopathologies, it has been discovered that certain antibodies can recognize and activate membrane-bound receptors, thus mimicking the actions of endogenous hormones or neurotransmitters (for review, see Wynn et al., 1994). This is particularly surprising in the case of β-adrenoceptor antibodies in view of the size of a typical IgG antibody molecule (mol wt ≈ 150,000) compared with the small binding pocket in the receptor protein that is designed to accommodate catecholamines (mol wt ≈ 300).

The first approach to develop a β-adrenoceptor vaccine is based on the concept of producing antiidiotypic antibodies to a β-adrenoceptor drug. Strosberg (1989) describes the results of various studies to generate antiidiotypic antibodies to the β-antagonist alprenolol, indicating the difficulty of this method. Antiidiotypic antibodies were produced with low frequency, some of which blocked the β-adrenoceptor or activated the receptor, while others were themselves inactivated by the production of anti-antiidiotypic antibodies. Despite these pitfalls, other workers have successfully produced monoclonal antibodies to clenbuterol and demonstrated, at least in principal, that these can be used as the basis for a vaccine to cause the production of β_2-adrenoceptor-stimulating antibodies in sheep (Hoskinson, 1995).

While the antiidiotypic approach remains the method of choice to produce antibodies for receptors that are not fully characterized, a more direct strategy is available for receptors that have a known amino acid sequence. Magnusson et al. (1991) demonstrated that antibodies to the β_1-adrenoceptor are naturally produced in patients who suffer idiopathic cardiomyopathy. Furthermore, the epitope for these antibodies was mapped to a specific extracellular portion of the β_1-adrenoceptor, and a short peptide that represents this segment could be used to produce similar antibodies in experimental animals (Magnusson et al., 1989). We have adopted this approach to generate antibodies that stimulate bovine β_2-adrenoceptors, with some success in activating bovine muscle tissue *in vitro* (Hill, 1995). Experiments to measure production responses in cattle treated with our β_2-adrenoceptor vaccine are currently under way.

TRANSGENIC TECHNOLOGIES

Hormone Targets

Soon after genetic engineering had delivered a method for industrial-scale synthesis of recombinant ST, the concept of producing transgenic animals became a reality. Because relatively few genes had (and have) been cloned for economically important traits, it is no coincidence that the ST gene became a prime candidate for this work. First, mice were transfected with the ST gene with outstanding results (Palmiter et al., 1982), followed by sheep and pigs (Brem et al., 1985) with less desirable consequences. Other targeted genes have been for the production of SRF and IGF-I, and, together with detailed accounts of ST experiments, these ventures have been reviewed and published recently (Pursel and Rexroad, 1993).

The enormous potential of transgenesis is best illustrated in experiments with rats and mice. Studies of ST, involving selective (site-directed) mutation and molecular probing with MABs, suggest that different segments of the ST molecule are responsible for activating different processes, such as lactation, lipogenesis, and protein synthesis (see Burton et al., 1994). Furthermore, molecular genetic studies have confirmed that considerable polymorphism exists naturally among the genes that encode for ST (Cowan et al., 1989; Rocha et al., 1992; Zhang et al., 1993), leading to the tempting conclusion that transgenic animals could be engineered to produce not only ST, but ST analogs selective for a chosen attribute. In fact, this possibility has already been realized to a great extent through the production of various lines of transgenic mice (Knapp et al., 1994). These animals range in size from 60 to 200% of that of normal mice according to the gene inserted, which dictates whether they have decreased growth of protein and fat, increased growth of protein and fat, or increased growth of protein only. As well as showing selective applications of ST transgenesis, studies in mice have also illustrated the feasibility of exerting tight control over the expression of a foreign gene. Shanahan et al. (1989) showed that transgenic mice maintained normal body size for many weeks until a selected salt was included in their drinking water to induce expression of their ovine ST transgene. Once this had occurred, plasma ST levels increased markedly, as did body weight. Upon the return to normal drinking water, plasma ST concentrations fell to control values within 24 h. With such outstanding results in the areas of selectivity and control obtained in rodents, it is disappointing perhaps that limited progress has been made toward the production of transgenic livestock.

Foreign ST genes have been inserted into sheep (Brem et al., 1985) and pigs (Pursel and Rexroad, 1993), principally through the method of microinjection, which involves introducing the cloned DNA into the pronucleus of a fertilized recipient egg. Unfortunately, although the efficiency of gene insertion through microinjection is extremely low in livestock (0.1 to 4.5%), this technique has not yet been surpassed by emerging methods such as retroviral infection, stem cell insertion, or the use of sperm vectors (Pursel and Rexroad, 1993). The mechanism by which the foreign gene becomes integrated in the host DNA is not well understood and appears to be somewhat random. As a result, either no integration occurs, or multiple copies of the gene are inserted in various arrays.

Because of random gene insertion, host animals may express the gene in all their cells, they may fail to express it at all, or expression may occur in some cells and not in others. In the last case, animals are termed "mosaic" and can have unusual phenotypes, perhaps displaying alternate patches of thick and thin subcutaneous fat when transfected with the ST gene. Mosaicism within germ cells also leads to difficulty in predicting which progeny of a founder animal will inherit the transgene. Finally, once a transgenic animal has been successfully engineered, the problem over control of gene expression arises. Although, in principal, gene expression may be controlled through fusion to a promoter gene that is dependent on the supply of a specific substrate, such as metallothionein, in practice, uncontrolled gene expression has been a significant problem in transgenic livestock. Unrestricted expression of ST genes in sheep has resulted in severe health problems such as degenerative kidney disease, hyperglycemia, glycosuria, and joint abnormalities (Nancarrow et al., 1991), as well as thermogenesis and immunodeficiency (Hoskinson, 1995). Transgenic pigs have endured similar problems with joint pathology, as well as suffering gastric ulcers and infertility (Pursel and Rexroad, 1993).

Receptor Targets

Although most attention has been given to engineering transgenic animals that overproduce anabolic hormones, receptors can also be a target for this technology. Once a receptor gene has been sequenced, that gene can be cloned causing the receptor to be expressed in either prokaryotic or eukaryotic cells. The β_2-adrenoceptor was among the first receptors to be treated this way and remains one of the most widely studied. Through selective mutation of the β_2-adrenoceptor gene, it was soon confirmed which amino acid residues in the receptor protein are important in the molecular mechanisms of drug binding (Frielle et al.1989), and the potential of altering this property of the receptor became apparent. Thus, through site-directed mutagenesis, Dixon et al. (1991) developed a mutant form of the β_2-adrenoceptor that responds only weakly to natural catecholamines, but that can be activated by a specific range of designer drugs that have no effect on unmodified or "wild-type" β_2-adrenoceptors. Next the mutant receptor gene was introduced into livestock and linked to an actin promoter to ensure that its expression was restricted to skeletal muscle. The concept is that the transgenic animals that overexpress mutant β_2-adrenoceptors would appear normal until treated with the designer drug, which causes their growth rate to increase. Theoretically, residual quantities of the drug in meat should not be a barrier to human consumption, as unmodified β_2-adrenoceptors such as those found in man are unresponsive to the drugs.

This ingenious approach illustrates how seemingly impossible problems can be overcome, such as the desire to retain the oral activity of β-agonists for convenience in livestock use, while protecting the consumer from drug residues, and the need to circumvent the problem of unrestricted transgene expression. Furthermore, this strategy offers a significant advantage to pharmaceutical companies which can maintain financial returns from transgenic livestock through controlling the supply of the activating compound. However, despite the ingenuity of this approach, such companies are doubtless aware that their greatest challenge will lie in consumer acceptance. In the present climate where consumer lobbyists are fighting, and often succeeding, in keeping straightforward natural hormones and synthetic drugs out of animal production, any attempt to introduce drug-treated animals carrying mutated receptor genes would have to be described as courageous. For now, this strategy is probably best regarded as a technology ahead of its time.

SUMMARY

Somatotropin and β-agonists are both new technologies that could be used in a safe and effective way to reduce N excretion, while yielding an economic return to the producer derived from improved feed efficiency and carcass grades. The main limitation to the use of ST is the necessity for frequent, if not daily, dosing, although this may be improved by the introduction of 2-week slow-release formulations. In contrast, β-agonists could be conveniently administered as a feed additive, but the fact that they increase the toughness of meat is likely to exclude their use in animals that are destined for quality-sensitive markets. There is evidence to show that the combined use of ST and β-agonists in cattle can produce an additive effect (Maltin et al., 1990), but this would compound their advantages as well as their disadvantages. Furthermore, because of their mode of delivery, neither technology is yet within reach of the grazier.

Attempts to modify growth through immunological means have been numerous and diverse, with some success in demonstrating that, in principal, N retention can be improved by means of a somatotropic vaccine. Unfortunately, many approaches have demonstrated the need for a more sound physiological understanding of the target system and the mechanism by which antibodies act once they have become attached to their quarry. Those treatments that involve the passive transfer of antibodies to animals, either alone or in combination with ST, are likely to be short term, and their use, if any, will be restricted to intensive production systems. Active immunization may be even less attractive to intensive producers than hormonal injections, as this strategy is likely to fail in a proportion of cattle because of their inherently low immunoresponsiveness. Instead, the greatest benefit to be derived from the development of a somatotropic or β-agonist vaccine may be in putting these technologies within the reach of graziers and producers in Europe, where hormonal treatments are banned.

The production of transgenic animals that overexpress anabolic hormones or their receptors continues to be pursued, with outstanding possibilities demonstrated in small animal studies, but comparatively disappointing results in livestock. Significant advances will need to be made in understanding the mechanisms of transgene integration and the control of gene expression before the enormous potential of this technology can be realized.

REFERENCES

Anfinson, M. S., Davis, S. L., Christian, E., and Everson, D. O., Episodic secretion of growth hormone in steers and bulls: an analysis of frequency and magnitude of secretory spikes occurring in a 24-hour period, *Proc. West. Sec. Am. Soc. Anim. Prod.,* 26, 175, 1975.

Anonymous, How ketones spare protein in starvation, *Nutr. Rev.,* 47, 80, 1989.

Anonymous, Black market drugs in EC beef, *Beef Improve. News,* 10, Aug. 1991.

Arkins, S., Dantzer, R., and Kelley, K. W., Somatolactogens, somatomedins, and immunity, *J. Dairy Sci.,* 76, 2437, 1993.

Armstrong, J. D., Harvey, R. W., Stanko, R. L., Cohick, W. S., Simpson, R. B., Moore, K. L., Schoppee, P. D., Clemmons, D. R., Whitacre, M. D., Britt, J. H., Lucy, M. C., Heimer, E. P., and Campbell, R. M., Active immunisation against growth hormone releasing factor: effects on growth, metabolism and reproduction in cattle and swine, in *Vaccines in Agriculture,* Wood, P. R., Willadsen, P., Vercoe, J. E., Hoskinson, R. M. and Demeyer, D., Eds., CSIRO, Melbourne, Australia, 1994, chapter 12.

Asato, G., Baker, P. K., Bass, R. T., Bentley, T. F., Chari, S., Dalrymple, R. H., France, P. J., Gringher, P. E., Lences, B. L., Pascauge, J. J., Pensock, J. M., and Ricks, C. A., Repartitioning agents: 5-[1-Hydroxy-2-(isopropylamino)ethyl]-anthranilonitrile and related phenethanolmines; agents for promoting growth, increasing muscle accretion and reducing fat deposition in meat-producing animals, *Agric. Biol. Chem.,* 48, 2883, 1984.

Azain, M. J., Kasser, T. R., Sabacky, M. J., and Baile, C. A., Comparison of the growth-promoting properties of daily verses continuous administration of somatotropin in female rats with intact pituitaries, *J. Anim. Sci.,* 71, 384, 1993.

Baker, P. K., Dalrymple, R. H., Ingle, D. L., and Ricks, C. A., Use of a β-adrenergic agonist to alter muscle and fat deposition in lambs, *J. Anim. Sci.,* 59, 1256, 1984.

Ballard, F. J., Francis, G. L., Walton, P. E., Knowles, S. E., Owens, P. C., Read, L. C., and Thomas, F. M., Modification of animal growth with growth hormone and insulin-like growth factors, *Aust. J. Agric. Res.,* 44, 567, 1993.

Barash, H., Peri, I., Gertler, A., and Bruckental, I., Effects of energy allowance and cimaterol feeding during the heifer rearing period on growth, puberty and milk production, *Anim. Prod.,* 59, 359, 1994.

Bass, J. J., Gluckman, P. D., Fairclough, R. J., Peterson, A. J., Davis, S. R., and Carter, W. D., Effect of nutrition and immunisation against somatostatin on growth and insulin-like growth factors in sheep, *J. Endocrinol.,* 112, 27, 1987.

Beermann, D. H., Hogue, D. E., Fishell, V. K., Aronica, S., Dickson, H. W., and Schricker, B. R., Exogenous growth hormone-releasing factor and ovine somatotropin improve growth performance and composition of gain in lambs, *J. Anim. Sci.,* 68, 4122, 1990.

Boisclair, Y. R., Bauman, D. E., Bell, A. W., Dunshea, F. R., and Harkins, M., Nutrient utilisation and protein turnover in the hindlimb of cattle treated with bovine somatotropin, *J. Nutr.,* 124, 664, 1993.

Brazeau, P., Bohlen, P., Esch, F., Ling, N., Wehrenberg, W. B., and Guillemin, R., Growth hormone releasing factor from ovine and caprine hypothalamus: isolation, sequence analysis and total synthesis, *Biochem. Biophys. Res. Commun.,* 125, 606, 1984.

Breier, B. H., Funk, B., Surus, A., Ambler, G. R., Wells, C. A., Waters, M. J., and Gluckman, P. D., Characterization of ovine growth hormone (oGH) and ovine placental lactogen (oPL) binding to fetal and adult hepatic tissue in sheep: evidence that oGH and oPL interact with a common receptor, *Endocrinology,* 135, 919, 1994.

Brem, G., Brenig, B., Goodman, H. M., Selden, R. C., Graf, F., Kruff, B., Springman, K., Hondele, J., Meyer, J., Winnaker, E.-L., and Krausslich, H., Production of transgenic mice, rabbits and pigs by microinjection into pronuclei, *Zuchthygiene (Berlin),* 20, 251, 1985.

Brockway, J. M., MacRae, J. C., and Williams, P. E. V., Side effects of clenbuterol as a repartitioning agent, *Vet. Rec.,* 20, 381, 1987.

Burton, J. L., McBride, B. W., Block, E., Glimm, D. R., and Kennelly, J. J., A review of bovine growth hormone, *Can. J. Anim. Sci.,* 74, 167, 1994.

Byatt, J. C., Eppard, P. J., Veenhuizen, J. J., Sorbet, R. H., Buonomo, F. C., Curran, D. F., and Collier, R. J., Serum half-life and *in vivo* actions of recombinant bovine placental lactogen in the dairy cow, *J. Endocrinol.,* 132, 185, 1992a.

Byatt, J. C., Warren, W. C., Eppard, P. J., Staten, N. R., Krivi, G. G. and Collier, R. J., Ruminant placental lactogens: structure and biology, *J. Anim. Sci.*, 70, 2911, 1992b.

Car, M., Nidar, A., and Filipan, T., An effect of the treatment of young steers with STH (growth hormone) upon nitrogen retention in intensive feeding, *Polijopr. Znan. Smotra.,* 37, 183, 1976.

Chikhou, F. H., Moloney, A. P., Allen, P., Joseph, R. L., Tarrant, P. V., Quirke, J. F., Austin, F. H., and Roche, J. F., Long-term effects of cimaterol in Freisian steers: II. Carcass composition and meat quality, *J. Anim. Sci.,* 71, 914, 1993a.

Chikhou, F. H., Moloney, A. P., Allen, P., Quirke, J. F., Austin, F. H., and Roche, J. F., Long-term effects of cimaterol in Freisian steers: I. Growth, feed efficiency and selected carcass traits, *J. Anim. Sci.,* 71, 906, 1993b.

Choo, J. J., Horan, M. A., Little, R. A., and Rothwell, N. J., Anabolic effects of clenbuterol on skeletal muscle are mediated by β_2-adrenoceptor activation, *Am. J. Physiol.,* 263, E50, 1992 *(Endocrinol. Metab.* 26).

Cottam, Y. H., Blair, H. T., Gallaher, B. W., Purchas, R. W., Breier, B. H., McCutcheon, S. N., and Gluckman, P. D., Body growth, carcass composition, and endocrine changes in lambs chronically treated with recombinantly derived insulin-like growth factor, *Endocrinology,* 130, 2924, 1992.

Cowan, C. M., Dentine, M. R., Ax, R. L., and Schuler, L. A., Restriction length polymorphism associated with growth hormone and prolactin genes in Holstein bulls: evidence for novel growth hormone allele, *Anim. Genet.,* 20, 157, 1989.

Dahl, G. E., Chapin, L. T., Allen, M .S., Moseley, W. M., and Tucker, H. A., Comparison of somatotropin and growth hormone-releasing factor on milk yield, serum hormones and energy status, *J. Dairy Sci.,* 74, 3421, 1991.

Dalke, B. S., Roeder, R. A., Kasser, T. R., Veenhuizen, J. J., Hunt, C. W., Hinman, D. D., and Schelling, G. T., Dose-response effects of recombinant bovine somatotropin implants on feedlot performance in steers, *J. Anim. Sci.,* 70, 2130, 1992.

Daughaday, W. H., The anterior pituitary, in *Williams Textbook of Endocrinology*, Wilson, J. D. and Foster, D. W., Eds., W. B. Saunders, Philadelphia, 1985, chapter 18.

Davis, S. L., Garrigus, U. S., and Hinds, F. C., Metabolic effects of growth hormone and diethylstilbestrol in lambs. II. Effects of daily ovine growth hormone injections on plasma metabolites and nitrogen-retention in fed lambs, *J. Anim. Sci.,* 30, 236, 1970.

Dawson, J. M., Buttery, P. J., Gill, M., and Beever, D. E., Muscle composition of steers treated with the β-agonist cimaterol, *Meat Sci.*, 28, 289, 1990.

Dixon, R. A. F., Patchett, A. A., Strader, C. D., Sugg, E. E., and Sigal, I. S., European Patent Application 0 453 119 A1, 1991.

Dunshea, F. R., Factors affecting efficacy of β-agonists for pigs, *Pig News Inf.,* 12, 227, 1991.

Dunshea, F. R., King, R. H., and Campbell, R. G., Interrelationships between dietary protein and ractopamine on protein and lipid deposition in finishing gilts, *J. Anim. Sci.,* 71, 2931, 1993a.

Dunshea, F. R., King, R. H., Campbell, R. G., Sainz, R. D., and Kim, Y. S., Interrelationships between sex and ractopamine on protein and lipid deposition in rapidly growing pigs, *J. Anim. Sci.,* 71, 2919, 1993b.

Early, R. J., McBride, B. W., and Ball, R. O., Growth and metabolism in somatotropin-treated steers: I. Growth, serum chemistry and carcass weights, *J. Anim. Sci.,* 68, 4134, 1990.

Eisemann, J. H., Huntington, G. B., and Ferrell, C. L., Effects of dietary clenbuterol on metabolism of the hindquarters in steers, *J. Anim. Sci.,* 66, 342, 1988.

Elliott, C. T., Shortt, H. D., and McCaughey, W. J., *Detection and Control of Clenbuterol Abuse in Northern Ireland,* I.G.A.P.A., Research Meeting Abstracts, 1994, 81.

Emery, P. W., Rothwell, N. J., Stock, M. J., and Winter, P. D., Chronic effects of β_2-adrenergic agonists on body composition and protein synthesis in the rat, *Biosci. Rep.,* 4, 83, 1984.

Engelhardt, G., Structure-activity relationship of some new amino-halogen-substituted phenylethanolamines, *Arzheim-Forsch.,* 22, 869, 1972.

Enright, W. J., Prendville, D. J., Spicer, J. L., Stricker, P. R., Moloney, D. P., Mowles, T. F., and Campbell, R. M., Effects of growth hormone-releasing factor and (or) thyrotropin-releasing hormone on growth, feed efficiency, carcass characteristics, and blood hormones and metabolites in beef heifers, *J. Anim. Sci.,* 71, 2393, 1993.

Esch, F., Bohlen, N., Ling, P., Brazeau, P., and Guillemin, R., Isolation and characterisation of the bovine hypothalamic growth hormone releasing factor, *Biochem. Biophys. Res. Commun.,* 117, 772, 1983.

Etherton, T. D., Wiggins, J. P., Chung, C. S., Evock, C. M., Rebhun, J. F., and Walton, P. E., Stimulation of pig growth performance by porcine growth-hormone and growth hormone-releasing factor, *J. Anim. Sci.,* 63, 1389, 1986.

Etherton, T. D., Wiggins, J. P., Evock, C. M., Chung, C. S., Rebhun, J. F., Walton, P. E., and Steele, N. C., Stimulation of pig growth performance by porcine growth hormone: determination of the dose-response relationship, *J. Anim. Sci.,* 64, 433, 1987.

Evock, C. M., Etherton, T. D., Chung, C. S., and Ivy, R. E., Pituitary porcine growth hormone (pGH) and a recombinant pGH analog stimulate pig growth performance in a similar manner, *J. Anim. Sci.,* 66, 1928, 1988.

Evock-Clover, C. M., Steele, N. C., Caperna, T. J., and Solomon, M. B., Effects of frequency of recombinant porcine somatotropin administration on growth performance, tissue accretion rates, and hormone and metabolite concentrations in pigs, *J. Anim. Sci.,* 70, 3709, 1992.

Fabry, J., Claes, V., and Ruelle, L., Effects of growth hormone on heifer meat production, *Reprod. Nutr. Dev.,* 27, 591, 1987.

Fiems, L. O., Buts, B., Boucque, Ch. V., Demeyer, D. I., and Cottyn, B. G., Effect of a β-agonist on meat quality and myofibrillar protein fragmentation in bulls, *Meat Sci.,* 27, 29, 1990.

Flint, D. J., Immunomodulatory approaches for regulation of growth and body composition, *Anim. Prod.,* 58, 301, 1994.

Forsberg, N. E., Ilian, M. A., Ali-Bar, A., Cheeke, P. R., and Wehr, N. B., Effects of cimaterol on rabbit growth and myofibrillar protein degradation and on calcium-dependent proteinase and calpastatin activities in skeletal muscle, *J. Anim. Sci.,* 67, 3313, 1989.

Frielle, T., Caron, M. G., and Lefkowitz, R. J., Properties of the β_1 and β_2-adrenergic receptor subtypes revealed by molecular cloning, *Clin. Chem.,* 35, 721, 1989.

Gardner, M. J., Morrison, C. A., Stevenson, L. Q., and Flint, D. J., Production of anti-idiotypic antisera to rat GH antibodies capable of binding to GH receptors and increasing body weight gain in hypophysectomized rats, *J. Endocrinol.,* 125, 53, 1990.

Geesink, G. H., Smulders, F. J. M., van Laack, H. L. J. M., van der Kolk, J. H., Wesing, Th., and Breukink, H. J., Effects on meat quality of the use of clenbuterol in veal calves, *J. Anim. Sci.,* 71, 1161, 1993.

Gluckman, P. D. and Breier, B. H., The regulation of the growth hormone receptor, in *Biotechnology in Growth Regulation,* Heap, R. B., Prosser, C. G., and Lamming, G. E., Eds., Butterworths, Toronto, 1989, 27.

Guillemin, R., Brazeau, P., Bohlen, P., Esch, F., Ling, N., and Wehrenberg, W. B., Growth and hormone-releasing factor from a human pancreatic tumour that caused acromegaly, *Science,* 218, 585, 1982.

Hanrahan, J. P., Summary and conclusions, in *Beta-Agonists and Their Effects on Animal Growth and Carcass Quality*, Hanrahan, J. P., Ed., Elsevier Applied Science, New York, 1987, 193.

Hanrahan, J. P., Allen, P., and Sommer, M., Food intake, growth and carcass characteristics of lambs treated with cimaterol — effect of length of withdrawal period, in *Control and Regulation of Animal Growth,* Quirke, J. F. and Schmid, H., Eds., Pudoc, Wageningen, The Netherlands, 1988, 149.

Hanrahan, J. P., Quirke, J. F., Bowmann, W., Allen, P., McEwan, J., Fitzsimmons, J., Kotzain, J., and Roche, J. F., β-Agonists and their effects on growth and carcass quality, in *Recent Advances in Animal Nutrition,* Haresign, W., Ed., Butterworths, London, 1986, chapter 9, 125.

Hill, R. A., Development of a β_2-adrenoceptor vaccine, Ph.D. Thesis, Central Queensland University, Australia, 1995.

Hoey, A. J., Matthews, M. L., Badran, T. W., Pegg, G. G., and Sillence, M. N., Cardiovascular effects of clenbuterol are β_2-adrenoceptor mediated in steers, *J. Anim. Sci.,* 73, 1754, 1995.

Holder, A. T., Aston, R., Preece, M. A., and Iranyi, J., Monoclonal antibody-mediated enhancement of growth hormone activity *in vivo, J. Endocrinol.,* 107, R9, 1985.

Hoskinson, R., Personal communication, 1995.

Hovell, F. D. De B., Kyle, D. J., Reeds, P. J., and Beerman, D. H., The effect of clenbuterol and cimaterol on the endogenous nitrogen loss of sheep, *Nutr. Rep. Int.,* 39, 1177, 1989.

Humbel, R. E., Insulin-like growth factors I and II, *Eur. J. Biochem.,* 190, 445, 1990.

Jacks, T., Hickey, G., Judith, F., Taylor, J., Chen, H., Krupa, D., Feeney, W., Schoen, W., Ok, D., Fisher, M., Wyvratt, M., and Smith, R., Effects of acute and repeated intravenous administration of L-692,585, a novel non-peptidyl growth hormone secretagogue, on plasma growth hormone, IGF-I, ACTH, cortisol, prolactin, insulin, and thyroxine levels in beagles, *J. Endocrinol.,* 143, 399, 1994.

Jones, R. W., Easter, R. A., McKeith, F. K., Dalrymple, R. H., Maddock, H. M., and Bechtel, P. J., Effect of the β-adrenergic agonist cimaterol (CL 263,780) on the growth and carcass characteristics of finishing swine, *J. Anim. Sci.,* 61, 905, 1985.

Juskevich, J. and Guyer, C. G., Bovine growth hormone: human food safety evaluation, *Science,* 249, 875, 1990.
Kensinger, R. S., McMunn, L. M., Stover, R. K., Schriker, B. R., Maccecchini, M. L., Harpster, H. W., and Kavanaugh, J. F., Plasma somatotropin response to exogenous growth hormone releasing factor in lambs, *J. Anim. Sci.,* 64, 1002, 1987.
Kim, Y. S., Lee, Y. B., and Dalrymple, R. H., Effect of the repartitioning agent cimaterol on growth, carcass, and skeletal muscle characteristics in lambs, *J. Anim. Sci.,* 65, 1392, 1987.
Kim, Y. S., Sainz, R. D., Summers, R. J., and Molenaar, P., Cimaterol reduces β-adrenergic receptor density in rat skeletal muscle, *J. Anim. Sci.,* 70, 115, 1992.
Klasing, K. C., Wagner, W. C., and Kelley, K. W., Impact of metabolic modifiers on target animal health and environmental safety with emphasis on somatotropin, *J. Anim. Sci.,* 69 (Suppl. 2), 88, 1991.
Knapp, J. R., Chen, W. Y., Turner, N. D., Byers, F. M., and Kopchick, J. J., Growth patterns and body composition of transgenic mice expressing mutated bovine somatotropin genes, *J. Anim. Sci.,* 72, 2812, 1994.
Koea, J. B., Breier, B. H., Shaw, J. H. F., and Gluckman, P. D., A possible role for IGF-II: evidence in sheep for *in vivo* regulation of IGF-I mediated protein anabolism, *Endocrinology,* 130, 2423, 1992.
Lapierre, H., Pelletier, G., Petitclerc, D., Dubreuil, P., Morisset, J., Gaudreau, P., Couture, Y., and Brazeau, P., Effect of human growth hormone-releasing factor and (or) thyrotropin-releasing factor on growth, carcass composition, diet digestibility, nutrient balance, and plasma constituents in dairy calves, *J. Anim. Sci.,* 69, 587, 1991.
Lawrence, M. E., Schelling, G. T., Byers, F. M., and Greene, L. W., Improvement of growth and feed efficiency in cattle by active immunization against somatostatin, *J. Anim. Sci.,* 63 (Suppl. 1), 215, 1986.
Lindsay, D. B., Hunter, R. A., Gazzola, C., Spiers, W. G., and Sillence, M. N., Energy and growth, *Aust. J. Agric. Res.,* 44, 875, 1993.
Machlin, L. J., Effect of porcine growth hormone on growth and carcass composition of the pig, *J. Anim. Sci.,* 35, 794, 1972.
MacRae, J. C., Bruce, L. A., Hovell, F. D. De B., Hart, I. C., Inkster, J., Walker, A., and Atkinson, T., Influence of protein nutrition on the response of growing lambs to exogenous bovine growth hormone, *J. Endocrinol.,* 130, 53, 1991.
Magnusson, Y., Hoyer, S., Lengange, R., Chapot, M. P., Guillet, J.-G., Hjalmarson, A., Strosberg, A. D., and Hoebeke, J., Antigenic analysis of the second extra-cellular loop of the human beta-adrenergic receptors, *Clin. Exp. Immunol.,* 78, 42, 1989.
Magnusson, Y., Wallukat, G., Guillet, J.-G., Hjalmarson, A., and Haebeke, J., Functional analysis of rabbit anti-peptide antibodies which mimic autoantibodies against the β_1-adrenergic receptor in patients with idiopathic dilated cardiopathy, *J. Autoimmun.,* 4, 893, 1991.
Maltin, C. A., Delday, M. I., Hay, S. M., Innes, G. M., and Williams, P. E. V., Effects of bovine pituitary growth hormone alone or in combination with the β-agonist clenbuterol on muscle growth and composition in veal calves, *Br. J. Nutr.,* 63, 535, 1990.
McCutcheon, S. N., Kadim, I. T., Wickham, G. A., and Purchas, R. W., A prolonged change in body composition induced by endocrine manipulation of the neonate, *Proc. N. Z. Soc. Anim. Prod.,* 54, 51, 1994.
McLaughlin, C. L., Byatt, J. C., Hendrick, H. B., Veenhuizen, J. J., Curran, D. F., Hintz, R. L., Hartnell, G. F., Kasser, T. R., Collier, R. J., and Baile, C. A., Performance, clinical chemistry, and carcass responses of finishing lambs to recombinant bovine somatotropin and bovine placental lactogen, *J. Anim. Sci.,* 71, 3307, 1993.

McLaughlin, C. L., Hendrick, H. B., Veenhuizen, J. J., Hintz, R. L, Munyakazi, L., Kasser, T. R., and Baile, C. A., Performance, clinical chemistry, and carcass responses of finishing lambs to formulated sometribove (Methionyl bovine somatotropin), *J. Anim. Sci.,* 72, 2544, 1994.

McShane, T. M., Schillo, K. K., Boling, J. A., Bradley, N. W., and Hall, J. B., Effects of somatotropin and dietary energy on development of beef heifers. I. Growth and puberty, *J. Anim. Sci.,* 67, 2230, 1989.

Meyer, H. H. D. and Rinke, L. M., The pharmacokinetics and residues of clenbuterol in veal calves, *J. Anim. Sci.,* 69, 4538, 1991.

Miller, M. F., Garcia, D. K., Coleman, M. E., Ekeren, P. A., Lunt, D. K., Wagner, K. A., Procknar, M., Welsh, T. H., and Smith, S. B., Adipose tissue, longissmus muscle and anterior pituitary growth and function in clenbuterol-fed heifers, *J. Anim. Sci.,* 66, 12, 1988.

Min, S. H., Mackenzie, D. D. S., Breier, B. H., McCutcheon, S. N., and Gluckman, P. D., Growth-promoting and metabolic actions of recombinant ovine placental lactogen and bovine growth hormone in young lambs, *Proc. N. Z. Soc. Anim. Prod.,* 54, 59, 1994.

Moloney, A. P., Allen, P., Joseph, R. L., and Tarrant, P. V., Carcass and meat quality of finishing Freisian steers fed the β-adrenergic agonist L-644,969, *Meat Sci.,* 38, 419, 1994.

Moloney, A. P., Allen, P., Ross, D. B., Olson, G., and Convey, E. M., Growth, feed efficiency and carcass composition of finishing Freisian steers fed the β-adrenergic agonist L-644,969, *J. Anim. Sci.,* 68, 1269, 1990.

Moore, N. G., Pegg, G. G., and Sillence, M. N., Anabolic effects of the β_2-adrenoceptor agonist salmeterol are dependent on route of administration, *Am. J. Physiol.,* 267, E475, 1994, *(Endocrinol. Metab.* 30).

Morgan, D. J., Clinical pharmacokinetics of β-agonists, *Clin. Pharmacokinet.,* 18, 270, 1990.

Moseley, W. M., Krabill, L. F., and Olsen, R. F., Effect of bovine growth hormone administered in various patterns on nitrogen metabolism in the holstein steer, *J. Anim. Sci.,* 55, 1062, 1982.

Moseley, W. M., Paulissen, J. B., Goodwin, M. C., Alaniz, G. R., and Caflin, W. H., Recombinant bovine somatotropin improves growth performance in finishing beef steers, *J. Anim. Sci.,* 70, 412, 1992.

Nancarrow, C. D., Marshall, J. T. A., Clarkson, J. D., Murray, R. M., Millard, C. M., Shanahan, C. M., Wynn, P. C., and Word, K. A., Expression and physiology of performance regulating genes in transgenic sheep, *J. Reprod. Fertil.,* 43 (Suppl.), 277, 1991.

O'Connor, R. M., Butler, W. R., Hogue, D. E., and Beermann, D. H., Temporal patterns of skeletal muscle changes in lambs fed cimaterol, *Domest. Anim. Endocrinol.,* 8, 549, 1991.

Palmiter, R. D., Brinster, R. L., Hammer, R. E., Trumbauer, M. E., Rosenfeld, M. G., Birnberg, N. C., and Evans, R. M., Dramatic growth of mice that develop from eggs microinjected with metallothionein-growth hormone fusion genes, *Nature,* 300, 611, 1982.

Parkins, J. J., Taylor, L. M., Reid, J., and Baggott, D. G., Nitrogen balance and digestibility studies conducted with calves given the repartitioning agents clenbuterol and bitolterol, *J. Anim. Physiol. Anim. Nutr.,* 62, 253, 1989.

Pell, J. M. and Aston, R., Active immunisation with a synthetic peptide region of growth hormone: increased lean tissue growth, *J. Endocrinol.,* 131, R1, 1991.

Pell, J. M., Elcock, C., Harding, R. L., Morrell, D. J., Simmonds, A. A., and Wallis, M., Growth, body composition, hormonal and metabolic status in lambs treated long-term with growth hormone, *Br. J. Nutr.,* 63, 431, 1990.

Pell, J. M., Johnsson, P., Morrell, D. J., Hart, I. C., Holder, A. T., and Aston, R., Potentiation of growth hormone activity in sheep using monoclonal antibodies, *J. Endocrinol.,* 120, R15, 1989.

Penny, R. H. C., Guise, H. J., Rolph, T. P., Tait, J. A., Johnston, A. M., Kempson, S. A., and Gettinby, G., Influence of the β-agonist salbutamol on claw horn lesions and walking soundness in finishing pigs, *Vet. Rec.,* 135, 374, 1994.

Perez-Llamas, F., Lopez, J. A., and Zamora, S., The digestive and metabolic utilization of the dietary protein: effect of clenbuterol and protein level, *Arch. Int. Phys. Biochem. Biophys.*, 100, 27, 1992.

Pringle, T. D., Calkins, C. R., Koohmaraie, M., and Jones, S. J., Effects over time of feeding a β-adrenergic agonist to wether lambs on animal performance, muscle growth, endogenous muscle proteinase activities, and meat tenderness, *J. Anim. Sci.,* 71, 636, 1993.

Pursel, V. G. and Rexroad, C. G., Jr., Status of research with transgenic farm animals, *J. Anim. Sci.,* 71 (Suppl. 3), 10, 1993.

Reeds, P. J., Hay, S. M., Dorwood, P. M., and Palmer, R. M., Stimulation of muscle growth by clenbuterol: lack of effect on muscle protein biosynthesis, *Br. J. Nutr.,* 56, 249, 1986.

Ricks, C. A., Dalrymple, R. H., Baker, P. K., and Ingle, D. L., Use of a β-agonist to alter fat and muscle deposition in steers, *J. Anim. Sci.,* 59, 1247, 1984.

Rivier, J., Spiess, J., Thorner, M., and Vale, W., Characterization of a growth hormone-releasing factor from a pancreatic islet tumour, *Nature,* 300, 276, 1982.

Rocha, J. L., Womack, J. E., and Baker, J. F., New allelic fragment for the growth hormone-Tag I marker in cattle, *Anim. Genet.,* 23, 480, 1992.

Rothwell, N. J., Stock, M. J., and Sudera, D. K., Changes in tissue blood flow and β-receptor density of skeletal muscle in rats treated with the β_2-adrenoceptor agonist clenbuterol, *Br. J. Pharmacol.,* 90, 601, 1987.

Sacca, L., Cittadini, A., and Fazio, S., Growth hormone and the heart, *Endocrinol. Rev.*, 15, 555, 1994.

Schalla, C. L., Roberge, S., Wilkie, B. N., McBride, B. W., and Walton, J. S., Production of anti-idiotypic antibodies resembling bovine somatotropin by active immunization of lactating cows, *J. Endocrinol.,* 141, 203, 1994.

Shanahan, C. M., Rigby, N. B., Murray, J. D., Marshall, J. T., Tounrou, C. A., Nancarrow, C. D., and Ward, K. A., Regulation of expression of a sheep metallothionein la-sheep growth hormone fusion gene in transgenic mice, *Mol. Cell. Biol.,* 9, 5473, 1989.

Sillence, M. N., Unpublished data, 1995.

Sillence, M. N., Hunter, R. A., Pegg, G. G., Brown, L., Matthews, M. L., Magner, T., Sleeman, M., and Lindsay, D. B., Growth, nitrogen metabolism, and cardiac responses to clenbuterol and ketoclenbuterol in rats and underfed cattle, *J. Anim. Sci.,* 71, 2942, 1993.

Sillence, M. N, Matthews, M. L., Spiers, W. G., Pegg, G. G., and Lindsay, D. B., Effects of clenbuterol, ICI118551 and sotalol on the growth of cardiac and skeletal muscle and on β_2-adrenoceptor density in female rats, *Naunyn-Schmiedeberg's Arch. Pharmacol.,* 344, 449, 1991.

Sillence, M. N., Reich, M. M., and Thomson, B. C., Sexual dimorphism in the growth response of entire and gonadectomized rats to clenbuterol, *Am. J. Physiol.,* 268, E1077, 1995 *(Endocrinol. Metab.* 31).

Singh, K., Ambler, G. R., Breier, B. H., Klempt, M., and Gluckman, P. D., Ovine placental lactogen is a potent somatogen in the growth hormone (GH)-deficient rat: comparison of somatogenic activity with bovine GH, *Endocrinology,* 130, 2758, 1992.

Smith, D. J. and Paulson, G. D., Growth characteristics of rats receiving ractopamine hydrochloride and the metabolic disposition of ractopamine hydrochloride after oral or intraperitonial administration, *J. Anim. Sci.,* 72, 404, 1994.

Smith, V. G. and Kasson, C. W., Growth performance and carcass characteristics of pigs administered recombinant porcine somatotropin during 30 to 110 kilogram live weight, *J. Anim. Sci.,* 68, 4109, 1990.

Spencer, G. S. G., Garssen, G. J., and Bergstrom, P. L., A novel approach to growth promotion using auto-immunisation against somatostatin II. Effects on appetite, carcass composition, and food utilisation in lambs, *Livestock Prod. Sci.,* 10, 469, 1983.

Spencer, G. S. G. and Williamson, E. D., Increased growth in lambs following autoimmunization against somatostatin, *Anim. Prod.,* 32, 376, 1981.

Stewart, C. E. W., Bates, P. C., Calder, T. A., Woodall, S. M., and Pell, J. M., Potentiation of insulin-like growth factor-I (IGF-I) activity by an antibody: supportive evidence for enhancement of IGF-I bioavailability *in vivo* by IGF binding proteins, *Endocrinology,* 133, 1462, 1993.

Strosberg, A. D., Anti-idiotypic antibodies that interact with β-adrenergic catecholamine receptor, *Methods Enzymol.,* 178, 265, 1989.

Thornton, R. F., Tume, R. K., Payne, G., Larsen, T. W., Johnson, G. W., and Hohenhaus, M. A., The influence of the β_2-adrenergic agonist, clenbuterol, on lipid metabolism and carcass composition of sheep, *Proc. N. Z. Soc. Anim. Prod.,* 45, 97, 1985.

Toutain, P. L., Schams, D., Laurentie, M. P., and Thomson, T. D., Pharmacokinetics of a recombinant bovine growth hormone and pituitary bovine growth hormone in lactating dairy cows, *J. Anim. Sci.,* 71, 1219, 1993.

Trout, W. E. and Schanbacher, B. D., Growth hormone and insulin-like growth factor-I responses in steers actively immunised against somatostatin or growth hormone-releasing factor, *J. Endocrinol.,* 125, 123, 1990.

Vestergaard, M., Sejersen, K., and Klastrup, S., Growth, composition and eating quality of longissimus dorsi from young bulls fed the β-agonist cimaterol at consecutive developmental stages, *Meat Sci.,* 38, 55, 1994.

Vicini, J. L., Chak, J. H., Hurley, W. L., and Bahr, J. M., The effect of immunization against somatostatin on growth and concentrations of somatotropin in plasma of Holstein calves, *Domest. Anim. Endocrinol.,* 5, 35, 1988.

Wallace, D. H., Hedrick, H. B., Seward, R. L., Baurio, C. P., and Convey, E. M., Growth and efficiency of feed utilisation of swine fed a beta-adrenergic agonist (L-644,969), in *Beta-Agonists and Their Effects on Animal Growth and Carcass Quality,* Hanrahan, J. P., Ed., Elsevier Applied Science, New York, 1987, 143.

Warriss, P. D., Brown, S. N., Rolph, T. P., and Kestin, S. C., Interactions between the beta-adrenergic agonist salbutamol and genotype on meat quality in pigs, *J. Anim. Sci.,* 68, 3669, 1990.

Warris, P. D., Kestin, S. C., and Brown, S. N., The effect of beta-adrenergic agonists on carcass and meat quality in sheep, *Anim. Prod.,* 48, 385, 1989.

Wellenreiter, R. H. and Tonkinson, L. V., Effect of ractopamine hydrochloride on growth performance of turkeys, *Poultry Sci.,* 69 (Suppl. 1), 142, 1990.

Westbrook, S. L. and McDowell, G. H., Immunization of lambs against somatotropin release inhibiting factor to improve productivity: comparison of adjuvants, *Aust. J. Agric. Res.,* 45, 1693, 1994.

Williams, P. E. V., Pagliani, L., Innes, G. M., Pennie, K., Hams, C. I., and Garthwaite, P., Effects of a β-agonist (clenbuterol) on growth, carcass composition, protein and energy metabolism of veal calves, *Br. J. Nutr.,* 57, 417, 1987.

Williams, G. D., Tonkinson, L. V., and Veenhuizen, E. L., The effects of elevated levels of ractopamine hydrochloride on finishing swine, *J. Anim. Sci.,* 67 (Suppl. 1), 260, 1989.

Wood, J. D., Enser, M., and Warris, P. D., Reducing fat quantity: implications for meat quality and health, in *Animal Biotechnology and the Quality of Meat Production,* Fiems, L. O., Cottyn, B. G., and Demeyer, D. I., Eds., Elsevier, Amsterdam, 1991, 69.

Wynn, P. C., Behrendt, R., Jones, M. R., Rigby, R. D. G., Bassett, J. R., and Hoskinson, R. M., Immuno-modulation of hormones controlling growth, *Aust. J. Agric. Res.,* 45, 1091, 1994.

Zainur, A. S., Hoskinson, R. M., and Kellaway, R. C., Production responses of lambs immunised against somatostatin, *Anim. Prod.,* 53, 339, 1991.

Zapf, J., Schmid, Ch., and Froesch, E. R., Biological and immunological properties of insulin-like growth factors (IGF) I and II, *Clin. Endocrinol. Metab.,* 13, 3, 1984.

Zhang, H. M., Brown, D. R., De Nise, S. K., and Ax, R. L., Rapid communication: polymerase chain reaction-restriction fragment length polymorphism analysis of the bovine somatotropin gene, *J. Anim. Sci.,* 71, 2276, 1993.

CHAPTER 10

Evaluation of Sources of Protein and Their Contributions to Amino Acid Adequacy in the Lactating Cow

Carl E. Polan

INTRODUCTION

Ruminants and especially lactating dairy cows have been heralded as efficient producers of animal protein from sources not competitive with human food. Potentially, this is true when considering utilization of roughages, waste, byproducts, and nonprotein nitrogen sources.

With increased milk potential per cow, recommended amounts of dietary protein have increased. Furthermore, because of the wide adoption of feeding cows complete diets in groups, dietary protein may likely exceed the requirements of many cows. Dietary crude protein exceeds 18% in a large proportion of Virginia's top dairy herds. In general, increments in dietary protein result in less efficient utilization of nitrogen (N) and greater quantities go into the waste stream. Providing optimal N utilization for rumen microbial growth and complementing protein (amino acid) needs for milk protein with a ruminally undegraded protein that would supply the right balance of amino acids may reduce N waste from the lactating cow.

A number of alternative feed sources have been improved on for use in dairy cow diets. Some of these are the result of specific processing and heating, such as certain soy products. In other cases, quality control of raw materials has improved and/or fat content has been reduced as in ruminant-grade fishmeal. Thus a variety of protein sources have the potential to supply specifically needed amino acid(s) to the animal. Included with the proper diet, less dietary N may be required to support the same amount of milk yield. As a result, demand for the products from the slaughter and rendering industry has resulted in more economic value for these products. Furthermore, this has resulted in better utilization (recycling) of these products, which would seem to have a positive environmental effect (less waste). However, this may not be true on a net basis if we consider the trends in protein feeding over the past several years.

In this chapter, several factors will be addressed that deal with the efficiency of use of dictary protein and considerations in the use of dietary ruminally degraded protein (RDP) and ruminally undegraded protein (RUP) with a goal for

1-56670-199-6/96/$0.00+$.50

providing the amino acid requirements for lactation. Where possible, efficiency of N utilization will be considered.

PROTEIN REQUIREMENTS FOR LACTATING COWS

Recommendations for protein feeding have paralleled the continuous increase in milk yield per cow over the years. In the 1966 edition of *Nutrient Requirements for Dairy Cattle* (NRC, 1966), 13 to 14% crude protein (CP) in the dietary dry matter (DM) was recommended. In 1971 (NRC, 1971), CP recommendations ranged from 14 to 16% and changed little in the later edition (NRC, 1978). A quotation from 1971 states, "The amino acid composition of the dietary protein is not critical because microorganisms in the ruminoreticulum synthesize the needed amino acids from lower quality proteins and nonprotein sources of nitrogen." Certainly, we know now that this is an oversimplification. In fact, because of the dynamic aspects of the rumen, protein nutrition of the lactating cow is quite complex.

How much protein is enough? Van Horn et al. (1979) summarized data from 14 experiments in which various amounts of soybean meal (SBM) were mixed in the diet and then dietary CP percentage was compared to milk yield. As observed in Figure 1, the response curve of milk yield to protein increment is one of diminishing returns. Dry matter intake follows a similar pattern, but is nearly level after 14% CP. When adjusted for difference in DM intake, the milk yield response to protein is smaller at low protein feeding.

Satter and Roffler (1975) showed that utilization of nonprotein nitrogen (NPN) in the rumen increased with an increase in total digestible nutrient (TDN) content of the diet, but added NPN was not utilized when dietary CP reached 13 or 14%. Crude protein percentage varied in an experiment with either SBM, urea, or combinations of the two so that a performance response curve could be estimated with either 0, 10, 20, 30, or 40% of CP coming from urea (Polan et al., 1976). Results presented in Figure 2 show that CP alone is unsatisfactory for evaluating protein nutrition of lactating cows.

Satter and Roffler (1975) stated that 15 to 16% CP was needed in early lactation for milk production, but that 12 to 12.5% was adequate for the last third of lactation. An ammonia (NH_3) concentration of 5 mg of NH_3 N/100 ml rumen fluid was considered adequate (Satter and Slyter, 1974), but Mehrez et al. (1977) suggested that 20 to 22 mg of NH_3 N/100 ml rumen fluid was required to maximize barley DM fermentation. So probably the need for more protein degradation or ruminal NH_3 depends on available carbohydrate and fermentation rate in the rumen.

Currently, 17 to 18% CP of the dietary DM are commonly recommended, and often even greater amounts are fed to cows in early lactation. Certainly, this is excessive for maximal ruminal microbial growth. Therefore, the excess protein would have to provide an RUP portion with a proper amino acid (AA) balance to supply the productive needs of the cow, if it is to be efficiently utilized and not result in excessive N losses. In fact, with the correct dietary content of RDP and

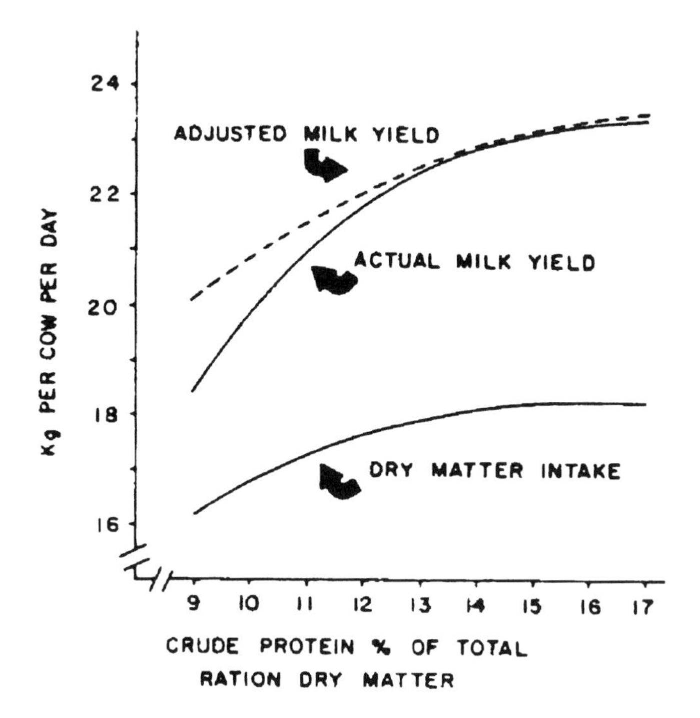

Figure 1 Responses of milk yield and feed intake to increasing ration protein with soybean meal. Represents data from 14 experiments as analyzed by Van Horn et al. (1979).

RUP (with AA content considered), it should be possible to reduce dietary protein. Much has been learned in recent years about protein nutrition in dairy cattle, but understanding how concepts can be utilized to provide specific AA needs is in the infancy stage.

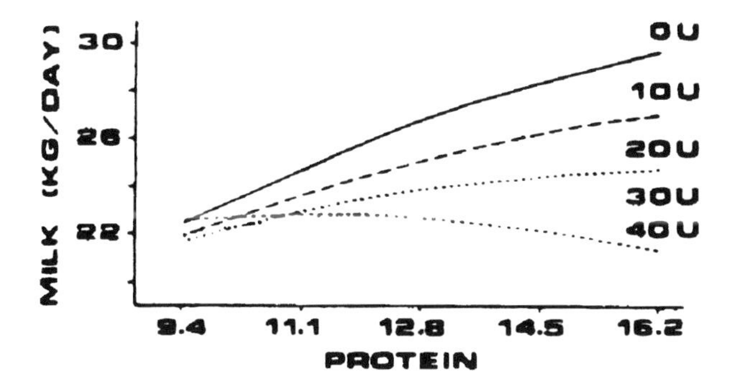

Figure 2 Milk production response curves to varying dietary protein and urea. Curves were developed from responses to diets arranged in a central composite design with protein ranging from 9.4 to 16.2% and substitution of urea (and corn to balance) on an equal N basis for SBM from 0 to 40% (Polan et al., 1976).

MICROBIAL PROTEIN CONTRIBUTION

In order to anticipate the need for absorbable AA directly from dietary sources offered as RUP, it is necessary to estimate the protein and AA contribution of rumen microorganisms. Intake of fermentable feedstuffs directly drives microbial protein production (NRC, 1985). Bacterial crude protein (BCP) can be calculated from consumption of either net energy lactation (NEL) or TDN, both with an r^2 of 0.77. This implies that the equation for BCP accounts for 77% of the variation in rumen microbial synthesis. Therefore, 23% variation is unexplained. For example, soluble protein or peptides may be inadequate to provide N forms required for BCP synthesis. Dietary fat is also a contributor to dietary energy, but contributes nothing to microbial growth.

Some rumen microorganisms require or are stimulated by forms of N more complex than NH_3. Some species require AA or short peptides. Those bacteria that ferment structural carbohydrate utilize only NH_3 as an N source, but those fermenting nonstructural carbohydrate (NSC) can utilize either NH_3 or peptides as an N source (Russell et al., 1992).

Stokes et al. (1991b) demonstrated by continuous culture that bacterial yield was enhanced when NSC replaced corn cobs in the diet. Bacterial efficiency was optimized when NSC increased from 25 to 37% of dietary DM. Maximum bacterial efficiency occurred when the ratio of NSC to RDP was less than 3. Bacterial CP synthesis was maximized in cannulated lactating cows with NSC and RDP at 31 and 12.9% of DM (Stokes et al., 1991a); ratios of less than 2.6 were implied as optimal.

Passage of microbial N per unit of organic matter fermented in the rumen is usually greatest in lactating cows when large quantities of readily fermentable carbohydrates are consumed. Therefore, data derived from animals eating smaller quantities should not be applied to animals with large intakes such as lactating cows (Clark, 1993). A computer disk that calculates BCP is included with current recommendations (NRC, 1989). A recent computer development is the Cornell model (CNCPS) (Russell et al., 1992). The authors claim an advantage over the NRC prediction of BCP because the model considers fermentable carbohydrate rather than TDN, changes in microbial growth efficiency with fermentation rates and rumen turnover, increase in yield of NSC-utilizing bacteria as available peptides increase, and partition of the microbial population according to metabolite activity. The relationship (r^2) between observed flows of microbial protein from the rumen and flows predicted by the model is 0.88.

ESTIMATING RUMEN DEGRADABILITY AND ESCAPE PROTEIN

Both NRC (1989) and Cornell models estimate that microbial protein supplies 50 to 65% of the AA needs of the lactating cow in practical feeding situations. It is desirable to optimize BCP production in the rumen because the essential AA (EAA) supplied by BCP is similar to that of milk protein (Table 1).

Table 1 A Comparison of the EAA Profiles of Body Tissue and Milk with that of Ruminal Bacteria, Protozoa, and Common Feeds[a]

	Arg	His	Ile	Leu	Lys	Met	Phe	Thr	Trp	Val	EAA
				% of total							
Lean tissue	16.8	6.3	7.1	17.0	16.3	5.1	8.9	9.9	2.5	10.1	—
Milk	7.2	5.5	11.4	19.5	16.0	5.5	10.0	8.9	3.0	13.0	—
Bacteria	10.6	4.3	11.6	15.5	17.3	4.9	10.0	11.0	2.6	12.2	40.0
Cell wall	10.2	4.7	10.7	15.8	15.0	6.4	11.3	8.8	4.3	12.6	37.3
Noncell wall	13.2	5.1	11.2	14.3	15.6	5.1	9.8	10.6	3.1	11.7	52.5
Protozoa	10.9	5.2	10.9	18.4	11.1	3.8	12.2	10.6	3.4	13.5	40.7
Feeds											
Alfalfa	10.9	5.2	10.9	18.4	11.1	3.8	12.2	10.6	3.4	13.5	40.7
Corn silage	6.4	5.5	10.3	27.8	7.5	4.8	12.0	10.1	1.4	14.1	—
Corn, yellow	10.8	7.0	8.2	29.1	7.0	5.0	11.3	8.4	1.7	11.5	42.3
Brewers grain	8.9	6.4	10.6	17.6	11.4	4.8	10.3	11.4	3.0	15.6	46.3
Corn gluten meal	6.9	4.7	9.3	36.4	3.8	5.5	13.8	7.5	1.5	10.7	44.2
Corn DDG[b] w/solubles	7.7	7.2	9.8	26.3	6.2	5.2	11.1	10.3	2.7	13.4	37.7
Cottonseed meal	25.4	6.0	7.7	13.9	9.6	3.8	12.2	7.7	2.9	10.8	43.1
Soybean meal	16.3	5.7	10.8	17.0	13.7	3.1	11.0	8.6	3.0	10.6	47.6
Bloodmeal	7.6	11.2	2.1	22.8	15.7	2.1	12.3	8.1	2.7	15.4	49.4
Fishmeal	13.1	5.7	9.3	16.5	17.0	6.3	8.8	9.5	2.4	11.3	44.8
Meat and bone-meal	20.5	5.5	7.8	16.2	14.2	3.6	9.2	9.0	1.8	12.1	38.0

[a] Adapted from Schwab (1994), except cell wall and noncell wall of bacteria (O'Conner et al., 1993).

[b] Dried distillers grains.

Ruminant nutritionists are attempting to manipulate the EAA content of the RUP and complement the microbial AA supply in order to optimize milk protein synthesis.

Because proteins vary in RUP, it is important to quantify the postruminal contribution of dietary protein sources. Protein solubility was evaluated as a predictor of ruminal degradability of CP, but has proven unreliable (Crawford et al., 1978).

Using cannulated cows, direct measurement of dietary protein intake and total protein flow at the duodenum with indirect measurement of BCP is used to calculate RUP by the difference (i.e., RUP = duodenal protein – BCP). However, this does not account for endogenous protein flow in the intestine (Stern and Satter, 1982). Another method requires protein to be fed in incremental amounts above the lowest CP intake that presumably provides adequate N for microbial synthesis. It is assumed that BCP production is constant, and, thus, any intestinal increase in protein (AA) is due to undegraded test protein (Stern and Satter, 1982).

Both preceding procedures possess experimental worth to estimate protein degradation and BCP synthesis in the rumen. However, both procedures require intensive effort, which prohibits routine use. Suspension of feedstuffs in the rumen has been used by a number of laboratories to determine RUP, but this also is too cumbersome for routine application.

An understanding of protein degradability in the rumen was enhanced by Pichard and Van Soest (1977) when feed protein was classified into various fractions according to solubility or rates of degradability. Fraction A was soluble N, which included a large portion of NPN, and was rapidly available to ruminal microbes. Fraction B_1 was insoluble, but rapidly available. Fraction B_2 was slowly available, and Fraction C was unavailable. Some B fractions would be degraded at a rate competitive with fractional turnover of digesta (i.e., some portion of this fraction may not be degraded because it escaped the rumen before all of it could be degraded). Later, the rate of passage was included in the formula (Orskov and McDonald, 1979) to estimate RUP, whether it was determined by the *in situ* technique or by solubility combined with enzymatic methods.

Although the *in situ* technique utilizes the many advantages of rumen environment, feed is not subjected to mastication, rumination, and passage (Nocek, 1988). Data for feedstuffs such as soybean meal or ground corn may be fairly reliable. However, because many feeds must be finely ground in contrast to the form in which they are usually fed, the difference in physical composition raises serious questions about reliability of RUP estimates for such feedstuffs.

Semipurified proteolytic enzymes have been used to measure potential degradability (Menhaden et al., 1980; Nocek et al., 1983). To be useful, the technique must include a rate of degradability of major N fractions so that results have *in vivo* applications and can be used in a dynamic model that also considers ruminal turnover.

Chalupa et al. (1991) described a model-generated submodel of the CNCPS that estimates degradability and undegradability of feedstuffs. Systematic use of borate-phosphate buffer, neutral detergent, and acid detergent solutions separates fractions of protein into A, B_1, B_2, B_3, and C. Ruminal degradation (Kd) of specific proteins are based on literature values where feeds have been incubated either *in situ* or *in vitro* with proteolytic enzymes.

Sniffen et al. (1992), in the description of the CNCPS model for evaluating cattle diets, partitioned protein into fractions similar to those described by Chalupa et al. (1991). Protein degradation rates were estimated by incubating with *Streptomyces griseus* protease, and curve-peeling methods were used to determine degradation rates of each fraction. If the protein degradation rate is rapid, peptides can accumulate, resulting in some peptide escape from the rumen.

FLOW OF NITROGEN FRACTIONS TO THE DUODENUM

Monitoring the flow of N fractions into the duodenum makes it possible to characterize how dietary protein affects the contribution from dietary sources or from microbial protein production. In our first experience with ruminal and duodenal cannula (Cummins et al., 1983), 250-kg steers were fed 13% CP diets with two physical forms of 13 to 15% acid detergent fiber and one that was all concentrate. Peak microbial production of N compounds occurred at approximately 50% ruminal degradation of the feed protein. Microbial protein synthesis and protein degradation were in balance at approximately 54% protein degradation. Above that point, degraded N was probably wasted.

Zerbini et al. (1988) was among the first in the United States to study duodenal protein flow in dual cannulated lactating cows. Fishmeal or soybean meal were compared in diets that contained corn silage, corn grain, and orchardgrass hay. Recovery of N at the duodenum was 93.2 and 84.3% for fishmeal and soybean meal diets, respectively. Total flows of AA to the duodenum were similar for both diets, because ruminal escape of the fishmeal protein was counterbalanced by less BCP synthesis in the rumen.

Eight trials were summarized by Clark et al. (1992), in which fishmeal, bloodmeal, corn gluten meal (CGM), feathermeal, dried distillers grains, or combinations of bloodmeal and fishmeal, bloodmeal and feathermeal, or dehydrated alfalfa and CGM were fed as sources of supplemental protein and compared with SBM. The data of Zerbini et al. (1988) was included in the analysis. When passage of nonammonia, nonmicrobial N (NANMN) to the small intestine was set at 100% for SBM, supplements of low ruminal degradable protein increased passage of NANMN ($r^2 = 0.44$); however, it was not increased much until 35% or more of CP was supplied from slowly degradable sources (Figure 3). When proteins with low rumen degradability were fed, passage of BCP to the small intestine was reduced (relative to SBM). Clark et al. (1992) concluded that limited degradability of protein in the rumen limited AA nitrogen needed for maximal BCP growth. However, flow of methionine (Met) and lysine (Lys) did increase remarkably when these AA were in higher concentrations in the RUP. For example, Met increased by 140% when CGM was fed and Lys increased when bloodmeal or fishmeal were fed.

It is possible that ruminal NH_3 (4.9 vs. 8.7 mg/dl) (Zerbini et al., 1988) limited microbial growth when cows were fed fishmeal compared to SBM (8.7 mg/dl),

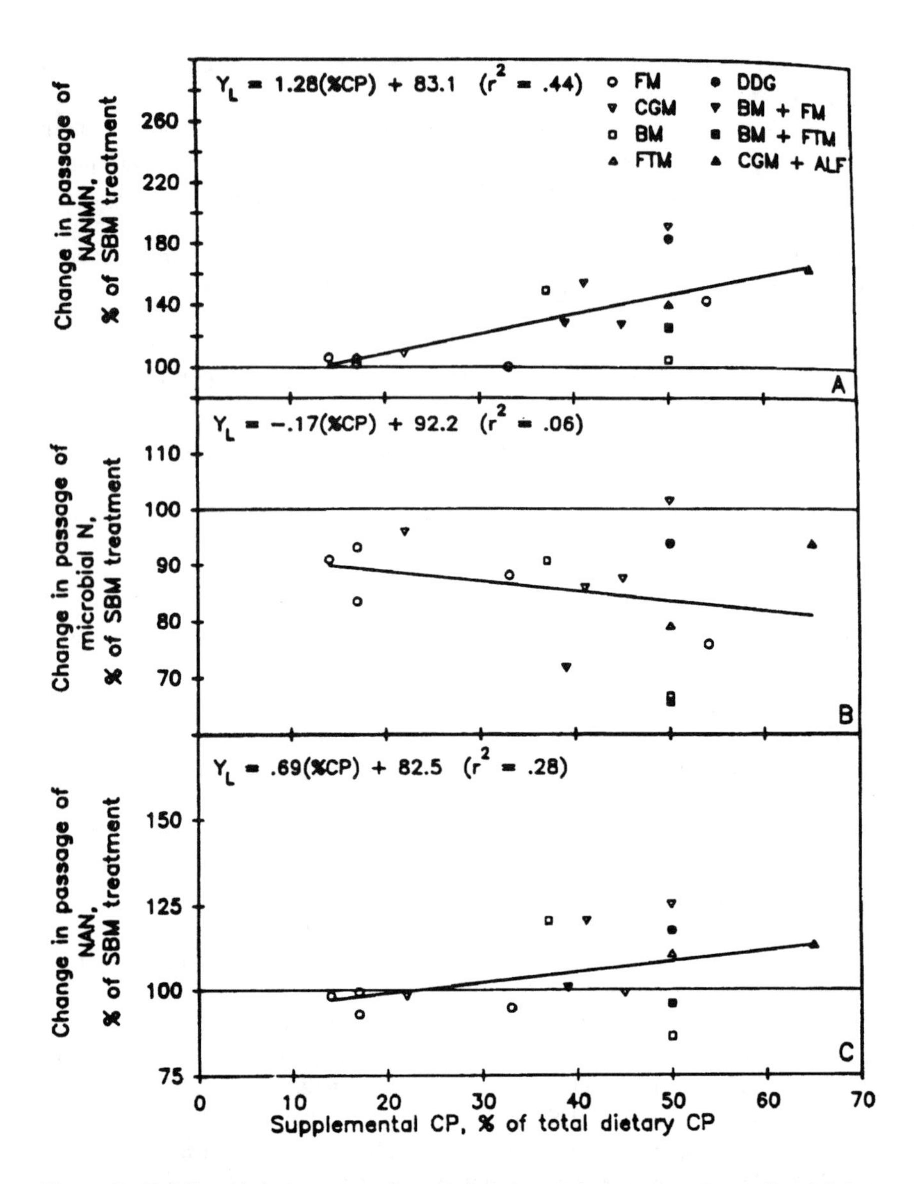

Figure 3 Relationship between supplemental crude protein as a percentage of total dietary crude protein and change in passage to the small intestine of NANMN (A), microbial N (B), and NAN (C) from feeding protein supplements with low ruminal degradability as a percentage of the SBM treatment. Data was taken from several sources. ALF = dehydrated alfalfa, BM = bloodmeal, CGM = corn gluten meal, DDG = dried distillers grains, FM = fishmeal, FTM = feathermeal. (Adapted from Clark et al., 1992.)

but it is more likely that peptide or amino N were limiting. Chapin (1986) fed soybean meal and fishmeal in a protein ratio of 2:1 and 1:2. When fishmeal was the major protein source, both microbial N flow (181 vs. 155 g/d) and total N flow (400 vs. 331 g/d) were greater ($p < 05$). Seymour et al. (1992), in our laboratory, compared diets with SBM to CGM plus SBM. Both microbial N flow (185 vs. 169 g/d) and protein N flow (226 vs. 180 g/d) were greater when CGM was in the diets. Similar studies in our laboratory with dried brewers grains (DBG) or bloodmeal compared to SBM have resulted in as much or more BCP and more total N flow when RUP sources were fed.

IMPROVING THE ABSORBABLE AMINO ACID POOL

Although it is uncertain that total protein flow to the duodenum will be increased extensively by RUP sources compared with SBM (Clark, 1992), sufficient evidence exists that the AA content of digesta can be changed. A comparison of the AA composition of potential RUP sources with those of microbial protein and milk protein is needed. Essential AA profiles show that the profile of rumen bacteria match well the profiles for milk protein (Table 1). Apparently, bacterial cell wall is unavailable (Sniffen et al., 1992) to proteolytic enzymes in the ruminant; so the important profile to consider is the noncell wall fraction which is 60% of BCP and is considered 100% digestible. Compared with AA in milk, SBM is as well balanced as any affordable dietary protein source. Various heat treatments enhance the bypass potential of soybean protein, and feeding heated soybean has enhanced milk production (Schingoethe et al., 1988). Corn gluten meal and fishmeal are good sources of Met. Bloodmeal and fishmeal are good Lys sources, but several other sources including soybean and alfalfa potentially contribute directly to the Lys pool.

Combinations of potential RUP sources can be selected to increase the supply of branched-chain AA; Leu, Ile, and Val. Branched-chain AA are slightly higher in milk protein than bacterial protein and most individual feed sources (Table 1). It is possible to select combinations of feed sources that potentially supply AA in proportion to requirements for milk protein synthesis. However, this may be a simplistic approach because of altered utilization in the rumen, disproportionate absorption or removal by the liver, and metabolic uses other than for milk synthesis.

It must be remembered that feedstuffs are composed of several proteins, each with different physical properties and varying resistance to microbial degradation in the rumen. Undoubtedly this alters the AA pattern of bypass protein relative to that of original feedstuff. Of AA that are released from protein in the rumen, probably only small amounts persist for absorption in the small intestine, although there is evidence of peptide absorption from the ruminant stomach (Webb et al., 1992). Romagnolo et al. (1990) quantified *in situ* ruminal degradation of subfractions of soy protein relative to the total protein when separated by SDS gel electrophoresis. Amino acid composition of protein fractions more resistant to ruminal degradation dictates AA available postruminally.

Mantysaari et al. (1989) measured AA composition of undegraded feed protein after *in vitro* incubation with protease from *S. griseus*. This allowed comparison of AA profiles before and after incubation. Changes were different for different feedstuffs, but usually the residue was higher in Leu than found in the original feed. Except for soybean meal, residue content of phenylalanine (Phe) tended to increase in all feeds tested. Valine increased in all feeds except CGM. The CNCPS (O'Conner et al., 1993) takes into account the AA profile of the rumen unavailable protein when predicting AA available for absorption.

LIMITING AA

The supply of rate-limiting AA available for absorption varies, depending on fermentability of the diet (extent of BCP synthesis) and AA contribution of RUP sources. Milk production responded to supplementation of ruminally protected Lys, but not to Met when cows were fed corn-based diets with CGM as the protein supplement (Polan et al., 1991). Even then, milk production was less than when cows were supplemented with SBM, so some other factor was limiting. A partial explanation may be that intake of CGM diets was less than that for SBM diets.

Diets high in corn and corn silage and supplemented primarily with bloodmeal, CGM, or cottonseed meal were compared on the basis of AA profiles available for digestion and metabolism (King et al., 1990). Based on mammary extraction coefficients, Met, Lys, and threonine (Thr) were apparently limiting AA for milk synthesis. Schwab (1989) tabulated a number of studies and concluded Lys, Met, or both were rate limiting for milk production. He suggested that a ratio of Lys to Met in the RUP fraction should fall between 2.5:1 and 3:1. This complements the ruminal microbial ratio which is between 3.4:1 and 4:1. Subsequently, Schwab (1994) indicated milk protein content and protein yield would respond optimally when intestinal flow of Lys and Met was 15.0 and 5.3% of the essential AA, respectively.

Schingoethe (1991) used milk protein scores (limiting AA in feed protein relative to milk) to compare milk protein sources and predict limiting AA. Using alfalfa, corn silage, and corn diets, first-, second-, and third-limiting AA for soybean meal were Met, histidine (His), and Val; for bloodmeal, they were Ile, Met, and Val; and for CGM, they were Lys, His, and Val.

Chandler (1989) used the classic approach of chemical score and resulting essential AA index which was based on milk protein instead of whole egg protein. He further applied Cornell-derived metabolic utilization efficiency coefficients for absorbed AA. Rumen BCP had superior biological value (82) over potential sources of RUP. The RUP sources with high scores were SBM (71), DBG (67), and fishmeal (66). Based on scores, branched-chain AA were the most limiting. Chandler (1989) proposed using a combination of supplements that would supply escape branched AA as well as Met and Lys.

The CNCPS (O'Conner et al., 1993) is now the most advanced published program to estimate adequacy and limitations of AA. The authors claim that in addition to improving milk and milk protein production, less dietary protein will be required, and therefore efficiency of utilization will increase. This model brings together a number of quantifiable factors to predict protein and AA status. Factors include fermentable carbohydrate and protein in the rumen, bacterial noncell wall fraction AA content, and AA composition of RUP for predicting AA available for absorption. The AA requirement of the cow was based on protein synthesis and turnover in tissue, milk, and keratin to account for endogenous and metabolic fecal losses. Metabolic utilization of individual AA varied, depending on physiological function.

The equations of the Cornell AA model were applied to several diets used in experiments in our laboratory or in conjunction with others (Wu et al., 1995). In every case, a bypass source was provided, either as RUP, abomasal infusion, or RUP accompanied by supplementation of protected Met or Lys. All diets were based on corn silage, but one contained 14% alfalfa.

Observed and permitted (limited by first-limiting AA) milk yields, milk protein, and calculated absorbable and usable AA according to the CNCPS for two experiments from our laboratory are shown in Table 2. In the first experiment, Met was first limiting when SBM was fed. When CGM was included in the diet, milk yield increased, but Lys became limiting such that only 42 g of Met could be used. When DBG was in the diet, milk and milk protein yield increased, and usable Met increased to 43 g, even though less was actually available for absorption when compared to the CGM diet. So we infer that our response in yield could be explained by available Met, but is still limited by Lys. Digestibility for dietary RUP sources is not considered here, so it is anticipated that available AA in the CGM diet is overestimated because of the difficulty of digesting zein protein.

In the second experiment, Met was limiting in the SBM control diet (18% CP). Again, milk and protein yield increased when DBG was included in the diet and Lys became limiting. However, the wet brewers grain diet resulted in similar protein yield, but usable Met was slightly less than in the case where Met was limiting. Therefore, this response cannot be explained by the projected limiting AA.

Researchers from five universities conducted a field study (35 commercial herds) to evaluate the value of a commercial animal-marine protein blend (AMPB) for stimulating milk yield (Beede et al., 1994). An animal-marine protein blend replaced an equal weight of CP in the control diet. For early lactation cows [<40 days in milk (DIM) initially], milk yield response to AMPB was 2.64 kg/d. In early lactation, 26 herds responded positively, 12 herds were nonresponders, and 2 herds responded negatively. The average response was less when all stages of lactation were considered — 1.2 kg/d. Milk yield increased in most herds when the dietary RUP increased due to AMPB compared with control. Other dietary influences could not be detected to explain responses. Unfortunately, intake was not known in this experiment. A highlight of this experiment was the development of an analysis to measure response for the treatment test period relative to expectations in the normal untreated lactation curve. These methods would increase the opportunity for field studies because herds would not have to be divided into control or experimental groups. This was accomplished by simulation of lactation curves and regression methods to estimate typical changes in milk yield over time and recording differences from zero responses.

Table 2 Treatment Means for Observed Milk, Milk Protein, Permitted Predicted Milk Yields, Absorbable and Usable AA Based on Limiting AA

		Milk (kg/d)				AA for (g/d[c])	
Ref.	Dietary treatment[a]	Observed	Permitted[b]	Milk protein (g/d)	Limiting AA for treatment	Abs	Usab
Cozzi and Polan (1994)	Protein source						
	18.3% SBM[d], control	30.8	32.2	924	Met	39	39
	9.4% SBM, 7% CGM[d]	31.7	34.5	951	Lys	52	42
	11.0% SBM, 16% DBG	35.0	36.6	1015	Lys	46	43
Polan et al., (1985)	Diet CP = 14%						
	6.5% SBM, control	26.8	28.2	879	Lys	127	127
	11.0% DBG[d]	28.0	33.3	932	Lys	150	150
	13.0% WBG[d]	28.4	28.3	963	Lys	128	128
	Diet CP = 16%						
	10.5% SBM, control	25.3	26.7	883	Lys	126	126
	17.4% DBG	29.1	30.3	928	Lys	129	129
	20.6% WBG	27.1	27.1	1003	Lys	145	145
	Diet CP = 18%						
	14.6% SBM, control	26.5	28.8	935	Met	44	44
	24.2% DBG	31.0	38.2	1029	Lys	60	52
	29.0% WBG	31.2	30.5	1055	Lys	49	42

[a] Diets contained corn silage, ground corn, and protein supplement shown (dry basis). Alfalfa (14%) was included in diets of Polan et al.

[b] Permitted milk is quantity possible by first-limiting AA based on Cornell model (O'Conner et al., 1993).

[c] Quantity represents the AA listed as first limiting the control diet of each set (O'Conner et al., 1993). Absorbable (Abs) and usable (Usab) does not change if the same AA remains first limiting, but usable AA is limited by the new limiting AA.

[d] SBM = soybean meal, CGM = corn gluten meal, DBG = dried brewers grains, and WBG = wet brewers grains.

CONCLUSIONS

Response to undegraded protein in diets of lactating cows has been variable in both milk protein and milk yield. Many of the studies over the past several years were not based on a balanced AA pool for absorptive needs. Others have not provided suitable conditions for acceptable rumen microbial fermentation or microbial protein yields. New modeling programs should continue to improve the predictability of meeting production needs at the tissue level. However, the greatest factor affecting predictability of AA supply is intake. When using the CNCPS, if observed feed intake substantially exceeds that which is predicted, limiting AA supply often exceeds the requirement, even without providing special RUP sources. However, if feed intake is near or less than the model predicts, then certain AA may become limiting.

Since intake of fermentable feeds may be a major factor in determining the need for an RUP source with a balanced AA profile, then the most likely situations for limiting AA can be anticipated. We suggest the most likely need for bypass AA is as follows: early lactation, because intake has not reached a maximum; fat addition to the diet, because it increases milk production without providing energy for BCP production, often inhibiting microbial growth; introduction of BST, because milk production increases in advance of intake; and other factors that reduce intake, such as diet changes, illnesses, etc. Consideration of these factors when formulating diets should make it possible to significantly reduce N intake and waste from lactating cows.

ACKNOWLEDGMENT

The John Lee Pratt Animal Nutrition Program for partial support of research since its inception.

REFERENCES

Beede, D. K., Ferguson, J. D., Shaver, R. D., Huber, J. T., and Polan, C. E., Study examines animal-marine protein blend use in cow diets, *Feedstuffs,* 66, 12, 1994.

Chalupa, W., Sniffen, C. J., Fox, D. G., and Van Soest, P. J., Model generated protein degradation nutritional information, in *Proceedings of the Cornell Nutrition Conference,* Cornell University, Ithaca, NY, 1991, 44.

Chandler, P. T., Achievement of optimum amino acid balance possible, *Feedstuffs,* 61, 14, 1989.

Chapin, C. A., Protein partition and digesta flow in lactating Holsteins fed 2:1 and 1:2 soybean meal:fish meal, M.S. Thesis, Virginia Polytechnic Institute and State University, Blacksburg, 1986.

Clark, J. H., Optimizing Ruminal Fermentation to Maximize Amino Acid Availability, in *Proceedings of the 4th Annual Florida Ruminant Nutrition Symposium,* University of Florida, Gainesville, January 13–14, 1993.

Clark, J. H., Klusmeyer, T. H., and Cameron, M. R., Microbial protein synthesis and flows of nitrogen fractions to the duodenum of dairy cows. *J. Dairy Sci.,* 75, 2304, 1992.

Cozzi, G. and Polan, C. E., Corn gluten meal and dried brewers grain as partial replacement for soybean meal in the diet of Holstein cows, *J. Dairy Sci.,* 77, 825, 1994.

Crawford, R. J., Jr., Hoover, W. H., Sniffen, C. J., and Crooker, B. A., Degradation of feedstuff nitrogen in the rumen vs. nitrogen availability in three solvents, *J. Anim. Sci.,* 46, 1768, 1978.

Cummins, K. A., Nocek, J. E., Polan, C. E., and Herbein, J. H., Nitrogen degradability and microbial protein synthesis in calves fed diets of varying degradability defined by the bag technique, *J. Dairy Sci.,* 66, 2356, 1983.

King, K. J., Huber, J. T., Sadik, M., Bergen, W. G., Grant, A. L., and King, V. L., Influence of dietary protein sources on amino acid profiles available for digestion and metabolites in lactating cows, *J. Dairy Sci.,* 3, 3208, 1990.

Mantysaari, P. E., Sniffen, C. J., and O'Conner, J. D., Protein fractions in animal byproduct meal cause variability, *Feedstuffs,* 61, 18, 1989.

Mehrez, A. Z., Orskov, E. R., and McDonald, I., Rates of rumen fermentation in relation to ammonia concentration, *Br. J. Nutr.,* 38, 447, 1977.

Menhaden, S., Erfle, J. D., and Sauer, F. D., Degradation of soluble and insoluble proteins by Bacteriodes amylophilus protease and by rumen microorganisms, *J. Anim Sci.,* 50, 723, 1980.

National Research Council (NRC), *Nutrient Requirements of Dairy Cattle,* 3rd rev. edition, National Academy of Science, Washington, DC, 1966.

National Research Council (NRC), *Nutrient Requirements of Dairy Cattle,* 4th rev. edition, National Academy of Science, Washington, DC, 1971.

National Research Council (NRC), *Nutrient Requirements of Dairy Cattle,* 5th rev. edition, National Academy of Science, Washington, DC, 1978.

National Research Council (NRC), *Ruminant Nitrogen Usage,* National Academy of Science, Washington, DC, 1985.

National Research Council (NRC), *Nutrient Requirements of Dairy Cattle,* 6th rev. ed., National Academy of Science, Washington, DC, 1989.

Nocek, J. E., In situ and other methods to estimate ruminal protein and energy digestibility: review, *J. Dairy Sci.,* 71, 2051, 1988.

Nocek, J. E., Herbein, J. H., and Polan, C. E., Total amino acid release rates of soluble and insoluble protein fractions and concentrate feedstuffs by Streptomyces griseus, *J. Dairy Sci.,* 66, 1663, 1983.

O'Conner, J. D., Sniffen, C. J., Fox, D. G., and Chalupa, W., A net carbohydrate and protein system for evaluating cattle diets: IV. Predicting amino acid adequacy, *J. Anim. Sci.,* 71, 1298, 1993.

Orskov, E. R. and McDonald, I., The estimation of protein degradability in the rumen from incubation measurements weighted according to rate of passage, *J. Agric. Sci. Camb.,* 92, 499, 1979.

Pichard, G. and Van Soest, P. J., Protein solubility of ruminant feed, in *Proceedings of the Cornell Nutrition Conference,* Cornell University, Ithaca, NY, 1977, 91.

Polan, C. E., Cummins, K. A., Sniffen, C. J., Muscato, T. V., Vicini, J. L., Crooker, B. A., Clark, J. H., Johnson, D. G., Otterby, D. E., Guillaume, B., Muller, L. D., Varga, G. A., Murray, R. A., and Peirce-Sandner, S. B., Responses of dairy cows to supplemental rumen-protected forms of methionine and lysine, *J. Dairy Sci.,* 74, 2997, 1991.

Polan, C. E., Herrington, T. A., Wark, W. A., and Armentano, L. E., Milk production response to diets supplemented with dried brewers grains, wet brewers grains, or soybean meal, *J. Dairy Sci.,* 68, 2016, 1985.

Polan, C. E., Miller, C. N., and McGilliard, M. L., Variable dietary protein and urea for intake and production in Holstein cows, *J. Dairy Sci.,* 59, 1910, 1976.

Romagnolo, D., Polan, C. E., and Barbeau, W. E., Degradability of soybean meal protein fractions as determined by sodium dodecyl sulfate-polyacrylamide gel electrophoresis, *J. Dairy Sci.,* 73, 2379, 1990.

Russell, J. B., O'Conner, J. D., Fox, D. G., Van Soest, P. J., and Sniffen, C. J., A net carbohydrate and protein system for evaluating cattle diets: I. Ruminal fermentation, *J. Anim. Sci.,* 70, 3551, 1992.

Satter, L. D. and Roffler, R. E., Nitrogen requirement and utilization in dairy cattle, *J. Dairy Sci.,* 58, 1219, 1975.

Satter, L. D. and Slyter, L. L., Effect of ammonia concentration on rumen microbial protein production in vitro, *Br. J. Nutr.,* 32, 199, 1974.

Schingoethe, D. J., Protein quality, amino acid supplementation in dairy cattle explored, *Feedstuffs,* 63, 11, 1991.

Schingoethe, D. J., Casper, D. P., Yong, C., Illg, D. J., Sommerfeld, J., and Mueller, C. R., Lactational response to soybean meal, heated soybean meal and entruded soybeans with ruminally protected methionine, *J. Dairy Sci.,* 71, 173, 1988.

Schwab, C. G., Amino Acids in Dairy Cow Nutrition, in Proceedings of the Rhone-Poulenc Animal Nutrition Technical Symposium, Fresno, CA Rhone-Poulenc Animal Nutrition, Atlanta, GA, 1989, 75.

Schwab, C. G., Amino acid requirements of lactating dairy cows, in Proceedings of the 55th Minnesota Animal Nutrition Conference and Roche Technical Symposium, University of Minnesota, St. Paul, 1994.

Seymour, W. M., Polan, C. E., and Herbein, J. H., In vivo degradation of protein in diets formulated for two degradabilities, *J. Dairy Sci.,* 75, 2447, 1992.

Sniffen, C. J., O'Conner, J. D., Van Soest, P. J., Fox, D. G., and Russell, J. B., A net carbohydrate and protein system for evaluating cattle diets: II. Carbohydrate and protein availability, *J. Anim. Sci.,* 71, 3562, 1992.

Stern, M. D. and Satter, L. D. In vivo estimation of protein degradability in the rumen, in *Protein Requirements for Dairy Cattle: Symp.,* Owens, F. N., Ed., MP-109, Oklahoma State University, Stillwater, 1982.

Stokes, S. R., Hoover, W. H., Miller, T. K., and Blauweikel, R., Ruminal digestion and microbial utilization of diets varying in type of carbohydrate and protein, *J. Dairy Sci.,* 74, 871, 1991a.

Stokes, S. R., Hoover, W. H., Miller, T. K., and Manski, R. P., Impact of carbohydrate and protein levels on bacterial metabolism in continuous culture, *J. Dairy Sci.,* 74, 860, 1991b.

Van Horn, H. H., Zometa, C. A., Wilcox, C. J., Marshall, S. P., and Harris, B., Jr., Complete rations for dairy cattle VIII. Effect of percent and source of protein on milk yield and ration digestibility, *J. Dairy Sci.,* 62, 1086, 1979.

Webb, K. E., Jr., Matthews, J. C., and DiRienzo, D. B., Peptide absorption: a review of current concepts and future perspectives, *J. Anim. Sci.,* 70, 3248, 1992.

Wu, Z., Polan, C. E., Fisher, R. J., and McGilliard, M. L., Amino acid adequacy in diets fed to lactating dairy cows, *J. Dairy Sci.,* submitted, 1996.

Zerbini, E., Polan, C. E., and Herbein, J. H., Effect of dietary soybean meal and fishmeal on protein digesta flow in Holstein cows during early and midlactation, *J. Dairy Sci.,* 71, 1248, 1988.

CHAPTER 11

Pending Advances in Understanding Nitrogen and Carbon Metabolism by Ruminal Microorganisms

Mark Morrison

INTRODUCTION

Ruminant nutrition has been the subject of intense research for many years, primarily to determine the nutrient requirements for maximal growth and production. In temperate, developed regions of the world, livestock production systems have evolved to become more intensive and, in some instances, dependent on diet formulations that are not consistent with the nutritional strategy that has evolved in ruminants. Feedlot cattle and high-producing dairy cows are now fed as little as 10 to 15% and 40 to 50% dry matter as forage, respectively, primarily to ensure proper ruminal function and digesta propulsion. This replacement of structural carbohydrates with rapidly degradable starch in ruminant diets has improved nitrogen retention in meat and milk (Huber and Herrera-Saldana, 1994), and, in feedlot cattle, the use of "protected" proteins and ionophores have also resulted in productive alterations of ruminal metabolism and animal performance. Today, the United States has a relatively cheap and abundant supply of beef, milk, and other dairy products as a result.

Even with these advances, there is still substantial opportunity to improve nutrient retention by ruminant animals, increase productivity, and further limit environmental impact(s). For example, protein digestion in forage-fed cattle can be "wasteful": the protein is degraded too rapidly in the rumen, excreted into the environment in animal waste, and a potential contaminant of groundwater and surface water. Relative to the organic matter digestibility of the diet, between 20 and 50% of the daily nitrogen intake may be released into the environment, and estimates of the costs from this waste approach $5 billion annually (Russell, 1993). Grain-based diets also predispose the dairy cow to ruminal acidosis, milk fat depression, lameness, and other health problems. In feedlot cattle, subclinical acidosis can result from prolonged feeding of grain-based finishing diets, causing feed intake variation, reductions in efficiency of gain, and, therefore, nutrient retention. Improvements in these areas could be facilitated by (1) altering in the kinetics of polysaccharide degradation, (2) altering the end product profile of ruminal fermentations, and (3) increased understanding of rumen microbial ecology in feedlot and forage-fed animals. Improvements in fiber digestion, especially

1-56670-199-6/96/$0.00+$.50

with grazing and forage-fed animals, would likely improve microbial growth, increasing nutrient supply and retention by the host animal.

Given the examples outlined here, greater research emphasis needs to be directed toward understanding the *fundamental principles* that control the uptake and metabolism of nutrients in ruminant tissues, and how these mechanisms might be manipulated to optimize nutrient retention. Rumen microbiology definitely fits into this category. The recombinant DNA and molecular biology techniques integral to today's version of "biotechnology" research also provide the means to advance our fundamental understanding of rumen microbial physiology and ecology. This chapter presents some of the current research activity that offers the promise for advancing our understanding in these areas. Such knowledge should help expedite the development of novel, productive manipulations of rumen microbial activity and nutrient retention by ruminants.

NITROGEN METABOLISM

Peptide Transport and the Smugglin Concept

It has long been established that despite adequate ruminal ammonia nitrogen to support microbial growth, feed proteins are extensively degraded in the rumen unless they are chemically pretreated or naturally resistant to microbial attack. The bacteria *Butyrivibrio fibrisolvens* and *Prevotella ruminicola* have a quantitatively important role in ruminal protein and peptide metabolism with a wide range of diets (for a recent review, see Wallace, 1994). Additionally, bacteria with active growth on peptides, plus rapid rates of ammonia production, have recently been isolated from laboratory animals (Russell et al., 1988; Chen and Russell, 1989). However, several of these bacteria are nonproteolytic, requiring that proteins first be degraded to peptides to permit growth. Therefore, the proteolytic enzymes of other rumen bacteria probably control the rate and extent of ammonia production by these peptide-fermenting bacteria. Although research has so far identified the major groups of rumen microorganisms involved with protein digestion, the enzymes involved have been only superficially characterized, and little success in manipulating microbial activity has been forthcoming.

Highly selective means of inhibiting microorganisms that hydrolyze proteins and peptides may provide the opportunity to manipulate successfully the kinetics of forage protein degradation. We have chosen to focus on *P. ruminicola* because its contribution to ruminal protein degradation is quantitatively important, and it is one of the species most amenable to molecular genetics-based studies. One approach would be to exploit the "smugglin" concept (Payne, 1972) also referred to as warhead delivery (Ringrose, 1972), and illicit transport (Ames et al., 1973). The approach is dependent upon peptides carrying amino acid mimetics, or other toxic molecules, which are accumulated intracellularly via peptide permease(s). Only those organisms with the appropriate peptide permease(s) would be capable of toxin accumulation, and thus only their growth would be inhibited. A number of synthetic smugglins have been devised and evaluated primarily with bacteria of clinical importance for humans (Payne and Smith, 1994). We have recently begun similar studies with *P. ruminicola* (Madeira et al., 1995). Strain $B_1$4 was tested by disk diffusion assays for its susceptibility to synthetic smugglins containing either chloroalanine, ethionine, tri-ornithine, or oxalysine. Chloroalanine did not inhibit growth of *P. ruminicola* within the range tested (0 to 250 μg). While tri-ornithine per se was not inhibitory to growth, a heptapeptide containing tri-ornithine was found to inhibit *P. ruminicola* $B_1$4. Pentapeptides containing oxalysine or ethionine were also inhibitory to strain $B_1$4 growing on ammonia, causing zones of clearing of 1.8 cm. The inhibition by these last two oligopeptides could be prevented by including peptone (0.2% w/v) in the growth medium, suggesting that the same transport system(s) may be used for the uptake of both synthetic and peptone peptides. These findings are consistent with early studies that showed oligopeptides ranging from four amino acid residues up to 2000 Da were utilized for growth, while growth with free amino acids, di-peptides, and tripeptides was limited (Pittman and Bryant, 1964). Although some Gram-positive bacteria also appear to possess a transport system for "large" peptides (Kunji et al., 1993; Tynkkynen et al., 1993), the selectivity demonstrated by *P. ruminicola* appears novel.

Studies with *P. ruminicola* (Pittman et al., 1967) and the colonic bacterium *Bacteroides fragilis* (Gibson and MacFarlane, 1988) suggested that peptide uptake in these bacteria probably involved an ATP-dependent transport system similar to the *Escherichia coli* oligopeptide permease *(opp)* system. We have recently constructed oligonucleotide probes with homology to DNA encoding highly conserved ATP-binding domains of the *E. coli oppD* gene, and have used these to screen a plasmid library of *P. ruminicola* $B_1$4 DNA. Several clones have been isolated, and the DNA sequences share homology with other membrane-bound translocator proteins (Peng and Morrison, 1995). These clones suggest the presence of *oppD*-like gene(s) in *P. ruminicola* and are currently being investigated in more detail to delineate further the mechanisms of peptide transport in these bacteria. Ethionine-containing oligopeptides are also being used to select for mutants defective in peptide transport. While these studies are still very fundamental, the smugglin concept appears to offer a powerful tool to study the molecular bases of peptide transport in *P. ruminicola.* The research also has the potential to design highly selective inhibitors of other peptidolytic microorganisms. Effective, yet highly selective means of manipulating rumen microbial activity is the ultimate objective of the research.

Protozoal digestion of particulate protein and rumen bacteria is also thought to be a major contributing factor to ammonia production and intraruminal nitrogen recycling. Defaunation can reduce ammonia concentration and increase protein flow to the small intestine (Van Nevel and Demeyer, 1988). Although chemical-based strategies designed to eliminate protozoa have been developed, they are best suited for experimental purposes. A novel approach would be to adapt the smugglin concept for long-term control of protozoal activity. Recombinant DNA technology now offers the potential to develop rumen bacteria as "smugglins" by having them express a recombinant protein toxic to the protozoan, either on bacterial cell surface or intracellularly. A bacterial delivery system for antiprotozoal compound(s) has several merits. It should prove highly selective for protozoa, compared with the current detergent-based methods of defaunation. Also, provided that the recombinant DNA construct is stably maintained in the host bacterium (which also is stable in the rumen), protozoal inhibition is likely to be long term, and the problems associated with reinoculation of the rumen with protozoa would be circumvented. Future success with this type of approach requires study of protozoa biology and physiology, as well as continued development of recombinant DNA strategies for use with rumen bacteria, such as those outlined by Flint (1994). Another intriguing example appears to be the development of an immunization protocol which effectively targets the rumen protozoa, presumably by protozoal engulfment of secretory antiprotozoal antibodies present in the saliva. However, little published literature is available regarding the development and efficacy of the technique (Gnanasampanthan et al., 1994).

Proteinases and Peptidases of Rumen Bacteria

With the possible exception of *Butyrivibrio fibrisolvens, P. ruminicola* is thought to be the most numerous proteolytic bacterium in the rumen. Hazlewood and Edwards (1981) suggested that at least three different classes of proteinase activity were present in *P. ruminicola* strain R8/4, and, based upon inhibition characteristics, the proteolytic activity was similar to that of rumen contents. Wallace and Brammall (1985) arrived at similar conclusions using several different isolates of *P. ruminicola* and a greater variety of substrates and proteinase inhibitors. Hazlewood et al. (1981) also found that the proteolytic activity of *P. ruminicola* strain R8/4 was affected by the type of protein used as a nitrogen source. More recently, Griswold and Mackie (1993) have found that soybean protein was utilized more efficiently than casein as a nitrogen source by *P. ruminicola* B_14. Dipeptides were found to accumulate in cultures, probably the result of dipeptidyl-aminopeptidase Type I activity produced by *P. ruminicola* strains. This type of activity also seems to be the major form of measurable peptidase activity in ruminal contents (Wallace, 1994). At least some strains of *P. ruminicola* possess two dipeptidyl aminopeptidases, and preliminary studies indicate that the peptidases differ in terms of substrate specificity and sensitivity to oxygen (Wallace et al., 1994). It will prove interesting to determine whether these peptidases possess inhibition patterns similar to those determined for proteinase(s).

Molecular biology and recombinant DNA technology offer a new dimension to the study and potential manipulation of the proteolytic and peptidolytic enzyme(s) of *P. ruminicola*. The construction of *P. ruminicola* mutant strains with diminished (or no) proteolytic and/or peptidolytic activity is a plausible and attractive research objective. Growth and development of *P. ruminicola* mutants could then be evaluated *in vitro* using a variety of nitrogen sources, including plant material, to determine whether proteolytic activity is required by *P. ruminicola* to effect significant degradation of plant tissue. Genetics-based studies can also be used to evaluate the cellular location and role(s) of peptidase(s) in oligopeptide utilization in *P. ruminicola*, and of particular interest is defining the role(s) that peptidase(s) may play in oligopeptide uptake and efflux. Such knowledge will be critical, not only to define oligopeptide composition of smugglins specific for *P. ruminicola,* but also for the design of smugglins that *are not* specific for *P. ruminicola.*

Fermentation uncoupled from bacterial growth ("energy spilling," Russell et al., 1988) seems to occur in *P. ruminicola* in response to the form of available nitrogen. Peptide nitrogen not only blocks ammonia assimilation in *P. ruminicola,* but it also impacts microbial yield once peptide nitrogen is depleted and replaced with ammonia (Russell, 1983). Such a situation might arise in cattle fed low protein forages, where the rumen receives some protein nitrogen, but the microbiota is primarily reliant upon recycled, nonprotein nitrogen sources. It is unclear whether ammonia assimilation in *P. ruminicola* is modulated in response to the amount of peptide nitrogen available, intracellular concentrations of amino acids, or feedback inhibition on amino acid biosynthetic pathways. Mutant strains defective in protein and peptide utilization should prove valuable for developing a mechanistic understanding of ammonia assimilation and of how growth and development of *P. ruminicola* is affected by the type, amounts, and fluctuations in nitrogen. Such fundamental information of bacterial physiology is crucial if "energy spilling" in the rumen is to be minimized and microbial protein synthesis maximized.

CARBON UTILIZATION

Byproduct Feeding and Rumen Microbiology in Feedlot Cattle

Feedlot acidoses are primarily caused by rapid production of organic acids and endotoxins by rumen bacteria. The ruminal concentration of lactic acid is considered to serve as a "barometer" of the severity of acidosis (Britton, 1986). While the biochemical bases of subacute acidosis are not well defined, the feeding of ionophores has helped to minimize its negative impact upon feedlot profitability (Russell and Strobel, 1989; Larson, 1992). Grain starch is also used for the production of fuel ethanol, and the quantity of distillers by-products will continue to increase, providing a steady supply of macronutrients (e.g., protein, fat, and fiber) for potential use by the feedlot industry. Past research at the University of Nebraska has shown that both wet and dried distiller's by-products are good sources of bypass protein and energy for finishing cattle and may alter rumen microbial activity. Recent findings indicate that the current performance of feedlot cattle may still be well below optimum, because the feeding of "thin stillage" to calves and yearlings improved animal performance (feed to gain ratio) by as much as 17% (Larson, 1992; Larson et al., 1993).

We have recently completed a series of studies to evaluate how the soluble and low-density fractions of condensed distillers byproducts (CDB) affect rumen

Table 1 Effect of a Diet of Dry Rolled Corn (DRC) or DRC Plus 15% CDB on Rumen Biochemistry and Microbiology[a]

Parameter	DRC	DRC + 15% CDB
Ruminal pH	5.8	5.5
Total VFA (m*M)*	115.6	123.3
Acetate	57.1	56.4
Propionate	39.7	47.9
Butyrate	14.3	14.4
Acetate:propionate	1.6	1.3
Lactate disappearance rate (mmol/min)	0.16	0.38
Total culturable bacteria ($\times 10^{10}$)	126.3	273.0
Starch-degrading bacteria ($\times 10^{10}$)	13.4	43.3
Lactate-utilizing bacteria ($\times 10^{9}$)	5.6	35.4

[a] Ruminal pH and VFA data taken from Ham (1994); lactate disappearance rates and culturable counts of rumen bacteria taken from Fron et al. (1996). Data from Fron et al. (1996) are the calculated means from six animals fed DRC and three animals fed DRC + 15% CDB.

microbiology and metabolism (Fron et al., 1996). Analyses in our laboratory have identified fumarate (0.1 to 1 m*M),* malate (1.2 to 6.2 m*M),* and succinate (3 to 16 m*M)* in several different sources of these by-products. In addition, we have detected high concentrations of lactic acid similar to levels reported by Huhtahnen and Nasie (1992). Nisbet and Martin (1990, 1991) have previously reported that water-soluble extracts of yeast cultures, malate, or fumarate can improve both lactate fermentation and growth of *Selenomonas ruminantium*, a rumen bacterium present in high numbers in grain-fed animals. In our studies, the *in vitro* rate of lactic acid disappearance was not stimulated by direct addition of CDB to mixed rumen contents collected from a steer adapted to a high concentrate diet. However, if animals were fed CDB for several weeks, both rumen microbiology and fermentation were affected (Table 1). Culturable counts of lactic acid-utilizing bacteria increased, and this coincided with a twofold increase in the *in vitro* rate of lactic acid fermentation. The lactic acid (and probably other dicarboxylic acids) present in CDB appears to selectively increase the numbers of microorganisms that ferment this acid(s) and increase the propionate to acetate ratio. Therefore, distiller's by-products possess nutrients in addition to protein, fiber, and lipids that can affect rumen microbiology and animal performance.

Fiber-Degrading Enzymes and Adherence Mechanism(s) in Rumen Bacteria

During the last 10 to 15 years, the molecular biology of polysaccharide-degrading enzymes has been a major focus in rumen microbiology research, and advances have been made in our understanding of enzyme structure and function

(Forsberg et al., 1993; Flint, 1994). Studies have also shown that colonization and adherence by some rumen bacteria are specific in nature (Latham et al., 1978a, 1978b), and that adherence can be modified by nutrients and growth conditions (e.g., Stack et al., 1983; Morrison et al., 1990). The catalytic domains of rumen bacterial cellulases and xylanases share significant homology with known bacterial and fungal enzyme families (Beguin and Aubert, 1994; Flint, 1994). Although microscopic observations may suggest otherwise, there is currently no direct evidence of a cellulosome-integrating protein or other "scaffolding proteins" involved with enzyme aggregation in rumen bacteria (Flint, 1994). Cellulase enzymes from many microorganisms also possess cellulose-binding domains (CBD), which are generally located at either termini of the protein and separated from catalytic domain(s) by glycosylated, Pro-Thr-Ser-rich "linker segments" (Beguin and Aubert, 1994). Motifs typical of the CBDs present in other cellulolytic enzymes are also missing from virtually all the rumen bacterial enzymes cloned so far. Pell and Schofield (1993) stated that CBDs have only been identified in *Fibrobacter succinogenes* and *B. fibrisolvens* endoglucanase genes. Furthermore, of the 13 sequences so far isolated from Ruminococci, no consensus CBDs or hydrophobic clusters have been identified (White, 1995). *Prevotella ruminicola* is an acid-tolerant bacterium and produces an acid stable, β-(1,4) endoglucanase that possesses limited activity against crystalline cellulose. The gene encoding this enzyme has been cloned and sequenced (Matsushita et al., 1990), and has since been used in gene fusions with *Thermonospora fusca* DNA, encoding for a CBD (Maglione et al., 1992). The reconstructed endoglucanase possessed tenfold faster rates of crystalline cellulose digestion. While similar experiments conducted with *Ruminococcus albus* SY3 were not as encouraging (Poole et al., 1991), these studies offer the promise of converting an acid-resistant bacterium such as *P. ruminicola* into one capable of cellulose degradation at low pH.

Despite these developments, our understanding of the adherence of rumen bacteria and their enzymes to plant surfaces remains descriptive rather than mechanistic. We have recently utilized functional screens in an attempt to identify cellulose binding proteins in *R. albus* 8 (Pegden and Morrison, 1995). Proteins bound to cellulose or remaining in the supernatant (unbound) fraction were subjected to denaturing sodium dodecyl sulfate polyacrylamide gel electrophoresis (SDS-PAGE) and compared to suitable control reactions. Despite the presence of some background protein bands, a 22-kDa protein is clearly absent after incubation with cellulose, but was readily visible once the cellulose particles were washed and boiled in protein running buffer to remove bound protein(s). The functional assays have since been combined with staining procedures to identify glycosylated proteins. No less than four protein bands, all ranging in size between 21 and 31 kDa, possess affinity for cellulose with the assay conditions used (Figure 1). These glycosylated proteins seem to be excellent candidates for further investigation by a varicty of molecular-based approaches.

Considering that ruminal pH in high-producing dairy cattle may be less than 6.2 for 70 to 80% of the day, the potential impact from the studies outlined here

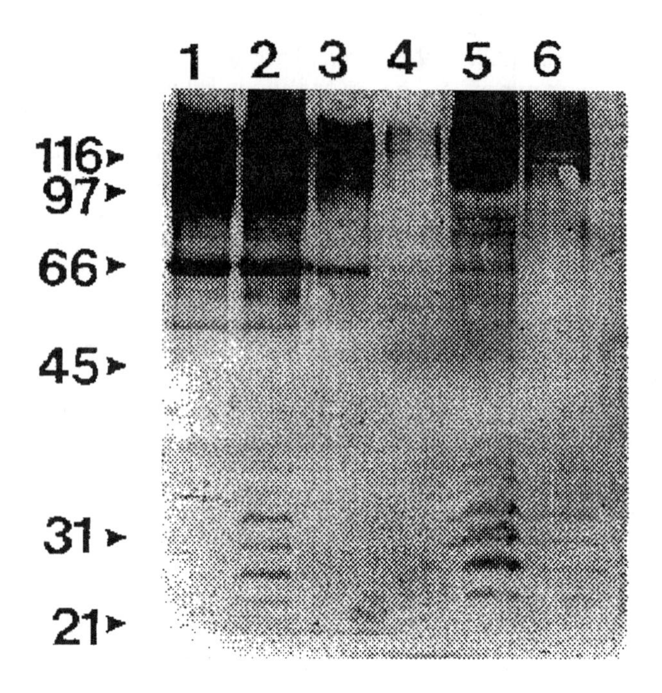

Figure 1 Identification of glycosylated proteins from *R. albus* 8 membrane fractions on the basis of their affinity for cellulose. Membrane fragments, with or without cellulose added, were incubated at room temperature for 1 h and then subjected to centrifugation. Note the disappearance of the glycosylated proteins between the 21 and 32 kDa molecular weight range in lane 1 (plus cellulose) compared to the control (lane 2, minus cellulose). Lanes 3 and 4 represent the supernatant fractions obtained after washing the test and control reactions, respectively, with phosphate buffer containing 0.05% (w/v) Triton X-100. The glycosylated proteins are virtually absent in these wash fractions, but are readily visible again in lane 5 after boiling the cellulose pellet in SDS-PAGE sample running buffer. Lane 6 represents the sedimentable protein present in control incubation.

could be far reaching. It may be possible to identify the "rate-limiting" ligand/receptor site(s) in either plant tissue or ruminal bacteria that affect adherence. Factors affecting the expression and(or) chemical "viability" of ligand/receptor sites (e.g., ruminal pH), and the relationship between these specific interactions and cellulase gene expression may ultimately be identified. Finally, the information gained may ultimately be used to model the effect of rumen environmental conditions and the adherence mechanism(s) on the kinetics of ruminal fiber digestion.

Regulation of Enzyme Activity

It is also possible that *colonization* of plant surfaces in the rumen is discrete from *adherence* and that the former process modulates the expression of at least some polysaccharide-degrading enzymes. Even in more intensively studied and better characterized bacterial systems, only superficial knowledge of the mechanisms involved with regulation of cellulase activity is currently available (Beguin and Aubert, 1994). Some evidence has now been obtained for differential protein expression and phosphorylation in *R. flavefaciens* FD-1 in response to growth with either cellobiose or cellulose (Vercoe et al., 1995). When the bacterium was grown with cellulose, no less than four new polypeptides appeared and five disappeared, as compared to cellobiose-grown cultures. "Induction experiments" were also performed by pulse labeling cells with γ^{32}P-ATP. After induction with cellulose, two high molecular weight polypeptides were found to lack radiolabel. The function(s) associated with these polypeptides, in addition to the mechanisms controlling their phosphorylation, remain to be determined. Now that some genes encoding fibrolytic enzymes have been cloned and characterized, advances in understanding whether and how ruminal conditions affect enzyme expression should soon be forthcoming. It will be interesting to determine whether the cellulase and xylanase genes readily cloned in the laboratory truly represent the major hydrolytic enzymes in the rumen and to what extent their activity is modulated by ruminal conditions.

Some Molecular Approaches to Understanding Bacterial Interactions in the Rumen

Microbial interactions in the rumen have long been recognized (Hobson, 1988), and the most notable of these was the identification of interspecies hydrogen transfer from studies by Bryant, Wolin, and colleagues. Various compounds have also been tested for their ability to modify ruminal fermentation (Van Nevel and Demeyer, 1988), and ionophores such as monensin have been widely utilized by the U.S. feedlot industry (Russell and Strobel, 1989). Recently, there has been heightened interest in ruminant methane emissions and their contribution to global warming. This interest has resulted in continued measurements of methane production, and, fortunately, there also seems to be an increasing commitment to microbiological-based approaches designed to further limit ruminant methane production. Evidence of this can be found in the Abstracts of Communications for the VIIIth International Symposium on Ruminant Physiology. No less than three invited papers addressed the well-established roles that rumen microbes and nutritional factors play in methane production. Additionally, several communications also described the effects of defined protozoal populations (Itabashi et al., 1994), L-cysteine, and archaebacterial supplements (Takahashi et al., 1994) upon methane production. An immunization protocol, similar to that developed with protozoa (Gnanasampanthan et al., 1994), is also being evaluated to inhibit methanogens selectively (Baker, 1995).

However, even if ruminal methanogens can be eliminated completely, alternative routes of hydrogen disposal need to be established. Acetogenic bacteria capable of growth using H_2 and CO_2 have frequently been identified as alternative "electron sinks," and research of rumen acetogens was recently reviewed (Mackie and Bryant, 1994). Recent observations that some acetogens preferentially use nitrate as an electron sink (and terminate acetate biosynthesis from H_2 and CO_2; Seifritz et al., 1993) and that one rumen acetogen shows diauxic growth in the presence of glucose, CO_2, and H_2 (Pinder and Patterson, 1994) merely reflects that ruminal acetogens may not simply inhabit the niche vacated by methanogenic bacteria if their removal from the rumen can be accomplished. Fundamental understanding of the ecology and physiology of acetogens, as well as other hydrogen-utilizing syntrophs (e.g., *Wolinella succinogenes),* must continue to be developed if progress is to be made in this area.

With the exceptions of protozoal engulfment of rumen bacteria and co-culture experiments with fibrolytic bacteria (Dehority, 1993), relatively little attention has been directed toward identifying negative interactions among rumen microorganisms. Microorganisms in most environments produce a diverse range of macromolecules that are antibacterial and(or) antifungal, and recent studies have shown that such compounds may also be produced by rumen bacteria. In mixed culture experiments, Odenyo et al. (1994) have observed that *R. albus* 8 effectively inhibited growth of *R. flavefaciens* FD-1, and it was subsequently shown that a proteinase-susceptible fraction of *R. albus* 8 culture fluid possessed bacteriocin-like properties. Bacteriocin production by numerous strains of *B. fibrisolvens* has since been demonstrated (Kalmokoff et al., 1995). Similar negative interactions between cellulolytic rumen bacteria and anaerobic fungi have also recently been identified (Stewart et al., 1992; Stewart, 1994). Further studies of this nature seem warranted and may offer a better understanding of microbial interactions in the rumen, as well as potential applications in manipulating specific microorganisms in the ruminal population.

Two limitations to more rapid advances from molecular studies of rumen microorganisms have been the reintroduction of genetic material into rumen bacteria, as well as the means to generate mutant strains of rumen bacteria (e.g., by transposon mutagenesis). Rumen bacteria have proven to be a relatively poor source of plasmid DNA and selective markers, but progress has been made utilizing broad host range systems from other bacteria (see Flint, 1994, for a review). Recently, plasmid vectors carrying genes encoding either a fluoroacetate-degrading enzyme (Gregg et al., 1994) or an endoglucanase (Whitehead and Cotta, 1995) have been stably reintroduced into rumen bacteria. Knowledge of the stability of these recombinant bacteria in the rumen, as well as their impacts upon rumen function and digestion, should soon be forthcoming. Similar studies are already being conducted with a recombinant strain of the colonic bacterium *Bacteroides thetaiotaomicron* bearing the xylanase genes from *P. ruminicola* 23 (strain BTX). The construction of this bacterium involved gene insertion into the chromosome using procedures developed by Salyers and Shoemaker (cited by Stevens et al., 1990). The xylanase gene in BTX has proven to be very stable, being maintained for over 60 generations in energy-limited continuous culture without antibiotic selection (Whitehead et al., 1991). While strain BTX is incapable of utilizing the degradation products of the recombinant enzyme, other ruminal microorganisms in co-culture are capable of rapid growth rates. The survival of strain BTX is currently being evaluated in rumen simulators, using both conventional plating procedures and 16S ribosomal RNA (rRNA)-based probes. Preliminary findings indicate that strain BTX persists in the dual-flow continuous cultures, although much of the enzyme activity remains trapped in the periplasm, thereby limiting the enzyme's access to substrate (Cotta, 1995). Increased effectiveness of such recombinant microorganisms probably will require efforts to ensure export of the recombinant proteins to the cell surface and continued emphasis upon strain improvement and selection.

The expansion of the 16S rRNA database, as well as advances in molecular-probe technology, offers the exciting opportunity to clearly document the dynamics of microbial fluctuations in the rumen that occur between animals and diets. Examples of the application of this technology include the reclassification of *Bacteroides succinogenes* as *Fibrobacter* genus nov. species nov. (Montgomery et al., 1988) and the *in vitro* studies of *Ruminococci* reported by Odenyo et al. (1994). Microorganisms not readily cultivated by conventional procedures may soon be identified and included in analyses, using primer pickup procedures similar to those described by Barns et al. (1994). In preliminary experiments, Sharp et al. (1995) have compared the microbial populations of rumen contents with "continuous batch cultures" and the Hoover/Stern dual-flow continuous culture system. Group (bacteria, archaea, eukaryotes) and genus *(Fibrobacter)* probes were used to monitor fluctuations. In the more complex dual-flow fermentors, bacterial 16S rRNA increased from 70 to 95% relative abundance, while eukaryotic rRNA decreased from 30 to 5% relative abundance. These changes coincided with a measurable decrease in protozoal concentrations in the fermentor. The proportions of archael and *Fibrobacter* rRNA remained constant; thus, active fibrolytic and methanogenic consortia were maintained. These techniques are also being used in attempts to conclusively identify, and monitor fluctuations in, the relative abundance of ammonia-producing bacteria isolated by Russell and co-workers. The isolates obtained from laboratory animals have been classified by 16S rRNA sequence analyses as *Clostridium sticklandii, C. aminophilum,* and *Peptostreptococcus anaerobius* (Paster et al., 1994). Molecular probe studies should help confirm or refute conclusions made from *in vitro* studies concerning the role of these bacteria in ruminal ammonia production.

CONCLUDING REMARKS

Priorities for ruminant nutrition research have and should continue to evolve, emphasizing both environmental impact and productivity. Manipulation of the rumen microbiota, resulting in changes in the degradation of dietary protein, offers the potential to improve nitrogen retention and animal productivity. The smugglin concept — the selective inhibition of peptide-degrading bacteria by the uptake of toxic/inhibitory compounds via peptide transport system(s) — offers the potential for very specific growth inhibition and/or fermentation of peptides by rumen bacteria. To determine the feasibility of this approach, mechanistic details of protein degradation and peptide transport must first be established. Future research should also continue to emphasize factors that affect ruminal degradation of cellulosic wastes and other byproducts (e.g., distillers byproducts), rumen microecology, and reduce greenhouse gas emissions from livestock and livestock wastes. While advances in these areas may continue to arise from traditional versions of ruminant nutrition and microbiology research, fundamental information of the biology of rumen microorganisms must also be gathered if advances are to be made expeditiously. An approach to rumen microbiology research that incorporates molecular biology techniques offers new and exciting opportunities, but it is important to emphasize that such studies need not be designed for, or justified by, creation of a "superbug." Molecular biology techniques also provide a powerful extension of our abilities to understand microbial physiology, diversity, and responses of microorganisms to nutrient fluctuations and environmental stresses. Advances in our understanding of ruminant nutrition and microbiology, as well as positive alterations of nutrient retention in production animals, most likely await those willing to change: those brave enough to accept and incorporate new directions and technology in their research without disregarding past achievements.

ACKNOWLEDGMENTS

I wish to express my thanks to Robert B. Hespell for encouraging my involvement in this symposium, as well as M. A. Cotta, R. I. Mackie, J. B. Russell, B. A. White, and T. R. Whitehead for helpful discussions and provision of both published and unpublished information included in the manuscript. Research in my lab has been supported by the Agricultural Research Division of the University of Nebraska and USDA Competitive Grants Program 9440481.

REFERENCES

Ames, B. N., Ames, G. F. L., Young, J. D., Isuchiya, D., and Lecocq, J., Illicit transport, the oligopeptide permease, *Proc. Natl. Acad. Sci. U.S.A.,* 70, 456, 1973.

Baker, S. K., Personal communication, 1995.

Barns, S. M., Fundyga, R. E., Jeffries, M. W., and Pace, N. R., Remarkable archael diversity detected in a Yellowstone National Park hot spring environment, *Proc. Natl. Acad. Sci. U.S.A.,* 91, 1609, 1994.

Beguin, P. and Aubert, J.-P., The biological degradation of cellulose, *FEMS Microbiol. Rev.,* 13, 25, 1994.

Britton, R. A., D-Lactic Acidosis Myth or Fact?, Elanco Products Company Technical Bulletin, 1986.

Chen, B. and Russell, J. B., More monensin sensitive ammonia producing bacteria from the rumen, *Appl. Environ. Microbiol.,* 55, 1052, 1989.

Cotta, M. A., Personal communication, 1995.

Dehority, B. A., Microbial ecology of cell wall fermentation, in *Forage Cell Wall Structure and Digestibility,* Jung, H. G., Buxton, D. R., Hatfield, R. D., and Ralph, J., Eds., Library of Congress Publications, 1993, 17.

Flint, H. J., Molecular genetics of obligate anaerobes from the rumen, *FEMS Microbiol. Lett.,* 121, 259, 1994.

Forsberg, C. W., Cheng, K.-J., Krell, P. J., and Phillips, J. P., Establishment of Rumen Microbial Gene Pools and Their Manipulation to Benefit Fibre Digestion by Domestic Animals, World Conference on Animal Production, 1993, 281.

Fron, M. J., Madeira, H. M. F., Richards, C. J., and Morrison, M., The Impact of Feeding Condensed Distillers Byproducts on Rumen Microbiology and Metabolism, *Anim. Feed Sci. Technol.,* 1996 (submitted).

Gibson, S. A. W. and MacFarlane, G. T., Studies on the proteolytic of *Bacteroides fragilis, J. Gen. Microbiol.,* 134, 19, 1988.

Gnanasampanthan, G., Hynd, P. J., and Mayrhofer, G., Effect of sheep antibodies on engulfment of bacteria by rumen protozoa, *Proc. Aust. Soc. Anim. Prod.,* 20, 389, 1994.

Gregg, K., Cooper, C. L., Schafer, D. J., Sharpe, H., Beard, C. E., Allen, G., and Xu, J., Detoxification of the plant toxin fluoroacetate by a genetically modified rumen bacterium, *Biotechnology,* 12, 1361, 1994.

Griswold, K. E. and Mackie, R. I., Degradation and utilization of isolated soybean protein and casein as nitrogen sources for growth of *Prevotella ruminicola* $B_1$4, *Abstr. 22nd Conf. Rumen Func.,* 1993, 36.

Ham, G. A., Corn byproducts for growing and finishing cattle, M.S. Thesis, University of Nebraska, Lincoln, 1994.

Hazlewood, G. P. and Edwards, R., Proteolytic activities of a rumen bacterium, *Bacteroides ruminicola* R8/4, *J. Gen. Microbiol.,* 125, 11, 1981.

Hazlewood, G. P., Jones, G. A., and Mangan, J. L., Hydrolysis of leaf fraction 1 protein by the proteolytic rumen bacterium *Bacteroides ruminicola* R8/4, *J. Gen. Microbiol.,* 123, 223, 1981.

Hobson, P. N., *The Rumen Microbial Ecosystem,* Elsevier Applied Science, London, U.K., 1988.

Huber, J. T. and Herrera-Saldana, R., Synchrony of protein and energy supply to enhance fermentation, *Principles of Protein Nutrition of Ruminants,* Asplund, J. M., Ed., CRC Press, Boca Raton, FL, 1994, 6.

Huhtahnen, P. J. and Nasie, J. M., Evaluation of feed fractions from integrated starch ethanol production from barley in diets of cattle, pigs and poultry, *Proc. Distillers Feed Conf.,* 47, 67, 1992.

Itabashi, H., Wahio, Y., Takenaka, A., Oda, S., and Ishibashi, T., Effects of a controlled faina on methanogenesis, fibre digestion, and rate of growth in calves, *Proc. Soc. Nutr. Physiol.,* 3, 179, 1994.

Kalmokoff, M. L., Forster, R. J., and Teather, R. M., Bacteriocin production among isolates of the rumen bacterium *Butyrivibrio fibrisolvens, Abstr. 95th Annu. Meet. Am. Soc. Microbiol.,* 318, 111, 1995.

Kunji, E. R. S., Smid, E. J., Plapp, R., Poolman, B., and Konings, W. N., Di-tripeptides and oligopeptides are taken up via distinct transport mechanisms in *Lactococcus lactis, J. Bacteriol.,* 175, 2052, 1993.

Larson, E. N., Corn by-products for finishing cattle, M.S. Thesis, University of Nebraska, Lincoln, 1992.

Larson, E. N., Stock, R. A., Klopfenstein, T. J., Sindt, M., Huffman, R. P., and Thompson, T., Wet distillers by-products for finishing cattle, *University of Nebraska Beef Cattle Report,* 43, 1993.

Latham, M. J., Brooker, B. E., Pettipher, G. L., and Harris, P. J., *Ruminococcus flavefaciens* cell coat and adhesion to cotton cellulose and to cell walls in leaves of perennial ryegrass, *Appl. Environ. Microbiol.,* 35, 156, 1978a.

Latham, M. J., Brooker, B. E., Pettipher, G. L., and Harris, P. J., Adhesion of *Bacteroides succinogenes* in pure culture and in the presence of *Ruminococcus flavefaciens* to cell walls in leaves of perennial ryegrass, *Appl. Environ. Microbiol.,* 35, 3146, 1978b.

Mackie, R. I. and Bryant, M. P., Acetogenesis and the rumen: syntrophic relationships, in *Acetogens,* Ferry, J. G. Ed., 1994, 12.

Madeira, H. M. F., Zhang, L., and Morrison, M., Use of the "smugglin concept" for the study of peptide transport in Prevotella ruminicola, *Abstr. 95th Annu. Meet. Am. Soc. Microbiol.,* K82, 550, 1995.

Maglione, G., Matsushita, O., Russell, J. B., and Wilson, D. B., Properties of a genetically reconstructed *Prevotella ruminicola* endoglucanase, *Appl. Environ. Microbiol.,* 58, 3593, 1992.

Matsushita, O., Russell, J. B., and Wilson, D. B., Cloning and sequencing of a *Bacteroides ruminicola* B_14 endoglucanase gene, *J. Bacteriol.,* 172, 3620, 1990.

Montgomery, L., Flesher, B., and Stahl, D. A., Transfer of *Bacteroides succinogenes* (Hungate) to *Fibrobacter* gen. nov. as *Fibrobacter succinogenes* comb. nov. and description of *Fibrobacter intestinalis* sp. nov, *Int. J. Syst. Bacteriol.,* 38, 430, 1988.

Morrison, M., Mackie, R. I., and Kistner, A., 3-Phenylpropanoic acid improves the affinity of *Ruminococcus albus* for cellulose in continuous culture, *Appl. Environ. Microbiol.,* 56, 3220, 1990.

Nisbet, D. J. and Martin, S. A., Effect of dicarboxylic acids and *Aspergillus oryzae* fermentation extract on lactate uptake by the ruminal bacteria *Selenomonas ruminantium, Appl. Environ. Microbiol.,* 56, 3515, 1990.

Nisbet, D. J. and Martin, S. A., Effect of a *Saccharomyces cerevisiae* culture on lactate utilization by the ruminal bacterium *Selenomonas Ruminantium, J. Anim. Sci.,* 69, 4628, 1991.

Odenyo, A. A., Mackie, R. I., Stahl, D. A., and White, B. A., The use of 16S rRNA-targeted oligonucleotide prodes to study competition between rumen fibrolytic bacteria: development of probes for *Ruminococcus* species, and evidence for bacteriocin production, *Appl. Environ. Microbiol.,* 60, 3688, 1994.

Paster, B. J., Dewhirst, F. E., Olsen, I., and Fraser, G. J., Phylogeny of *Bacteroides, Prevotella,* and *Porphyromonas* spp. and related bacteria, *J. Bacteriol.,* 176, 725, 1994.

Payne, J. W., Mechanisms of bacterial peptide transport, *Peptide Transport in Bacteria and the Mammalian Gut,* CIBA Foundation Symposium, Vol. 4., Associated Scientific Publishers, Amsterdam, 1972, 2.

Payne, J. W. and Smith, M. W., Peptide transport by microorganisms, *Adv. Microbial Physiol.,* 36, 1, 1994.

Pegden, R.S. and Morrison, M., Unpublished data, 1995.

Pell, A. N. and Schofield, P., Microbial adhesion and degradation of plant cell walls, *Forage Cell Wall Structure and Digestibility.,* Jung, H. G., Buxton, D. R., Hatfield, R. D., and Ralph, J., Eds., Library of Congress Publications, Washington, D.C., 1993, 18.

Peng, L.. and Morrison, M., Unpublished data, 1995.

Pinder, R. S. and Patterson, J. A., Glucose and hydrogen utilization by a non-mixotrophic ruminal acetogenic bacteria, *Proc. Nutr. Soc. Physiol.,* 3, 156, 1994.

Pittman, K. A. and Bryant, M. P., Peptides and other nitrogen sources for growth of *Bacteroides ruminicola, J. Bacteriol.,* 88, 401, 1964.

Pittman, K. A., Lakshmanan, S., and Bryant, M. P., Oligopeptide uptake by *Bacteroides ruminicola, J. Bacteriol.,* 93, 1499, 1967.

Poole, D. M., Durrant, A. J., Hazlewood, G. P., and Gilbert, H. S., Characterization of hybrid proteins consisting of catalytic domains of *Clostridium* and *Ruminococcus* endoglucanases, fused to *Pseudomonas* non-catalytic cellulose binding domains, *Biochem. J.,* 279, 787, 1991.

Ringrose, P. S., Peptides as antimicrobial agents, in *Microorganisms and Nitrogen Sources,* Payne, J. W., Ed., John Wiley, New York, 1972.
Russell, J. B., Rumen bacteria rob cattle of nutrients, *Agric. Res.,* 40, 17, 1993.
Russell, J. B., Strobel, H. J., and Chen, G., The enrichment and isolation of a ruminal bacterium with a very high specific activity of ammonia production, *Appl. Environ. Microbiol.,* 54, 872, 1988.
Russell, J. B. and Strobel, H. J., Effect of ionophores on ruminal fermentation, *Appl. Environ. Microbiol.,* 55, 1, 1989.
Seifritz, C., Daniel, S. L., Gobner, A., and Drake, H. L., Nitrate as a preferred electron sink for the acetogen *Clostridium thermoaceticum, J. Bacteriol.,* 175, 8008, 1993.
Sharp, R., Ziemer, C. J., Whitehead, T. R., Cotta, M. A., Stern, M. D., and Stahl, D. A., Molecular phylogenetic based comparisons of model rumens, *Abstr. 95th Annu. Meet. Am. Soc. Microbiol.,* N-43, 340, 1995.
Stack, R. J., Hungate, R. E., and Opsahl, W. P., Phenylacetic acid stimulation of cellulose digestion by *Ruminococcus albus* 8, *Appl. Environ. Microbiol.,* 46, 539, 1983.
Stevens, A. M., Shoemaker, N. B., and Salyers, A. A., The region of a *Bacteroides* conjugal chromosomal tetracycline resistance element which is responsible for production of plasmidlike forms from unlinked chromosomal DNA might be involved in transfer of the element, *J. Bacteriol.,* 172, 4271, 1990.
Stewart, C. S., Duncan, S. H., and Richardson, A. J., The inhibition of fungal cellulolysis by cell-free preparations from Ruminococcus, *FEMS Microbiol. Lett.,* 97, 83, 1992.
Stewart, C. S., Factors affecting fermentation and polymer degradation by anaerobic fungi and the potential for manipulation, *Proc. Soc. Nutr. Physiol.,* 3, 145, 1994.
Takahashi, J., Beneke, R. G., Aoki, M., Fukushima, M., Nakano, M., and Young, B. A., Effects of L-cysteine and archaebacterial supplementation on ruminal methanogenesis in sheep, *Proc. Soc. Nutr. Physiol.,* 3, 180, 1994.
Tynkkynen, S., Buist, G., Kunji, E., Kok, J., Poolman, B., Venema, G., and Haandrikman, A., Genetic and biochemical characterization of the oligopeptide transport system of *Lactococcus lactis, J. Bacteriol.,* 175, 7523, 1993.
Van Nevel, C. J. and Demeyer, D. I., Manipulation of rumen fermentation, *The Rumen Microbial Ecosystem,* Hobson, P. N., Ed., Elsevier Applied Science, London, 1988, 13.
Vercoe, P. E., Kocherginskaya, S. A., and White, B. A., Evidence for differential protein expression and phosphorylation in *Ruminococcus flavefaciens* FD-1 grown on cellobiose or cellulose, *Abstr. 95th Annu. Meet. Am. Soc. Microbiol.,* H-232, 532, 1995.
Wallace, R. J., Amino acid and protein biosynthesis, turnover, and breakdown by ruminal microorganisms, *Principles of Protein Nutrition of Ruminants,* Asplund, J. M., Ed., CRC Press, Boca Raton, FL, 1994, 5.
Wallace, R. J. and Brammall, M. J., The role of different species of bacteria in the hydrolysis of protein in the rumen, *J. Gen. Microbiol.,* 131, 821, 1985.
Wallace, R. J., McKain, N., Walker, N. D., and Newbold, C. J., Two dipeptidyl aminopeptidases of *Prevotella ruminicola, Proc. Soc. Nutr. Physiol.,* 3, 164, 1994.
White, B. A., Personal communication , 1995.
Whitehead, T. R., Cotta, M. A., and Hespell, R. B., Introduction of a *Bacteroides ruminocola* xylanase gene into the *Bacteroides thetaiotamicron* chromosome for the production of xylanase activity, *Appl. Environ. Microbiol.,* 57, 277, 1991.
Whitehead, T. R. and Cotta, M. A., Personal communication, 1995.

CHAPTER 12

Designer Proteins Through Biotechnology for Use As Feeds

Roger A. Kleese

INTRODUCTION

Our feed grains are treated as commodities. Corn is used not only as a feed, but as a food and for a variety of industrial uses. The Iowa Corn Promotion Board states that there are over 3000 uses for corn, and a vast majority of these come from the same yellow dent that is used to feed livestock (Anonymous, 1988). Because corn is used in many ways, the farmer generally has not been paid on the basis of composition, but on the basis of weight delivered to the elevator.

As a result, seed companies have heavily emphasized the development of hybrids with improved yield and productivity in terms of kilograms per hectare. Corn breeders have concentrated on developing hybrids that meet these performance criteria. Developing hybrids that are high yielding is a demanding job, and corn breeders generally have had their hands full focusing upon developing hybrids with high, stable yield. Increasing yield is what plant breeders have been doing, and it is what they have been good at.

Soybeans represent a similar picture. Although soybeans are grown only for protein and oil, almost none of the beans are sold off the farm on the basis of protein or oil content. High, stable yield (kilograms per hectare) is what the farmer seeks and gets paid for.

Because of the emphasis on yield, the seed industry has almost no knowledge of the feed value of current products. New hybrids and varieties are sold to farmers almost entirely on the basis of their improved agronomic characteristics. This is true for corn, sorghum, and soybeans.

Although the primary use of corn, sorghum, and soybean meal is for animal feed, they are not satisfactory feed ingredients if used alone. Corn and sorghum are good sources of energy, but have poor protein quality. A nutritionally balanced diet is possible when supplemented with a high-quality protein source. However, for monogastrics, excess nitrogen is introduced into the diet due to the nonessential amino acids in the cereals. This leads to wasted nitrogen, which shows up in the manure. In concentrated livestock operations, where disposal of manure can be a problem, high levels of nitrogen can create an environmental hazard. Is it possible to use plant genetics to manipulate protein quality or quantity to produce a more environmentally sound feed grain?

1-56670-199-6/96/$0.00+$.50

Let us look at an experiment that was begun in 1896 by Hopkins (1899) to select for oil and protein in corn (Dudley et al., 1974). There were four selection programs: high oil, low oil, high protein, and low protein. This experiment, which continues today, was summarized after 70 generations of selection. The high oil selection has been driven to nearly 18% oil, whereas the low oil selection is well below 0.5%. Oil content in U.S. hybrids is approximately 3.5% at 15% moisture. The high protein selection is near 28% protein, whereas the low protein selection is about 4%. Protein content in U.S. hybrids varies, but generally falls between 7.5 and 9.0% at 15% moisture. Clearly, it is possible to alter nutrient composition dramatically. However, it should be pointed out that none of these lines has been used directly as commercial lines, although high oil material has been a source of germplasm for developing high oil hybrids. The agronomic performance, including yield, of these lines is inferior to commercial U.S. dent corn hybrids, but this is to be expected since protein and oil content were the only traits under selection.

It is interesting to note the practical objectives of this research as listed in 1908 by Smith (1908), a researcher involved in the experiment at that time. The stated justifications for the research were

1. Increasing percent protein is desirable because corn does not contain sufficient protein for most feeding purposes.
2. Decreasing percent protein is desirable for manufacturers of products derived from the starch in corn.
3. Increasing percent oil is desirable because corn oil is pound-for-pound the most valuable constituent of the grain in commercial use.
4. Decreasing percent oil is desirable for feeding purposes because corn oil tends to produce soft fat in pork.

What then are the challenges and opportunities for genetically improving feed value? What are the improvements that should be made? Where is the need greatest? What is the value of each of these improvements? What improvements are technically possible? How do we market these improved products? How are the improvements measured quickly and inexpensively to determine value? Can nutritional enhancement compete with improved commodity grain productivity? How will we deal with the price of alternative feed ingredients, especially if those prices drop when improved feed grains are marketed? I will not attempt to answer most of these questions, but I raise them to illustrate the breadth and scope of the issues raised as we consider genetically modifying feed value of grains and oilseeds.

I have been asked to focus my attention on protein, and, more specifically, I will address protein quality. Many of my comments will focus on corn because that is our company's core business and I am most familiar with corn. Also, I will be thinking of corn grain for monogastrics. Generally, genetic technology will not be considered as a constraint, even though it is today. However, today's constraints may become routine tomorrow.

GENERAL CONSIDERATIONS

The following are general considerations that underlie any strategy to improve protein quality.

- We are interested in amino acids from the animal nutrition standpoint, but amino acids in plants come packaged as proteins. This protein packaging constrains strategies for enhancing amino acids. Generally, levels of free amino acids in grain or meal are nominal.

- A plant genetic strategy to address protein quality should at least acknowledge that an animal diet typically has more than one source of protein. If we manipulate one plant species, what assumptions do we make about the other source(s) of protein in the diet? If we manipulate soybean protein, we can make a fairly good assumption today about the other source of protein in a diet formulated in the United States. The other source of protein will come from corn most of the time, but if the soybean meal is fed in Europe that assumption is not sound.

- Do we manipulate the composition of corn or soybeans? Interestingly, most efforts to date have been on changing the composition of corn because that is the source of protein considered to have the greatest deficiency. This deficiency is largely related to its low lysine content. In fact, soybeans are a relatively poorer source of methionine than corn when evaluated on the basis of protein quality. Since soybeans already produce 40% protein compared to 7 to 8% protein in corn, it might make more sense to manipulate the species with greater capacity to synthesize amino acids and package them in proteins. Thus a rationale for increasing lysine might be more easily done in soybeans, even though the species already has a fairly high level of the amino acid.

- Synthetic sources of lysine, methionine, tryptophan, and threonine are now available as feed ingredients, although the cost of tryptophan and threonine may yet be beyond their wide-scale use in animal diets today. What are the expectations for these latter two amino acids to become increasingly cost competitive with other sources of protein? Are there other amino acids that could be synthesized and marketed to the livestock industry?

- Simply elevating total protein, particularly in corn, leads to lower-quality protein. Therefore, this is not a viable strategy for reducing nonessential nitrogen in the diet.

- What are the targeted improvements in protein quality? Amino acid requirements vary among livestock and poultry species as well as among the age and condition of the animals or birds. What is to be the target or are we to have multiple targets and, if so, how many? Furthermore, animal and poultry genetics are powerful technologies that have created major changes in the animals and birds and have altered the quantities and balance of nutrients needed to maximize the productivity of these species. How will these requirements change in the next 10 to 20 years?

STRATEGIES TO MANIPULATE PROTEIN QUALITY

Five strategies have been outlined for manipulating protein quality in plants. The first two strategies are supported by reports in the literature. The latter three are speculative, and I am not aware of laboratories working on them. Because there is commercial interest in this area, there is no doubt some work with which I am not familiar.

Altering the Amino Acid Biosynthetic Pathway

This approach allows one to focus on a particular limiting amino acid. Because amino acid biosynthesis is generally regulated, the objective is to deregulate the pathway to achieve some level of overproduction of the free amino acid. Hibberd et al. (1986) were successful in selecting for overproduction of tryptophan in corn by using the selection agent 5-methyl tryptophan in corn tissue culture. This approach mimics that used routinely in microbes to select for overproducers of a variety of metabolites. Levels of tryptophan in corn sufficient to meet dietary requirements of growing swine were achieved (Kirihara et al., 1991).

Glassman et al. (1993) used a molecular approach to select for overproducers of lysine. Biosynthesis of lysine in plants is feedback inhibited at two enzymatic steps. The first of these steps is common to the biosynthesis of four amino acids: isoleucine, methionine, threonine, and lysine. The second step is unique to lysine biosynthesis and is catalyzed by the enzyme dihydrodipicolinic acid synthase.

Lysine biosynthesis in *Escherichia coli* follows essentially the same pathway as in plants. There is some feedback sensitivity of the enzyme dihydrodipicolinic acid synthase to lysine, but the enzyme in *E. coli* is at least 200-fold less sensitive to lysine inhibition *in vitro* than the plant enzyme. By cloning the bacterial gene for this enzyme and transferring it to tobacco, Glassman et al. (1993) were able to elevate the expression of free lysine 200-fold in tobacco leaves.

Overexpression of a Gene Coding for a Protein with High Levels of the Desired Amino Acid

Certain proteins have very high levels of particular amino acids. Using gene transfer technology, it is feasible to engineer one of these proteins containing high levels of the desired amino acid into a feed grain or oilseed crop. But simply introducing the gene for the foreign protein might not significantly impact the total level of the desired amino acid. Manipulating the regulatory sequences associated with the gene could bring about the overexpression desired.

Kirihara et al. (1988) selected a zein in corn containing 22% methionine to manipulate for elevating methionine in corn. Since the zeins are endogenous corn proteins, the objective was to reengineer the promoter on the gene coding for this protein to bring about overexpression of the gene when reintroduced into corn.

There are several families of zein genes, some of which have multiple copies in corn. Furthermore, the expression level of these genes is known to vary.

Kirihara et al. (1988, 1991) chose a 27-kDa zein known to have a high expression level, removed the promoter sequence from this gene, and substituted it for the endogenous promoter on the 10-kDa, high methionine gene. This construct has been reintroduced into corn with the hope of achieving elevated levels of this high methionine protein (Kirihara et al., 1991).

Both of these strategies focus upon one amino acid at a time. This allows one to emphasize the amino acid in question, but does not accommodate a need to consider more than one amino acid simultaneously. These approaches may challenge the capacity of the plant to supply the necessary carbon skeletons if the targeted increase in the level of amino acid is very high. In addition, there could be problems of toxicity to the plant if the levels of certain free amino acids are excessive. This second strategy has a limited capacity to address more than one amino acid, depending upon the amino acid composition of the protein that is expressed.

Do either of these strategies impact total nitrogen going into the environment? If the strategy simply elevates the most limiting amino acid, the protein supplement that is normally added to cover only the requirement for that limiting amino acid could be reduced. If synthetic amino acids are being used to balance the diet, then it would not appear that there is any benefit to the environment.

Overexpression of an individual protein can raise concerns about allergenicity. The FDA (1992) has issued guidelines for "Foods Derived from New Plant Varieties," and potential allergenicity of overexpressed proteins may be considered a significant risk to humans depending upon the source of that protein. The guidelines do not make reference to specific levels of specific proteins in plants.

Manipulating Several Limiting Amino Acids Simultaneously

To date, most efforts to manipulate protein quality have focused on individual amino acids simply because of the state of the art in biotechnology. Clearly, there are needs to address more than one amino acid at a time. This probably requires choosing a suitable protein, preferably one that already has a significant number of the desired residues and that occurs at high levels in the seed. Nonetheless, some of the codons of the gene coding for this protein should probably be altered to further increase the number of desired residues of the target amino acids. If this is extensive, it is not clear what will happen to protein structure and the capacity of the plant to deposit high levels of this altered protein.

An alternative strategy could be to design the protein *de novo*. Enzymatic sites and receptor sites of pharmacologic proteins are both very active areas of research. Also, there is considerable effort to understand and predict three-dimensional structure based on amino acid sequence. But these efforts are experimental today and do not directly predict the design of a nutritionally enhanced protein that would fold and pack into the desired feed grain.

In either of the previous cases, the capacity of the plant to synthesize the desired levels of the amino acids may be an issue. It may be necessary to alter individual amino acid biosynthetic pathways as well.

This strategy may impact the environment if it reduces the need for protein supplement. If synthetic amino acids are used to balance the diet, this strategy has no impact on the environment.

Reducing Nonessential Amino Acids

If we are truly driven by environmental concerns, then it seems the focus should be on nonessential amino acids. Excess amounts of these amino acids in the diet represent nitrogen that is not utilized by the animal. However, these nonessential amino acids can serve as a source of nitrogen for transamination reactions. But assuming this is not a major need, specific nonessential amino acids in the diet represent excess nitrogen which ends up in the manure.

Is it possible to reduce this nitrogen load? Should we breed for lower protein content in corn? Would 4 to 5% protein in corn with significantly less nonessential amino acids be more environmentally friendly and make sense nutritionally?

It is possible to breed for lower protein content as we noted earlier, but we know little of the impact this would have on nonessential vs. more limiting amino acids. Molecular biology strategies to lower nonessential amino acids may be more complicated than the suggestions offered earlier to enhance the level of expression of a limiting amino acid.

In general, zeins in corn are very deficient in lysine and tryptophan. Techniques to down-regulate genes associated with zein production might be considered as a strategy to reduce nonessential amino acids. However, there are at least two issues to consider in such an effort. First, there are multiple copies of zein genes as well as several families of these genes, so manipulating as many as 100 genes is not presently attractive from a molecular biology standpoint. Second, zein proteins have been altered previously using the opaque-2 gene. Although protein quality was improved, grain yield and grain quality suffered tremendously.

Furthermore, total protein in the opaque-2 corn was similar to normal corn, indicating a compensatory affect. Thus, even though the lysine and tryptophan were increased, there were still significant levels of nonessential amino acids present.

Reduction of All Protein in Corn

An extension of the strategy to reduce nonessential amino acids might be to eliminate all protein. Corn is a starch crop and one of the most efficient in transferring light energy into chemical energy in the form of starch. Clearly, corn is fed primarily as an energy source, although a value is placed on the protein from corn entering a diet.

In the longer term, it makes more sense to maximize what corn seems to be "good at" (e.g., energy production in the form of starch). The question, of course, is whether greater total value can be created when breeding for kernels of starch vs. kernels with starch, protein, and oil. Germless mutants exist that would eliminate much of the protein and almost all of the oil. A genetic system to provide viable seed that gives rise to germless grain would have to be developed. There are molecular biology approaches that have the potential to manipulate such a system. This grain, when fed with significant levels of synthetic amino acids, would have the potential to substantially reduce nitrogen in the manure.

The last strategy really begins to address the issue of genetically altering protein, starch, and oil in plants to maximize their value for a particular end use. Crop species are quite resilient genetically and physiologically, and there are many examples illustrating altering both quantity and quality of protein, starch, and oil. It is also accurate to say that plant breeders have not emphasized these traits nearly as much as the agronomic traits. Therefore, for most plants and especially for the feed grains, there is not a good understanding of the genetic control of many of these traits nor the limits to which they can be pushed. End users requesting raw material ingredients from plants better suited for a specific end use will push plant geneticists to learn much more about these traits in the next 10 to 15 years.

COMPETITIVE FORCES

Ultimately, the issue is not if plants can be genetically manipulated, but whether or not it can be done cost effectively. A major feature of plant genetic improvements is that they are generally permanent for that hybrid or variety. An improvement in protein quality could lock in the improvement in that germplasm. This would provide a base upon which to build incrementally, as has been the pattern for agronomic traits. Animal breeding offers the same opportunity.

But a cost is associated with each improvement. Furthermore, there are alternative opportunities for investing research dollars. Some of these opportunities are related to feed value, but others are not. The question is not simply which protein quality strategy do we adopt and in which species, but rather do we choose to focus on feed value at all. Corn refiners would like greater starch yield per kilogram of corn. Ethanol producers want increased fermentable starch. The food industry is looking for more healthy vegetable oils with desired functional properties. Hard-textured, pure white corn is desired for corn chips, and so on.

Plant genetics is adding flexibility in herbicide choice by developing hybrids and varieties resistant to selected herbicides. Corn resistant to the European corn borer will be offered commercially in the next year or two, thus alleviating the need for insecticides to control this pest.

Each of these genetic manipulations is an opportunity, but each also has an incremental cost and associated benefit. Delivering these new technologies through plant genetics is attractive, but priorities have to be set and choices have to be made.

There are three sources of competition that we have to consider as we establish these priorities and decide which products to develop.

Commodity Grain

How can we compete with commodity grain productivity from improved hybrids and varieties? In corn, the current rate of improvement in yield is about 120 kg/ha/year. Technically, productivity is the main competition we face in developing corn hybrids with altered composition better suited to a feed end use. Anytime an additional trait is added to a plant breeder's program, the total effort is diluted. It is not possible in a plant breeding program to add a trait and maintain the same rate of improvement unless resources are added.

Many of the feed value traits have never been manipulated in a plant breeding program. We have very little information on genetic variability or environmental variation or their interaction on these traits. I assume that some of these traits are very complex genetically and therefore will be affected by the environment.

Other Plant Genetics Capabilities

The list of market needs is widely known. Both private and public concerns are active in pursuing these product opportunities. The choice by a competitor to pursue a product improvement, which we do not pursue, poses a risk to us. Yet we cannot be all things to all customers. In the past, plant genetics technology was largely in the public domain. A delay in adopting new technology could be expensive to the business, but generally did not close one out of the technology completely. Today that has changed. Utility patents, which restrict individual pieces of technology, have become the rule. Some technologies may be impossible to use during the life of the patent.

Alternatively, the patent does provide a vehicle for accessing technology even from one's competition. The technology is clearly defined in a patent, enabling one to place a value on it, thus facilitating its purchase or licensing.

Alternative Ingredients

The value of improving a particular trait is related to the cost of alternative feed ingredients. Understanding the business and technology factors that influence the price of these ingredients will be a substantial challenge. Synthetic amino acid production capacity continues to increase. This will continue to put pressure on our technology to be as efficient as possible. It is this issue of alternative ingredients as much as any other that reminds us of the need to understand the meat production industry. Our largest product lines are corn hybrids and soybean varieties, and the largest market area is the central United States. Corn and soybeans are the only two crops that many farmers in that area grow. Good alternatives do not exist.

In contrast, as you formulate a diet, you let the least-cost feed formulation model pull in those ingredients that are most cost effective. I recognize that the exercise involves other considerations as well, but you do not have to feed corn if other, less expensive alternatives with adequate nutritional value are available.

In theory, the same is true for the corn/soybean farmer in Iowa, except that the alternatives tend to be quite a bit less attractive.

In the seed industry, we will have to learn what it means to be an ingredient supplier of raw material if we are to improve grains and oilseeds for particular end use markets. With the increased use of raw materials will come a better appreciation of the competitive forces of alternative ingredients.

SUMMARY

Several strategies for designing plants with improved proteins are outlined. Some of these efforts currently are in progress. There is little doubt that the technologies for designing proteins for use as feeds will continue to evolve, thus opening the door to a variety of new feeding strategies.

The emphasis of this conference is on enhancing the environment. Balancing a diet due to environmental issues is a different concept than simply using the least-cost formulation of ingredients to balance for the nutrient needs of the animal. How is the environment to be weighted when balancing a diet? What is to be the relative emphasis placed on environment and nutrition for balancing a diet? We will be challenged to sort this out, for it will vary from state to state, region to region, and country to country. And finally, it will be critical to place a value on a targeted feed value improvement in order to prioritize it relative to a host of other plant genetic opportunities.

REFERENCES

Anonymous, *Corn, It's Your Business!* 1987/1988, Iowa Corn Promotion Board, Iowa Corn Growers Association, West Des Moines, 1988.

Dudley, J. W., Lambert, R. J., and Alexander, D. E., Seventy generations of selection for oil and protein concentration in the maize kernel, in *Seventy Generations of Selection for Oil and Protein in Maize,* Dudley, J. W., Ed., ASA, CSSA, Madison, WI, 1974.

Glassman, K. F., Barnes, L. J., and Pilacinski, W. P., Method of inducing lysine overproduction in plants, U.S. Patent No. 5,258,300, 1993.

Hibberd, K. A., Anderson, P. C., and Barker, M., Tryptophan over-producer mutants of cereal crops, U.S. Patent No. 4,581,847, 1986.

Hopkins, C. G., Improvement in the chemical composition of the corn kernel, *Ill. Agric. Exp. Stat. Bull.,* 55, 205–240, 1899.

Kirihara, J. A., Petri, J. B., and Messing, J., Isolation and sequence of a gene encoding a methionine-rich 10 kDa zein protein from maize, *Gene,* 71, 359–370, 1988.

Kirihara, J. A., Sandahl, G. A., and Kleese, R. A., Gene transfer to elevate methionine levels. Proceedings of the 46th Annual Corn and Sorghum Industry Research Conference, American Seed Trade Association, Washington, DC, 1991, 124.

Smith, L. H., Ten generations of corn breeding, *Ill. Agric. Exp. Sta. Bull.,* 128, 457, 1908.

U.S. FDA, Statement of Policy: Foods Derived from New Plant Varieties, U.S. Food and Drug Administration, *Fed. Reg.,* 57(104), 22984–23005, Friday, May 29, 1992.

CHAPTER 13

Environmentally Friendly Methods to Process Crop Residues to Enhance Fiber Digestion

G. C. Fahey, Jr., H. S. Hussein, K. Karunanandaa, and G. A. Varga

INTRODUCTION

In recent years, the senior author of this chapter has written extensively about processing of crop residues to improve their nutritive value. A chapter entitled "Postharvest Treatment of Fibrous Feedstuffs to Improve Their Nutritive Value" by G. C. Fahey, Jr., et al. (1993) was published in *Forage Cell Wall Structure and Digestibility* edited by H. G. Jung et al. This book was prepared in conjunction with the International Symposium on Forage Cell Wall Structure and Digestibility held in Madison, WI in October 1991. In April 1994, the National Conference on Forage Quality, Evaluation, and Utilization was held at the University of Nebraska, Lincoln. From the conference came a text entitled *Forage Quality, Evaluation, and Utilization* edited by G. C. Fahey, Jr., et al. (1994). A chapter in the text "Modification of Forage Quality After Harvest" by L. L. Berger et al. (1994) covered many of the same processing methodologies outlined in the above-mentioned chapter.

The words *environmentally friendly* at the beginning of this chapter title cause us to take a somewhat different approach to the topic than was taken in the two previously cited contributions. Indeed, to rupture the highly organized cell wall matrix of crop residues, strong reagents frequently are necessary. But certain of these have a negative environmental impact when treated substrates are fed to ruminants. In this chapter, we will examine treatment methods that have little negative impact on the environment. Additionally, we will discuss how certain already existing treatment methodologies might be modified to become more environmentally friendly.

FUNGI

History of the Origin of Biological Delignification

As early as 1863, it was documented that wild-type fungi were capable of decomposing wood (review by Blanchette, 1991). Later, in 1893, it was reported

1-56670-199-6/96/$0.00+$.50

that decomposed wood was used as feed for cattle and horses in southern Chile (Zadrazil et al., 1982). The ability of these fungi to selectively remove lignin or cellulose from wood was first reported in 1878 (Blanchette, 1991). When lignin was selectively removed from wood, the decayed material turned white. Thus the name white rot was given to this particular group of fungi. All species of white-rot fungi (WRF) were grouped under basidiomycetous because of their ability to degrade lignin. Based on rate of removal of lignin and polysaccharides from substrates, WRF can be divided into three groups. The first group may remove lignin at a faster rate than polysaccharide-selective delignifiers. The second group may remove lignin and polysaccharides at an equal rate, and the third group may remove polysaccharides at a faster rate than lignin throughout the decay process (Zadrazil, 1984; Blanchette, 1991).

Knowledge in the area of molecular biology and biochemistry of enzymes produced by WRF has increased tremendously during the past two decades (Eriksson, 1993). However, progress on processes to treat lignocellulosic substrates with WRF to improve its feeding value is still in the preliminary stage. Most of the published reports on biological delignification to upgrade low-quality roughages with WRF deal with the ability of different species and strains of WRF to colonize different substrates. These studies are preliminary, but provide vital information. To date, considerable information is available on factors influencing solid-state fermentation (SSF) of substrates, changes in chemical composition of decayed material, and ruminal *in vitro* dry matter digestibility (IVDMD).

Solid-State Fermentation

Solid-state fermentation (SSF) is the most widely used method for growing WRF on lignocellulosic material. Substrate is moistened, but not saturated, with enough water to provide 65 to 85% moisture (Reid, 1989). The water-holding capacity of the substrates varies, depending on their type and particle size. Thus weight ratios of water to substrate for optimal SSF of lignocellulosics ranges between 1:1 and 10:1. During SSF, exchange of oxygen between the gaseous phase and the solid substrate is one of the crucial factors that determines the success of the fermentation process. The presence of excessive water can impede the exchange of gases and create an anaerobic condition, while insufficient water may prevent optimal fungal growth (Reid, 1989). Several studies have investigated culture criteria that influence optimal SSF of lignocellulosics by WRF. The effects of physical criteria (moisture, temperature, length of SSF) and nitrogen supplementation on SSF have been reported (Zadrazil and Brunnert, 1980, 1981, 1982; Leisola et al., 1983; Kahlon and Dass, 1987) and reviewed (Reid, 1989).

Mechanisms of Lignin Degradation During Solid-State Fermentation

Fungal growth and metabolism on lignocellulosics, from initial colonization to effective degradation of plant cell wall components, can be divided into two phases: (1) primary phase and (2) secondary phase. During the primary phase, initial growth of the fungus occurs on substrates by utilization of readily digestible polysaccharides present in the plant cell wall and in cell contents. The primary phase is characterized by an increase in fungal respiratory activity, a decrease in free sugars, no lignin degradation, and reduced IVDMD (Valmaseda et al., 1991). At this stage, the fungus faces no deprivation of nutrients and has no need to secrete lignin-degrading enzymes to liberate plant cell wall polysaccharides. Lignin degradation takes place only when readily digestible nutrients are exhausted and substrate becomes limiting in an essential nutrient such as nitrogen, carbon, sulfate, or phosphate (Eriksson, 1993). The duration of this primary phase may vary depending upon fungal species as well as type of substrate. Valmaseda et al. (1991) reported that the duration of the primary phase for *Pleurotus ostreatus* and *Trametes versicolor* on wheat straw was 10 and 7 d, respectively. According to Eriksson (1993), it is advantageous if lignin is degraded during the primary phase because this will minimize organic matter loss and shorten the time required for the entire process.

When fungi become limiting in nutrients, they switch to the secondary phase of growth, during which degradation of lignin and polysaccharides takes place. Valmaseda et al. (1991) named this the "degradation phase," and it is characterized by the onset of lignolytic activity, polysaccharide degradation, and an increase in IVDMD and crude protein content of the substrate (Valmaseda et al., 1991). The increase in digestibility of the plant substrate is associated with the removal of lignin and the liberation of cell wall polysaccharides.

Fungi secrete a variety of extracellular enzymes that are responsible for the degradation of lignocellulosic materials. Based on their mode of action, these enzymes can be grouped into hydrolytic and oxidative types (Joseleau et al., 1994). Most of the polysaccharidase enzymes are of the hydrolytic type, except those of cellobiose:quinone oxidoreductase and cellobiose oxidase. These are involved in oxidation of cellobiose, reduction of quinones, and the production of hydroxy radicals that are involved in lignin degradation (Ander, 1994). In contrast, there are lignin-degrading enzymes that are oxidative in nature. The hydrolytic enzymes have a narrow range of specificity and their function is mediated by protein. However, lignolytic enzymes have a wide range of substrate oxidation capacities and their function is controlled by nonprotein mediators. Four kinds of lignin-degrading enzymes have been reported (Kirk, 1988): (1) ligninases (lignin peroxidases [LiP]), (2) manganese-dependent peroxidases (MnP), (3) phenol oxidases (laccase), and (4) hydrogen peroxide (H_2O_2)-producing enzymes (glyoxal oxidase). Of all the enzymes, ligninase is the major one and consists of several isoenzymes. Ligninase was first discovered in *Phanerochaete chrysosporium* by Tien and Kirk (1983) and Glenn et al. (1983). Lignin peroxidase and MnP are heme-containing glycoproteins. Lignin peroxidase catalyzes several oxidative reactions in lignin by the formation of cation free radicals. These reactions include C-C cleavage, loss of methoxyls, oxidation of benzylic hydroxyls to ketones, and ring opening (Tien, 1987). Laccase and MnP catalyze reactions which use free radical formations similar to LiP, but are capable only of oxidizing phenolic substrates (Kirk, 1988). Degradation of lignin and cellulose are interlinked through the H_2O_2 in WRF (Hatakka et al., 1989; Ander, 1994). The oxidative enzymes, cellobiose:quinone oxidoreductase and cellobiose oxidase, are related to the production of H_2O_2 and the conversion of Fe(III) to Fe(II) with oxidation of cellobiose (Ander, 1994). The presence of the combination of Fe(II) and H_2O_2 leads to the production of the hydroxy radical that is directly involved in the degradation of lignin. Thus degradation of lignin has been referred to as "enzymatic combustion" by Kirk and Farrell (1987), indicating a nonspecific enzyme-catalyzed burning of lignin. Hydrogen peroxide is necessary for the action of lignolytic enzymes, LiP, and MnP. Therefore, lignin degradation can link cellulose degradation pathways by regulating lignolytic and cellulolytic activities (Eriksson, 1978). Based on the enzyme production pattern of WRF, Hatakka (1994) grouped WRF into three groups: (1) lignin-Mn peroxidase group, (2) Mn peroxidase-laccase group, and (3) lignin peroxidase-laccase group. Of these three groups, the most efficient lignin degraders belong to those in group 1.

An array of enzymes produced by WRF during SSF have been implicated in lignin and polysaccharide degradation of lignocellulosic materials. Electron microscopy and biochemical studies indicated that the enzymes secreted by wood-rotting fungi such as lignin peroxidase, Mn-dependent peroxidase, laccases, and cellulases are too large to penetrate wood cell walls (reviewed by Evans et al., 1994). Therefore, it was suggested that the degradation of lignocellulose by fungal enzymes is a surface-enzyme interaction, and any degradation of cell wall components far from the fungal mycelium should be mediated through the production of diffusible low molecular weight mass agents such as H_2O_2, veratryl alcohol, oxalate, Fe(II)-Fe(III), and Mn(II)-Mn(III). In order for the enzyme to penetrate the cell wall, the cell wall must be partially degraded. This phenomenon of enzyme penetration has been documented by immunogold-cytochemical labeling techniques. The hypothesis put forward for the total degradation of the material by fungi was that first the diffusible agents penetrate into the cell wall and initiate decay, and once the pore size is large enough, the enzyme will penetrate and complete the degradation process (Evans et al., 1994). However, the involvement of diffusible agents related to the degradation of the cell wall has not yet been conclusively proven. There is an indirect involvement of these diffusible agents produced during lignin degradation that also has been implicated with cellulose degradation in wood cells (reviewed by Joseleau et al., 1994). It was observed using electron microscopy that when wood was colonized by fungal strains in which the cellulase system had been inhibited, a substantial degradation of cellulose occurred. These findings may further support other suggestions that lignin degradation is always associated with polysaccharide removal from lignocellulosic material (Kirk and Moore, 1972).

The WRF is the only known organism on earth that can cause extensive degradation of lignin present in lignocellulosics (Hatakka, 1994). Degradation of lignin by WRF in the plant cell is an inside-out process. Removal of lignin begins from the cell lumen, progresses toward the middle lamella, and ends up in the cell corners (Blanchette, 1991). In contrast, lignin deposition in plants during biosynthesis begins in the corners of the middle lamella and progresses toward the secondary layers in the direction of the cell lumen. Kirk and Moore (1972) reported that wood lignin degradation by WRF always took place with removal of polysaccharide, but was not correlated with removal of any single component of the polysaccharide. Hatakka et al. (1989) suggested that lignin degradation by WRF required an easily digestible cosubstrate. Wheat straw lignin degradation by *Cyathus stercoreus* and *Dichomitus squalens*, as estimated from the ^{14}C-lignin to $^{14}CO_2$ ratio, indicated that both fungi displayed qualitative differences in lignin degradation (Agosin and Odier, 1985). More lignin was degraded by *D. squalens* and most of the ^{14}C-lignin was converted to $^{14}CO_2$, while *C. stercoreus* degraded less lignin and converted relatively more ^{14}C-lignin to low molecular weight ^{14}C-solubles and thus produced less $^{14}CO_2$ from lignin. This study indicates that *C. stercoreus* produced more intermediate compounds (polymeric lignin) during lignin degradation and lacked the ability to further convert this compound to CO_2.

There are conflicting results reported on the extent of lignin degradation by WRF due to nitrogen availability in the medium. Zadrazil and Brunnert (1980, 1982) reported a negative effect of nitrogen, in the form of ammonium nitrate, on wheat straw lignin degradation by *Pleurotus erngii* and *Lentinus edodes*. However, in some species of WRF *(Stropharia rugosoannulata* and *Sporotrichum pulverulentum),* the same authors reported that a low level of nitrogen stimulated lignin degradation, while higher concentrations had either no effect, or an inhibitory effect, on lignin degradation. Similar results with urea on wheat straw lignin degradation by the alkaliphilic WRF, *Coprinus* sp., was reported by Yadav (1987). Kahlon and Dass (1987) investigated the effects of different sources and levels of nitrogen on rice straw degradability by five strains of WRF. They reported that nitrogen supplementation in the form of ammonium chloride at 0.08% was optimum for *Pleurotus ostreatus* and *S. pulverulentum* for improving IVDMD. However, in this study, the workers failed to find any inhibitory effect of nitrogen on lignin degradation. In contrast, more detailed enzyme kinetic studies reported that the availability of nitrogen in the medium during SSF depressed enzyme (ligninase) activity, thus reducing lignin degradation (Kirk et al., 1976). The most widely studied WRF for rapid lignin degradation is *Phanerochaete chrysosporium* (Agosin et al., 1987; Akin et al., 1993; Karunanandaa et al., 1992, 1995). Of the phenolic acids, ferulic acid was always preferentially degraded compared to *p*-coumaric acid. Degradation of phenolic compounds in lignocellulosics by phenol-oxidizing enzymes may lead to the production of quinones, which, in turn, have inhibitory effects on other enzymes (Hatakka et al., 1989). Additionally, low molecular weight phenolic compounds can inhibit cellulolytic bacteria in the rumen (Chesson et al., 1982). It appears that a single species of WRF exhibits great variation in lignin degradation within substrates and with time of incubation of different substrates, which makes generalization difficult in terms of lignin removal.

Chemical Composition of Fungal-Treated Substrates

Few studies have reported the changes in polysaccharide components of the plant cell wall due to fungal treatment (Agosin et al., 1987; Karunanandaa, 1995).

Biodegradation of wheat straw by three species of WRF, *Cyathus stercoreus*, *D. squalens*, and *P. chrysosporium*, revealed that the degradation of monomers of wheat straw polysaccharides varied depending on the species of WRF; all three fungi degraded arabinose, while xylose was degraded only by *P. chrysosporium*. Uronic acids accumulated during fungal degradation except for *P. chrysosporium*. In this study, a cell wall preparation was used for the monosaccharide analysis. Karunanandaa (1995) used a whole sample preparation of fungal-treated material for monosaccharide analysis and found differences in the pattern of degradation of monosaccharides from that of Agosin et al. (1987). The monomers of hemicelluloses and cellulose of rice leaf decayed by *C. stercoreus*, *Pleurotus pulmonarius* (previously named *Pleurotus sajor-caju*), and *Phanerochaete chrysosporium* were determined on whole sample as described by Blakeney et al. (1983). During the 30-d SSF of rice leaf, only glucose derived from cellulose was extensively degraded and metabolized by *P. chrysosporium*, while arabinose and xylose were intact. However, *Pleurotus pulmonarius* and *C. stercoreus* showed a preference for metabolizing arabinose and xylose over glucose.

In Vitro Evaluations of Fungal-Treated Substrates

Conversion of plant wastes such as sawdust, reed, rape, and sunflower straws, and rice husks into feed by four species of WRF for 30 and 60 d of SSF was reported by Zadrazil (1980). All fungi tested effectively colonized substrates used, but the IVDMD of the final decayed material varied from a negative effect with rice husks to a maximum improvement of 37 percentage units with reed decayed by *Pleurotus* sp. in Florida for 60 d. The author concluded that lignin degradation and the subsequent increase in IVDMD strongly depended upon the fungal species selected and type of substrate. Moreover, despite the large variations found in IVDMD, fungal colonization of all substrates increased the water-soluble substances after 60 d of SSF. The same author screened more than 200 strains of WRF on wheat straw for organic matter loss, lignin degradation, and changes in IVDMD with varying temperature and duration of incubation (Kamra and Zadrazil, 1988). Interestingly, the authors calculated the "process efficiency" as changes in IVDMD (percentage units) per unit dry matter loss. The variable process efficiency compensated for the dry matter loss incurred during SSF. On the basis of process efficiency, the highest improvement, a 2.27 percentage unit increase in IVDMD per unit loss of dry matter, was found with *Hymenochaete tabacina*. A selection of 75 strains of WRF was evaluated with wheat straw for their lignolytic capacity as well as for their selective removal of hemicelluloses and cellulose (Agosin et al., 1985a). The following species of WRF were best suited for their capacity to degrade lignin: *Pycnoporus cinnabarinus*, *Pleurotus ostreatus*, *D. squalens*, *Cyathus stercoreus*, *Vavaria effuscata*, and *Bjerkandera adusta*. However, the most widely studied fungus, *Phanerochaete chrysoporium* (previously named *Sporotrichum pulverulentum*), was found to be very aggressive in degrading hemicelluloses and cellulose along with the removal of lignin. Agosin et al. (1985c) reported that *C. stercoreus* and *Pycnoporus cinnabarinus* had the ability to selectively degrade hemicelluloses over cellulose during SSF. The selective degradation of hemicelluloses resulted in an increase in the final IVDMD to 45 and 40%, respectively, for *P. cinnabarinus* and *C. stercoreus*. The same authors reported that *Phanerochaete chrysosporium* degraded hemicelluloses and cellulose of wheat straw indiscriminately, and thus the increase in IVDMD was very low. A similar pattern of selective degradation of hemicelluloses was reported for *C. stercoreus* and *D. squalens* on wheat straw (Agosin and Odier, 1985). Since then many studies have reported selective degradation of components of cell wall polysaccharides from various substrates by WRF (Rolz et al., 1986; Karunanandaa et al., 1992). Rolz et al. (1986) screened 12 strains of WRF for their ability to improve the quality of lemon grass *(Cymbopogon citratus)* and citronella *(C. winterianus)* bagasse. This study reported the *in vitro* dry matter enzyme digestibility (IVDMED) as a percentage weight loss due to enzymatic action compared to the initial dry matter of the sample. Best results for both substrates were obtained with *Bondarzewia berkeleyi*.

Al-Ani and Smith (1988) pretreated sugarcane bagasse with various chemicals singly or in combination (NaOH, H_2SO_4, NH_3, $Ca(OH)_2$ + Na_2CO_3, or ethylenediamine [EDA]) before it was decayed by *P. chrysosporium* for 14 d. The highest IVDMD (59%) was reported for the 0.25 *M* NaOH treatment followed by fungal treatment. Yadav (1987) investigated varying levels of nitrogen (as urea), phosphorus + sulfur, and free carbohydrates (as molasses or whey) on SSF of wheat straw with alkaliphilic WRF *(Coprinus* sp.). The addition of urea stimulated fungal growth as well as lignin degradation under sterile conditions. The IVDMD also was increased with increasing levels of urea (2 g/100 g dry matter). This study, along with that of Al-Ani and Smith (1988), demonstrated that WRF has the ability to grow successfully on alkali-treated material. This is important because the growth of unwanted microorganisms can be suppressed by keeping the pH on the alkaline side.

In Situ Evaluations of Fungal-Treated Substrates

In situ ruminal degradation of cell wall components of fungal-treated wheat straw was reported by Agosin et al. (1986b). The rate of cell wall degradability of straw decayed by *P. chrysosporium* was adversely affected during the 96 h of ruminal incubation compared to the control (2.4 vs. $2.9\%/h^{-1}$). However, the potentially digestible fraction was higher in this treatment (54.6 vs. 48.2%), suggesting that *P. chrysosporium* converts only the major proportion of the cell wall into a soluble fraction (20%) and the rest of the cell wall becomes less digestible. The other two fungi, *Cyathus stercoreus* and *D. squalens*, increased the rate of degradation of the cell wall by 1.5-fold (4.4 and 4.0 vs. $2.9\%/h^{-1}$, respectively). Increased cell wall degradation has been attributed to increased cellulose degradation (by 50%) by ruminal microorganisms.

Continuous Culture Evaluations of Fungal-Treated Substrates

Continuous and semicontinuous culture fermenters have been employed in the evaluation of fungal-treated straw. Jalc et al. (1994) reported the effects of fungal-treated *(Polyporus cilatus* or *Lentinus tigrinus)* wheat straw on ruminal microbial activity. The diets consisted of treated or untreated straw and barley (80:20) with a crude protein content of 13%. These two treatments were compared with a positive control (hay) diet. Fungal treatments of wheat straw, except for *L. tigrinus,* resulted in increased digestibilities of dry matter, organic matter, and acid detergent fiber (ADF) compared to the untreated wheat straw (48.2 vs. 42.2%; 49.5 vs. 43.1%; 37.8 vs. 33.1%, respectively). However, the hay diet resulted in the highest values for all of these criteria. An inhibitory effect was reported with *L. tigrinus*-treated wheat straw when compared to the control straw. The fermentation pattern of diets containing treated or untreated straws showed no differences in the production of total volatile fatty acids (millimoles per day, mmol/d). Interestingly, the energetic efficiency of ruminal volatile fatty acid production was significantly higher for wheat straw treated with *P. cilatus* compared to the control straw or hay diet. This was mainly due to the increased molar proportion of propionate found for the fungal-treated diet. A similar trend for increased molar proportions of propionate was reported in other studies under *in vitro* conditions (Akin et al., 1993; Karunanandaa, 1995). A higher molar proportion of propionate leads to a decreased production of H_2 and, consequently, lower methane production (Jalc et al., 1994) in fungal-treated diets compared to the control or hay diets. Although, in the studies of Jalc et al. (1994), the efficiency of fermentation was increased by fungal treatment, no differences were found in the efficiency of microbial protein synthesis. This may suggest that the availability of carbohydrate and protein to the microbes was not synchronized during the fermentation of the diet containing the fungal-treated straw.

Nutrient utilization and fermentation characteristics of rice straw colonized by *C. stercoreus* during continuous culture fermentation were reported by Karunanandaa (1995). Experimental diets consisted of fungal-treated rice straw (FRS-diet) or untreated rice straw:concentrate (75:25, dry matter basis). Both diets were formulated to be isonitrogenous (11% crude protein), and soybean meal was the major protein supplement. Fungal treatment significantly increased dry matter and organic matter digestibilities (44 vs. 35% and 51 vs. 42%) compared with the control. Cellulose digestibility was significantly higher in the FRS-diet compared to the control (61.1 vs. 48.8%). This increased digestibility of cellulose was further evident from an increased glucose digestibility. No improvement was found in the digestibility of the major monomers (arabinose and xylose) of hemicellulloses. Total volatile fatty acid production was significantly higher for the FRS-diet, and this was due to increased molar proportions of propionate and butyrate (28.8 vs. 24.2%; 13.1 vs. 7.9%, respectively). Compared with the control diet, however, production of total isoacids and ammonia nitrogen was significantly lower for the FRS-diet (0.19 vs. 2.6%; 0.77 vs. 22.3 mg/dl, respectively). Decreased production of ammonia nitrogen and reduced digestibility of dietary crude protein of the FRS-diet resulted in an increased daily flow of dietary nitrogen from the fermenter. Unusually high concentrations of bacterial purines found with the FRS-diet may suggest that fungal treatment influenced the composition and population of ruminal microbes. This study, along with others, clearly demonstrated that fermentable energy was not a limiting factor in fungal-treated material; however, other factor(s) such as degradability of crude protein may limit the optimal digestibility of carbohydrates and(or) microbial protein synthesis.

In Vivo Evaluations of Fungal-Treated Substrates

Lack of knowledge regarding large-scale treatment of lignocellulosics with WRF has resulted in a limited number of studies on the evaluation of feeding fungal-treated substrates to animals. Zadrazil et al. (1990) first demonstrated large-scale treatments (1500 kg) of wheat straw by SSF with *Pleurotus* spp. in a bioreactor. The IVDMD of straw varied depending on the site of sampling in the bioreactor; layers proximate to the surface were more digestible compared to the middle layers. In the past, fungal-treated straws utilized in feeding trials were obtained from two sources: (1) decayed straws, those obtained after SSF and with no fruiting bodies (mushrooms) formed during the treatment period, and (2) spent straw, defined as the decomposed straw obtained after the harvest of fruiting bodies. However, with the formation of fruiting bodies, the content of α-cellulose and pentosan of the compost (spent straws) has been shown to decrease drastically, leaving a very recalcitrant substrate (Gerrits, 1969). Calzada et al. (1987) determined the feeding value for lambs of spent wheat straw, which was obtained after the harvest of fruiting bodies of *Pleurotus sajor-caju*. The chemical composition of untreated straw and decayed straw indicated no differences in total carbohydrate or lignin content due to treatment. The workers found no differences in dry matter intake, body weight change, or dry matter digestibility due to treatment. Bakshi et al. (1985) reported a decrease in digestibility of spent wheat straw compared with control straw when fed to male buffaloes at ad libitum intake. Inclusion of spent wheat straw, obtained from *P. ostreatus*, at four levels (0, 6.6, 10.4, and 13.7% of dry matter) in a diet for steers was evaluated along with ruminal fermentation pattern, daily intake, weight gain, and carcass characteristics (Henics, 1987). The 48-h *in situ* degradability of dietary components also was evaluated. Total volatile fatty acid production was significantly reduced, and ammonia nitrogen concentration of ruminal fluid increased as the ratio of spent straw increased in the diet. It was suggested that lignin degradation products released during fungal degradation selectively inhibited ruminal cellulolytic bacteria and thus reduced the incorporation of ammonia nitrogen into bacterial protein. Accordingly, the *in situ* cellulose degradability of spent straw also decreased as the amount of spent straw in the diet reached the highest level. However, no inhibition of cellulose degradation of spent straw was found in steers that received the diet consisting of the lowest level of spent straw. No differences were observed in average daily gain, dressing percentage, or percentage renal fat. Ahuja et al. (1986) showed that *Volvariella volvacea*–grown spent rice straw could be used as a basal roughage for sheep. No differences in dry matter intake (1.3 vs. 1.3 kg/d), dry matter digestibility (63 vs. 61.6%), or intake of digestible crude protein (95.4 vs. 106.7 g/d) were found when diets consisting of control rice straw and spent rice straw were fed to crossbred male sheep. In conclusion, these studies involving feeding of spent straw indicated that fungal treatment neither improved the feeding value of straw nor adversely affected the performance of animals.

The following studies used fungal-treated straw obtained from SSF where no fruiting bodies were formed during the incubation period. Chandra et al. (1991) evaluated the effects of fungal-treated rice straw on nutrient utilization when supplemented at 50% of the total diet and fed to sheep. Rice straw was decayed by *Trichoderma viride*, *Aspergillus niger*, and a mixture of both *T. viride* and *A. niger* for 30 d. The voluntary dry matter intake and water intake were higher on diets containing treated straws. However, digestibility coefficients for dry matter, organic matter, and crude protein were higher for control straw. Total ash and acid-insoluble ash content of rice straw significantly increased after the fungal treatment and may have had an inhibitory effect on digestibility. Rice straw, in contrast to other cereal straws, is very high in silica (Karunanandaa, 1995). Fungal treatment does not remove silica present in rice straw, and the relative concentration of silica increases after fungal treatment. Chandra et al. (1991) also indicated that fungi were not able to utilize lignin. In this study, no information was reported on the effect of these fungi on chemical composition (hemicelluloses, cellulose, and lignin) and IVDMD.

Walli et al. (1991) compared nutrient utilization in male calves (Tharparkar × Holstein Friesian) that received urea-treated or fungal-treated wheat straw. Fungal treatment was carried out with *Corprinus fimetarius* incubated at 35 to 45°C for 2 weeks in 6-kg capacity, perforated polyethylene bags. Chemical composition of fungal-decayed wheat straw indicated that only cellulose was slightly consumed by the fungus during the treatment process. No information was available on lignin degradation by this fungus. Total dry matter intake and the digestibilities of dry matter and ADF were lower in fungal-treated straw compared to urea treated straw (2.6 vs. 2.7 kg/100 kg of body weight; 46.6 vs. 49.6%; 39.8 vs. 44.4%, respectively). However, there was a tendency for higher nitrogen retention in calves fed fungal-treated straw. A similar finding was reported for nitrogen retention by Ahuja et al. (1986) and Chandra et al. (1991) when fungal-treated rice straw was fed to sheep.

Mpofu and Ndlovu (1994) investigated a different approach to the treatment of lignocellulosic materials with WRF. A culture extract from the SSF of a WRF, *Armillaria heimii,* and yeast granules were used as feed additives either singly or in combination to improve fiber digestibility of veld hay consumed by mature wethers. This study also investigated the digestibility of fiber components under *in vitro* conditions. During the *in vitro* study, all of the microbial treatments increased the digestibility of neutral detergent fiber (NDF) compared to the control treatment, with the combination of yeast plus fungi being the highest (30.5 vs. 59.6%). Fungal treatment alone significantly increased NDF digestibility of veld hay compared to yeast treatment (50.9 vs. 45.8%). A similar trend was noted *in vivo*. Of all the treatments, the combination of yeast plus fungal treatment gave the highest value for dry matter intake, metabolizable energy, and NDF digestibility for veld hay compared to the control (0.889 vs. 0.742 kg/d; 6.69 vs. 3.55 MJ/d; 58 vs. 37.4%, respectively). Values for fungal treatment alone are 0.748 kg/d, 4.64

MJ/d, and 49.2%, respectively, and the values for yeast are 0.786 kg/d, 4.47 MJ/d, and 43.8%, respectively. This is the only study that has demonstrated the significant effect of fungal treatment in improving fiber digestibility *in vitro* and *in vivo*. Furthermore, results of this study suggest that in the future we may not need to use live fungi, but rather a culture extract from the fungus. However, this study needs to be repeated with several other fungi that have shown potential for improving roughage digestibility under *in vitro* conditions.

Future Applications

Molecular technology can be expected to rapidly expand our knowledge of fungal enzymes. Research is currently under way to create strains of *Phanerochaete chrysosporium* that produce large amounts of ligninases (Holzbaur et al., 1991). Normally, ligninases are induced only during secondary metabolism, but molecular biology may allow the creation of strains of WRF that produce greater amounts of ligninase during primary growth. If these enzymes could be harvested and applied to lignocellulosics, a faster and more extensive biological delignification may be possible. The ligninase genes of WRF have been cloned into *Escherichia coli*. Unfortunately, the enzyme produced by *E. coli* often lacks the heme group and is insoluble, thus requiring costly downstream processing prior to use (Holzbaur et al., 1991).

ENZYMES

Enzymes have been used to treat fibrous feeds in an effort to improve their quality (e.g., silage) and/or digestibility (e.g., crop residues). Enzymes play their greatest role in the ensiling process (1) when there are limited soluble sugars available in the substrate (e.g., crop residues) and (2) where there is an adequate level of lactic acid bacteria on the crop at the time of ensiling.

Purified Enzymes

Enzymes used in treating crop residues to enhance digestibility include commercial enzyme preparations (microbial byproducts having enzymatic activities harvested from one or several microorganisms) or single enzymes. The commercial enzyme preparations (polysaccharidases) contain a broad spectrum of activities and degrade two (cellulose and hemicelluloses; Vanbelle and Bertin, 1989) or more (cellulose, hemicelluloses, and pectin; Nakashima et al., 1988; Nakashima and Ørskov, 1989) components of the plant cell wall. Enzymes also have been isolated from different fungal species to attack specific fiber moieties. These include cellulases from *Aspergillus niger* (Ben-Ghedalia and Marcipar, 1979; Ben-Ghedalia and Miron, 1981) and *T. viride* (Ben-Ghedalia and Marcipar, 1979; Ben-Ghedalia and Miron, 1981; Morrison, 1988, 1991), hemicellulases from *T. viride* (Morrison, 1988, 1991), and ligninases from *P. chrysosporium* (Khazaal et al., 1990).

Mode of Action

Addition of cellulases to forages prior to storage improves silage characteristics (Autrey et al., 1975; Jorgensen and Cowan, 1989; van der Meer and Ketelaar, 1989). This is a result of the rapid decrease in silage pH by the lactic acid bacteria in response to increased availability of readily fermentable carbohydrates by hydrolysis of cellulose (Henderson and McDonald, 1977). Treatment of crop residues with cellulases or hemicellulases also increased the soluble fraction of cell wall components and therefore increased rate of digestibility (Ben-Ghedalia and Miron, 1981; Nakashima et al., 1988; Nakashima and Ørskov, 1989). A partial hydrolysis of lignin also was achieved by a fungal ligninase (Khazaal et al., 1990). It would be expected that physical treatments that provide greater surface area for attack by enzymes will improve the efficacy of the enzymatic treatment.

Evaluation of Enzymatic Treatments

In vivo digestibility of dry matter was improved from 17.3 to 30.6% when rice hulls were treated with a fungal cellulase (Daniels and Hashim, 1977). *In vitro* dry matter digestibility also was increased from 29.8 to 40.2% for capim *(Panicum maximum* Jacq.) and from 64.8 to 76.5% for alfalfa when these forages were treated with several commercial cellulases containing significant hemicellulolytic activity (Vanbelle and Bertin, 1989). However, other trials showed either a decrease (Morrison, 1988, 1991) or no response (Nakashima et al., 1988) in digestibility due to enzymatic treatment. Morrison (1988, 1991) treated barley straw, ryegrass, and alfalfa with either cellulase or hemicellulase and observed decreases in IVDMD. This was attributed to the loss of sugars (from cell wall origin) that were solubilized by enzymatic hydrolysis and removed prior to *in vitro* evaluation. Nakashima et al. (1988) observed that the extent (72 h) of *in situ* dry matter disappearance of rice straw was not affected by polysaccharidase (containing a broad spectrum of activities) treatment. However, the soluble fraction was increased from 15.0 to 21.0% of dry matter, and the rate of disappearance was increased from 0.05 to $0.08/h^{-1}$. Similar increases in the soluble fraction of barley straw were reported (Nakashima and Ørskov, 1989) when treated with the polysaccharidase used by Nakashima et al. (1988). The enzyme was found to be more active on fibers of leaf blades and leaf sheaths than nodes or internodes of barley straw.

The sequential (chemical and enzymatic) treatment of forage was evaluated by Ben-Ghedalia and Marcipar (1979), who observed small increases in *in vitro* organic matter digestibility (IVOMD) of wheat straw treated with cellulases from *A. niger* or *T. viride* following a pretreatment with NaOH or NH_3 gas at 5% of dry matter. In another trial (Ben-Ghedalia and Miron, 1981), these cellulases increased rate without affecting the extent (48 h) of IVDMD of wheat straw treated chemically (NaOH, O_3, or SO_2).

Khazaal et al. (1990) treated barley straw with a ligninase enzyme from *Phanerochaete chrysosporium* and measured the effects on composition and degradability. A modest reduction in the lignin content of the straw was observed (from 116 to 105 g/kg of organic matter). However, no significant improvements in digestibility were observed. The lack of response to this enzyme may be because lignin degradation is a free radical reaction that requires other enzymes (i.e., cellobiose:quinone oxidoreductase) to shift the equilibrium of the reactions involving degradation and repolymerization (Ander, 1990).

Future Applications

Molecular biology may allow the large-scale production of enzymes targeted to attack specific fiber moieties. Considerable research probably will be required before enzymes are used on a practical basis.

BACTERIA

Use of bacterial inoculants in ruminant feeding is designed to preserve and/or enhance the quality of forages (e.g., silage). In recent years, bacterial inoculants have been used as feed additives to enhance ruminal fiber digestion and/or to maintain stable ruminal fermentation. Most contain live cultures of lactate-producing bacteria from the genus *Lactobacillus, Pediococcus,* or *Streptococcus faecium.* Many different species exist within each genus, and it is estimated that there may be as many as 5000 different strains within each species (Pioneer Hi-Bred International, 1990). These strains differ in their ability to ferment various substrates and their ability to grow at various moistures and temperatures. Different strains of bacteria that belong to the genus *Lactobacillus* and *Enterococcus* have been used more recently as feed additives because of their probiotic potential (Wallace and Newbold, 1992).

Seale (1986) reviewed the relationship between bacterial inoculants and silage quality and indicated that moisture is the main factor affecting quality. The benefits of bacterial inoculants in improving silage characteristics decrease when the dry matter content of the forage is 60% or greater. Therefore, the lower moisture content of crop residues could be one of the main limitations in improving silage quality of crop residues when bacterial inoculants are added. An approach to dealing with the low moisture content of crop residues is ensiling them (after enzymatic or microbial treatment) with temperate forages that are high in moisture content.

In contrast to forages, crop residues (e.g., straws, tops, and stovers) are low-quality fibrous feeds that have low digestibility and promote low feed intake (Flores, 1989). This is because of their nutritional limitations (high degree of lignification and low content of readily fermentable carbohydrates, N, and S). These limitations also would minimize the responses to bacterial inoculation during ensiling. Therefore, additional treatment and(or) supplementation (Bolsen, 1978) of such fibrous feed before ensiling is an essential step to benefit ruminal fermentation (Sukanto and Soedomo, 1982; Holmes et al., 1987; van Eys et al.,

1987; Wan Mohammed, 1987). Environmentally friendly chemical treatments such as $Ca(OH)_2$ and urea (to generate ammonia and heat) have been used to break down the lignin-carbohydrate complexes and cause swelling of the fiber to facilitate microbial access (Dunlop et al., 1976). Supplementation of crop residues with molasses (to provide readily fermentable carbohydrates and S) and urea (to provide N) to stimulate bacterial fermentation in the silo and subsequent ruminal digestion was suggested (Leng, 1985; Holmes et al., 1987). Fitzsimons and O'Connell (1994) evaluated two starch-degrading *Lactobacillus* strains and concluded that *L. amylovorus* was suitable as a silage inoculant for crops low in water-soluble carbohydrates but that contain starch, which is unavailable to most conventional silage inoculants. Therefore, starch as a carrier of inoculants containing these bacterial strains may be useful to stimulate lactate production when ensiling crop residues. Glucose as an inoculant carrier produces a theoretically high lactate yield among lactate-producing bacteria and has produced better results with temperate silage inoculants (Edwards and McDonald, 1978; Beck, 1978).

Future Applications

Any enhancement of the quality of crop residues by bacterial inoculation during ensiling will require manipulation of both the substrate and the bacterial inoculant. Manipulation of bacterial inoculants should include screening for those strains of bacteria that can proliferate on substrates with low moisture content and low water-soluble carbohydrates while maintaining their high lactate productivity. This should be followed by genetic manipulation (recombination and mutation) of selected strains to boost lactate production in these silages and to stabilize against clostridial proliferation during the ensiling process. Sharp et al. (1992) showed that two recombinant forms of *L. plantarum* proliferated in grass silage and induced a rapid decline in pH similar to that induced by the parent strain. The advantage, however, was that one of the two strains contained a shuttle vector containing the *Clostridium thermocellum* cellulase gene. The survival and proliferation of the bacterial strain containing the cellulase gene is a sign of potential success in cloning and expressing other genes (cellulases and ligninases) that can improve the quality of crop residue silages.

CHEMICAL TREATMENTS

Numerous chemical treatment methods have been developed in the past century. The two major treatment categories involve hydrolytic and oxidative agents. Common hydrolytic agents are sodium hydroxide, other alkali metal hydroxides, ammonia, and urea. Common oxidants are ozone, sulfur dioxide, H_2O_2, and various other delignifying agents such as chlorite, peracetic acid, and permanganate. Combinations of hydrolytic agents and oxidants also have been studied, but to a lesser extent. Each of these treatment methods improve either intake potential or fiber digestion by ruminants, preferably both. However, each has its own problems regarding environmental pollution. For example, one of the more effective hydrolytic agents is sodium hydroxide. For optimal treatment to occur, it must be added at approximately 5% of the substrate dry matter. The amount of sodium in 5% NaOH will cause animals to produce more urine at more frequent intervals. Ruminants fed lignocellulosics treated with sodium hydroxide have increased sodium concentrations in both urine and feces, which could increase both soil and water salinity if the manure is used as fertilizer. This is not a problem if the manure is applied to the same area of land from which the substrate is collected. Only when manure is concentrated on smaller land areas does sodium accumulation occur. Another example — ozone — is a very efficacious treatment method, yet it holds little promise as a practical means of producing large quantities of treated material for animal production purposes because, among other things, ozone at ground level is a pollutant. Of the other oxidants, sulfur dioxide and H_2O_2 appear to have few negative environmental consequences at the concentrations needed for treatment purposes. Hydrogen peroxide is actually consumed in the reaction with lignocellulosic substrates by being converted to oxygen and water.

Perhaps the most efficacious chemical treatment to date is alkaline H_2O_2. The methodology involves use of 5% sodium hydroxide and 2% H_2O_2 (as a percentage of the substrate dry matter). Ruminants fed treated substrates at high concentrations (50 to 60% of diet dry matter) exhibit high levels of feed intake (2 to 4.5% of body weight) and high nutrient digestibilities (60 to 75%). These responses translate into excellent rates of gain, feed efficiencies, and milk production. Alkaline H_2O_2-treated lignocellulosics have an energy value that is 60 to 70% that of corn. Details regarding this treatment process can be obtained in Fahey et al. (1993). A potential means of reducing the environmental problem posed by sodium is to define precisely the ratio of sodium hydroxide to other alkali sources that would allow cell wall modification to occur while at the same time minimizing the amount of sodium in treated substrates. Hydroxides such as calcium hydroxide, potassium hydroxide, and ammonium hydroxide would cause few if any environmental problems as all supply plant and animal nutrients. Sodium hydroxide is essential in the alkaline H_2O_2 treatment reaction as a pH of 11.5 ± 0.2 in the final product is mandatory for treatment efficacy. Most of the other hydroxide compounds will not achieve this pH when used alone.

Ammoniation is perhaps the most widely used of the chemical treatments on a practical scale. The mechanism of action of ammoniation is assumed to be similar to that of sodium hydroxide. However, several features distinguish treatment of roughages with ammonia from treatment with sodium hydroxide. Increases in digestibility of roughages are generally not as dramatic as those observed with sodium hydroxide. But in addition to increasing structural carbohydrate digestibility, ammonia treatment is an effective means of decreasing the amount of supplemental N needed in diets containing treated residues (Sundstøl and Coxworth, 1984). Ammoniation also is an effective means of increasing the intake of crop residues such as corn stover (Morris and Mowat, 1980) and wheat straw

(Saenger et al., 1983). The amount of ammonia (optimal level = 20 to 30 kg/metric ton), temperature, length of treatment time, water content (optimal level = 293 kg/ metric ton), and type and quality of material being treated are factors affecting the efficacy of the process (Sundstøl and Coxworth, 1984).

Little disturbance to the environment comes about as a result of ammonia use. When the bags are open and the treated material is fed to animals, approximately 30% of the ammonia is volatilized into the air. The remaining portion is incorporated into the substrate matrix. The manure excreted by ruminants fed ammonia-treated substrates might have slightly more N content.

Regarding future applications of chemical treatment technology, most barriers related to environmental concerns could be overcome once the proper ratio of chemicals required are defined. It will be important to strike a proper balance between treatment efficacy and sound environmental management practices.

CONCLUSIONS

Crop residues continue to be vastly underutilized resources, particularly in the United States and other developed countries. Grain prices have never increased to the point where it has become expedient for ruminant animal producers to consider using processed crop residues as an integral component of their feeding regimen. Clearly, treatment methods exist that greatly enhance the digestion and ultimate utilization of the fiber component of these residues. The more efficacious treatment methods are not necessarily "environmentally friendly," although certain components of the process might be changed such that the processed substrate and any waste stream resulting from its preparation are less environmentally unfriendly.

A significant improvement in fiber digestibility with the use of culture extract from WRF alone or in combination with yeast has opened new opportunities for fungal treatment (Mpofu and Ndlovu, 1994). The coexistence of yeast with WRF in wood decay (Gonzalez et al., 1989) and in the removal of soil pollutants (Sasek et al., 1993) already has been reported. The mode of action of yeast and WRF in improving fiber digestibility seems similar, but needs to be investigated further. Wallace and Newbold (1992) emphasized the importance of screening for effective yeast strains such as those having the ability to stimulate one of the major cellulolytic bacterial species *(Fibrobacter succinogenes)* as illustrated by Dawson (1990). Wallace and Newbold (1992) anticipated the development of three generations of yeast cultures. Products of the first generation (available now) can be screened for their bacterial-stimulating activity. The second generation will be products selected from the first generation to fit specific dietary conditions (e.g., certain strains of *Saccharomyces cerevisiae* stimulate cellulolytic bacteria more effectively than others, while others enhance the growth of the lactate utilizers). The third generation will be the utilization of yeasts as a vehicle for implementing the benefits of recombinant DNA technology to ruminal fermentation, given that ecological problems exist in modifying and maintaining indigenous rumen bacteria

(Wallace and Newbold, 1992). The genetics of *S. cerevisiae* are much better known than any species of ruminal microorganism. The microbial cultures are fed daily, so no selection pressure is necessary. The technology exists for their large-scale production, and, as food products already, they are perceived to be safe. Products from the third generation, therefore, could produce enzymes such as cellulases.

Biological treatments of crop residues, as an alternative to nonbiological treatments, have the potential to improve substrate quality without having certain disadvantages that are related to chemical pretreatments. Future studies concerning their efficacy should be conducted in a systematic, detailed, definitive manner so that more progress might be made in this area than has occurred in the past.

REFERENCES

Agosin, E. and Odier, E., Solid-state fermentation, lignin degradation and resulting digestibility of wheat straw fermented by selected white-rot fungi, *Eur. J. Appl. Microbiol. Biotechnol.,* 21, 397, 1985.

Agosin, E., Daudin, J.-J., and Odier, E., Screening of white-rot fungi on (^{14}C) lignin-labelled and (^{14}C) whole-labelled wheat straw, *Appl. Microbiol. Biotechnol.,* 22, 132, 1985a.

Agosin, E., Monties, B., and Odier, E., Structural changes in wheat straw components during decay by lignin-degrading white-rot fungi in relation to improvement of digestibility for ruminants, *J. Sci. Food Agric.,* 36, 925, 1985b.

Agosin, E., Tollier, M. T., Brillouet, J. M., Thivend, P., and Odier, E., Fungal pretreatment of wheat straw: effects on the biodegradability of cell walls, structural polysaccharides, lignin and phenolic acids by rumen microorganisms, *J. Sci. Food Agric.,* 36, 925, 1985c.

Agosin, E., Tollier, M. T., Heckmann, E., Brillouet, J. M., Thivend, P., Monties, B., and Odier, E., Effect of fungal treatment of lignocellulosics on biodegradability, in *Proceedings of the Workshop on Degradation of Lignocellulosics in Ruminants and in Industrial Processes,* van der Meer, J. M., Rijkens, B. A., and Ferranti, M. P., Eds., Elsevier Applied Science, New York, 1987, 35.

Ahuja, A. K., Kakkar, V. K., Garcha, H. S., and Makkar, G. S., Spent paddy straw as a basal roughage for sheep, *Indian J. Anim. Sci.,* 56, 285, 1986.

Akin, D. E., Sethuraman, A., Morrison, W. H., III, Martin, S. A., and Ericksson, K., Microbial delignification with white-rot fungi improves forage digestibility, *Appl. Environ. Microbiol.,* 59, 4274, 1993.

Al-Ani, F. and Smith, J. E., Effect of chemical pretreatments on the fermentation and ultimate digestibility of bagasse by *Phanerochaete chrysosporium, J. Sci. Food Agric.,* 42, 19, 1988.

Ander, P., The cellobiose-oxidizing enzymes CBQ and CbO as related to lignin and cellulose degradation — a review, *FEMS Microbiol. Rev.*, 13, 297, 1994.

Ander, P., The use of white-rot fungi and their enzymes for biopulping and biobleaching, in *Advances in Biological Treatment of Lignocellulosic Materials,* Coughlan, M. P. and Collaco, M. T. A., Eds., Elsevier Applied Science, London, 1990, 287.

Autrey, K. M., McCaskey, T. A., and Little, J. A., Cellulose digestibility of fibrous materials treated with *Trichoderma viride* cellulase, *J. Dairy Sci.,* 58, 67, 1975.

Bakshi, M. P. S., Gupta, V. K., and Langar, P. N., Acceptability and nutritive evaluation of *Pleurotus* harvested spent wheat straw in buffaloes, *Agric. Wastes,* 13, 51, 1985.

Beck, Th., The microbiology of silage fermentation, a review, in *Fermentation of Silage,* McCullough, M. E., Ed., National Feeds Ingredients Association, West Des Moines, IA, 1978, 181.

Ben-Ghedalia, D. and Marcipar, A., The effect of chemical pretreatments and subsequent enzymatic treatments on the organic matter digestibility in vitro of wheat straw, *Nutr. Rep. Int.,* 19, 499, 1979.

Ben-Ghedalia, D. and Miron, J., The effect of combined chemical and enzyme treatments on the saccharification and in vitro digestion rate of wheat straw, *Biotechnol. Bioeng.,* 23, 823, 1981.

Berger, L. L., Fahey, G. C., Jr., Bourquin, L. D., and Titgemeyer, E. C., Modification of forage quality after harvest, in *Forage Quality, Evaluation, and Utilization,* Fahey, G. C., Jr., Collins, M., Mertens, D. R., and Moser, L. E., Eds., American Society of Agronomy, Crop Science Society of America, and Soil Science Society of America, Madison, WI, 1994, 922.

Blakeney, A. B., Harris, P. J., Henry, R. J., and Stone, B. A., A simple and rapid preparation of alditol acetates for monosaccharide analysis, *Carbohydr. Res.,* 113, 291, 1983.

Blanchette, R. A., Delignification by wood-decay fungi, *Annu. Rev. Phytopathol.,* 29, 381, 1991.

Bolsen, K. K., The use of aids to fermentation in silage production, in *Fermentation of Silage — A Review*, McCullough, M. E., Ed., National Feeds Ingredients Association, West Des Moines, IA, 1978, 181.

Calzada, J. F., Franco, L. F., de Arriola, M. C., Rolz, C., and Ortiz, M. A., Acceptability, body weight changes and digestibility of spent wheat straw after harvesting of *Pleurotus sajor-caju, Biol. Wastes,* 22, 303, 1987.

Chandra, S., Reddy, M. R., and Reddy, G. V. N., Effect of fungal treatment of paddy straw on nutrient utilization in complete diets for sheep, *Indian J. Anim. Sci.,* 61, 1330, 1991.

Chesson, A., Stewart, C. S., and Wallace, J. R., Influence of plant phenolic acids on growth and cellulolytic activity of rumen bacteria, *Appl. Environ. Microbiol.,* 44, 597, 1982.

Daniels, L. B. and Hashim, R. B., Evaluation of fungal cellulases in rice hull based diets for ruminants, *J. Dairy Sci.,* 60, 1563, 1977.

Dunlop, C. E., Thompson, J., and Lin, C. C., Treatment processes to increase cellulose microbial digestibility, *Am. Inst. Chem. Eng. Symp. Ser.,* 72, 58, 1976.

Edwards, R. A. and McDonald, P., The chemistry of silage, in *Fermentation of Silage — A Review,* McCullough, M. E., Ed., National Feeds Ingredients Association, West Des Moines, IA, 1978, 27.

Eriksson, K.-E. L., Enzyme mechanisms involved in cellulose hydrolysis by the white-rot fungus *Sporotrichum pulverulentum, Biotechnol. Bioeng.,* 20, 317, 1978.

Eriksson, K.-E. L., Concluding remarks: where do we stand and where are we going? Lignin biodegradation and practical utilization, *J. Biotechnol.,* 30, 149, 1993.

Evans, C. S., Dutton, M. V., Guillen, F., and Veness, R. G., Enzymes and small molecular mass agents involved with lignocellulose degradation, *FEMS Microbiol. Rev.,* 13, 235, 1994.

Fahey, G. C., Jr., Bourquin, L. D., Titgemeyer, E. C., and Atwell, D. G., Postharvest treatment of fibrous feedstuffs to improve their nutritive value, in *Forage Cell Wall Structure and Digestibility*, Jung, H. G., Buxton, D. R., Hatfield, R. D., and Ralph, J., Eds., American Society of Agronomy, Crop Science Society of America, and Soil Science Society of America, Madison, WI, 1993, 715.

Fitzsimons, A. and O'Connel, M., Comparative analysis of amylolotic lactobacilli and *Lactobacillus plantorum* as potential silage inoculants, *FEMS Microbiol. Letters*, 116, 137, 1994.

Flores, D. A., The application of recombinant DNA to rumen microbes for the improvement of low quality feeds utilization, *J. Biotechnol.*, 10, 95, 1989.

Gerrits, J. P. G., Organic compost constituents and water utilized by the cultivated mushroom during spawn run and cropping, *Mushroom Sci.*, 7, 111, 1969.

Glenn, J. K., Morgan, M. A., Mayfield, M. B., Kuwahara, M., and Gold, M. H., An extracellular H_2O_2-requiring enzyme preparation involved in lignin biodegradation by the white-rot basidiomycete *Phanerochaete chrysosporium*, *Biochem. Biophys. Res. Commun.*, 114, 1077, 1983.

Gonzalez, A. E., Martinez, A. T., Almendros, G., and Grinbergs, J., A study of yeasts during the delignification and fungal transformation of wood into cattle feed in chilean rain forest, *Antonie van Leeuwenhoek*, 55, 221, 1989.

Hatakka, A. I., Mohammadi, O. K., and Lundell, T. K., The potential of white-rot fungi and their enzymes in the treatment of lignocellulosic feed, *Food Biotechnol.*, 3, 45, 1989.

Hatakka, A., Lignin-modifying enzymes from selected white-rot fungi: production and role in lignin degradation, *FEMS Microbiol. Rev.*, 13, 125, 1994.

Henderson, A. R. and McDonald, P., The effect of cellulase preparations on the chemical changes during the ensilage of grass in laboratory silos, *J. Sci. Food Agric.*, 28, 486, 1977.

Henics, Z., Wheat straw upgraded by *Pleurotus ostreatus*, *World Rev. Anim. Prod.*, 23, 55, 1987.

Holmes, J. H. G., Egan, A. R., and Doyle, P. T., Intrinsic systems based on crop residues and cultivated fodders, in *Small Ruminant Production Systems in South and Southeast Asia*, Devendra, C., Ed., International Development Research Centre, Ottawa, Canada, 1987, 118.

Holzbaur, E. L. F., Andrawis, A., and Tien, M., Molecular biology of lignin peroxidases from *Phanerochaete chrysosporium*, in *Molecular Industrial Mycology, Systems and Applications for Filamentous Fungi*, Leong, S. A. and Berka, R. M., Eds., Marcel Dekker, New York, 1991, 197.

Jalc, D., Zitnan, R., and Nerud, F., Effect of fungus-treated straw on ruminal fermentation in vitro, *Anim. Feed Sci. Technol.*, 46, 131, 1994.

Jorgensen, O. B. and Cowan, D., Use of enzymes in feed and in ensiling, in *Enzyme Systems for Lignocellulose Degradation*, Coughlan, M. P., Ed., Elsevier Applied Science, London, 1989, 347.

Joseleau, J.-P., Gharibian, S., Comtat, J., Lefebvre, A., and Ruel, K., Indirect involvement of ligninolytic enzyme systems in cell wall degradation, *FEMS Microbiol. Rev.*, 13, 255, 1994.

Kahlon, S. S. and Dass, S. K., Biological conversion of paddy straw into feed, *Biol. Wastes*, 22, 11, 1987.

Kamra, D. N. and Zadrazil, F., Microbiological improvement of lignocellulosics in animal feed production: a review, in *Treatment of Lignocellulosics with White-Rot Fungi*, Zadrazil, F. and Reiniger, P., Eds., Elsevier Applied Science, New York, 1988, 56.

Karunanandaa, K., Colonization of rice straw *(Oriza sativa* L.) by white-rot fungi: effects on structure, chemical composition, cell wall components and nutrient digestibility *in vitro*, Ph.D. Thesis, The Pennsylvania State University, University Park, 1995.

Karunanandaa, K., Fales, S. L., Varga, G. A., and Royse, D. J., Chemical composition and biodegradability of crop residues colonized by white-rot fungi, *J. Sci. Food Agric.*, 60, 105, 1992.

Khazaal, K. A., Owen, E., Dodson, A. P., Harvey, P., and Palmer, J., A preliminary study of the treatment of barley straw with ligninase enzyme: effect on in vitro digestibility and chemical composition, *Biol. Wastes,* 33, 53, 1990.

Kirk, T. K., Biochemistry of lignin degradation by *Phanerochaete chrysosporium,* in *FEMS Symposium: Biochemistry and Genetics of Cellulose Degradation,* Aubert, J.-P., Beguin, P., and Millet, J., Eds., Academic Press, New York, 1988, 315.

Kirk, T. K., Connors, W. J., and Zeikus, J. G., Requirement of a growth substrate during lignin decomposition by two wood rotting fungi, *Appl. Environ. Microbiol.,* 32, 192, 1976.

Kirk, T. K. and Farrell, R. L., Enzymatic "combustion": the microbial degradation of lignin, *Annu. Rev. Microbiol.,* 41, 465, 1987.

Kirk, T. K. and Moore, W. E., Removing lignin from wood with white-rot fungi and digestibility of resulting wood, *Wood Fiber,* 4, 72, 1972.

Leisola, M., Ulmer, D., and Fiechter, A., Problem of oxygen transfer during degradation of lignin by *Phanerochaete chrysosporium, Eur. J. Appl. Microbiol. Biotechnol.,* 17, 113, 1983.

Leng, R. A., The key role of urea and protein supplementation in increasing production by ruminants fed crop residues for pastures, in *Proceedings of the FAO Expert Consultation of the Substitution of Imported Concentrate Feeds in Animal Production Systems in Developing Countries,* Sansoucy R., Preston, T. R., and Leng, A., Eds., Bangkok, Thailand, FAO-UN, Rome, Italy, 1985, 50.

Morris, P. J. and Mowat, D. N., Nutritive value of ground and(or) ammoniated corn stover, *Can. J. Anim. Sci.,* 60, 327, 1980.

Morrison, I. M., Influence of chemical and biological pretreatments on the degradation of lignocellulosic material by biological systems, *J. Sci. Food Agric.,* 42, 295, 1988.

Morrison, I. M., Changes in the biodegradability of ryegrass and legume fibres by chemical and biological pretreatments, *J. Sci. Food Agric.,* 54, 521, 1991.

Mpofu, I. D. T. and Ndlovu, L. R., The potential of yeast and natural fungi for enhancing fibre digestibility of forages and roughages, *Anim. Feed Sci. Technol.,* 48, 39, 1994.

Nakashima, Y., Ørskov, E. R., Hotten, P. M., Ambo, K., and Takase, Y., Rumen degradation of straw, 6. Effect of polysaccharidase enzymes on degradation characteristics of ensiled rice straw, *Anim. Prod.,* 47, 421, 1988.

Nakashima, Y. and Ørskov, E. R., Rumen degradation of straw, 7. Effects of chemical pretreatment and addition of propionic acid on degradation characteristics of botanical fractions of barley straw treated with a cellulase preparation, *Anim. Prod.,* 48, 543, 1989.

Pioneer Forage Manual — A Nutritional Guide, Pioneer Hi-Bred International, Inc., Des Moines, IA, 1990.

Reid, I. D., Solid-state fermentation for biological delignification, *Enzyme Microbial. Technol.,* 11, 786, 1989.

Rolz, C., Leon, R., Arriola, M. C., and Cabrera, S., Biodelignification of lemon grass and citronella bagasse by white-rot fungi, *Appl. Environ. Microbiol.,* 52, 607, 1986.

Saenger, P. F., Lemenager, R. P., and Hendrix, K. S., Effects of anhydrous ammonia treatment of wheat straw upon in vitro digestion, performance and intake by beef cattle, *J. Anim. Sci.,* 56, 15, 1983.

Sasek, V., Volfova, O., Erbanova, P., Vyas, B. R. M., and Matucha, M., Degradation of PCBs by white rot fungi, methylotrophic and hydrocarbon utilizing yeasts and bacteria, *Biotechnol. Letters,* 15, 521, 1993.

Seale, D. R., Bacterial inoculants as silage additives, *J. Appl. Bacteriol.,* 9S(Symp. Suppl.), 1986.

Sharp, R., O'Donnell, A. G., Gilbert, H. G., and Hazlewood, G. P., Growth and survival of genetically manipulated *Lactobacillus plantarum* in silage, *Appl. Environ. Microbiol.*, 58, 2517, 1992.

Sukanto, L. and Soedomo, R., Low-cost feed rations: the prospect for substitution, in *Livestock in Asia, Issues and Policies*, Fine, J. C. and Lattimore, R. G., Eds., International Development Research Centre, Ottawa, 1982, 79.

Sundstøl, F. and Coxworth, E. M., Ammonia treatment, in *Straw and Other Fibrous By-Products as Feed,* Sundstøl, F. and Owen, E., Eds., Elsevier Science Publishers, Amsterdam, 1984, 196.

Tien, M. and Kirk, T. K., Lignin-degrading enzyme from the hymenomycete *Phanerochaete chrysosporium* Burds, *Science,* 221, 661, 1983.

Tien, M., Properties of ligninase from *Phanerochaete chrysosporium* and their possible applications, *CRC Crit. Rev. Microbiol.,* 15, 141, 1987.

Valmaseda, M., Martinez, M. J., and Martinez, A. T., Kinetics of wheat straw solid-state fermentation with *Trametes versicolor* and *Pleurotus ostreatus* — lignin and polysaccharide alteration and production of related enzymatic activities, *Appl. Microbiol. Biotechnol.,* 35, 817, 1991.

Vanbelle, M. and Bertin, G., Screening of fungal cellulolytic preparations for application in ensiling processes, in *Enzyme Systems for Lignocellulose Degradation,* Coughlan, M. P., Ed., Elsevier Applied Science, London, 1989, 357.

van der Meer, J. M. and Ketelaar, R., Evaluation of enzymatically changed feeds, in *Enzyme Systems for Lignocellulose Degradation,* Coughlan, M. P., Ed., Elsevier Applied Science, London, 1989, 383.

van Eys, J. E., Rangkuti, M., and Johnson, W. L., Feed resources and feeding systems for small ruminants in south and southeast Asia, in *Small Ruminant Production Systems in South and Southeast Asia,* Devendra, C., Ed., International Development Research Centre, Ottawa, 1987, 52.

Wallace, R. J. and Newbold, C. J., Probiotics for ruminants, in *Probiotics: The Scientific Basis,* Fuller, R., Ed., Chapman and Hall, London, 1992, 12.

Walli, T. K., Rai, S. N., Gupta, B. N., and Singh, K., Influence of fungal treated and urea treated wheat straw on nutrient utilization in calves, *Indian J. Anim. Nutr.,* 8, 227, 1991.

Wan Mohammed, W. E., Integrated small ruminants with rubber and oil palm cultivation in Malaysia, in *Small Ruminant Production Systems in South and Southeast Asia,* Devendra, C., Ed., International Development Research Centre, Ottawa, 1987, 239.

Yadav, J. S., Influence of nutritional supplementation on solid-substrate fermentation of wheat straw with an alkaliphilic white-rot fungus *(Coprinus* sp.), *Appl. Microbiol. Biotechnol.,* 26, 474, 1987.

Zadrazil, F., Conversion of different plant waste into feed by Basidiomycetes, *Eur. J. Appl. Microbiol. Biotechnol.,* 9, 243, 1980.

Zadrazil, F., Microbial conversion of lignocellulose into feed, in *Straw and Other Fibrous By-Products as Feed,* Sundstøl, F. and Owen, E., Eds., Elsevier Science Publishers, Amsterdam, 1984, 276

Zadrazil, F. and Brunnert, H., The influence of ammonium nitrate supplementation on degradation and in vitro digestibility of straw colonized by higher fungi, *Eur. J. Appl. Microbiol. Biotechnol.,* 9, 37, 1980.

Zadrazil, F. and Brunnert, H., Investigation of physical parameters important for the solid-state fermentation of straw by white-rot fungi, *Eur. J. Appl. Microbiol. Biotechnol.,* 11, 183, 1981.

Zadrazil, F. and Brunnert, H., Solid-state fermentation of lignocellulose containing plant residues with *Sporotrichum pulverulentum* Nov. and *Dichomitus squalens* (Karst.) Reid., *Eur. J. Appl. Microbiol. Biotechnol.,* 16, 45, 1982.

Zadrazil, F., Grinbergs, J., and Gonzalez, A., "Palo podrido" — decomposed wood used as feed, *Eur. J. Appl. Microbiol. Biotechnol.,* 15, 167, 1982.

Zadrazil, F., Janssen, H., Diedrichs, M., and Schuchardt, F., Pilot-scale reactor for solid-state fermentation of lignocellulosics with higher fungi: production of feed, chemical feedstocks and substrates suitable for biofilters, in *Proceedings of Workshop Advances in Biological Treatment of Lignocellulosic Materials,* Coughlan, M. P. and Collaco, M. T. A., Eds., Elsevier Applied Science, New York, 1990, 31.

CHAPTER 14

Potential for Recycling Animal Wastes by Feeding To Reduce Environmental Contamination

J. P. Fontenot, G. A. Ayangbile, and V. G. Allen

INTRODUCTION

Animal wastes were regarded as valuable sources of plant nutrients in the early part of this century and were used extensively as fertilizer. Following the introduction of intensive confinement systems during the past 40 to 50 years animal wastes became a liability rather than an asset. This change in perceived value of waste resulted from the availability of low-cost commercial fertilizer and the high cost of handling and transporting the wastes. Animal wastes were identified as a concern in the control of water, soil, and air pollution (Freeman and Bennett, 1969).

Estimates vary concerning the amount of animal waste produced annually, but the amount is substantial. In the United States, over 100 million tons of animal waste dry matter (DM) are produced per year (Fontenot and Ross, 1980). About 50% of the wastes are collectable, and, in some enterprises such as poultry, virtually all of the wastes are collectable. In large, concentrated animal production enterprises most of the wastes can be collected.

Recycling of animal wastes refers to making further use of the nutrients in the waste. Options available for recycling nutrients in animal wastes include (1) sources of plant nutrients, (2) feed ingredients for farm animals, (3) substrate for methane generation, and (4) substrates for microbial and insect protein synthesis. Although it is technically feasible to use animal waste to produce microbial and insect protein, the practice is not economically feasible (Calvert, 1979). Likewise, methane generation from animal waste is technically feasible (Smith et al., 1979), but wastes possess low monetary value for this purpose (Table 1; Fontenot et al., 1983). Thus the most feasible methods of recycling animal wastes are as sources of nutrients for plants and animals.

Renewed interest has been shown in the use of animal wastes as fertilizer because of the increased cost of fossil fuel used in manufacturing fertilizer. However, although collection of wastes from large, concentrated animal production enterprises is possible, in some areas use of all of the wastes as fertilizer may not be possible because of insufficient land area in close proximity to the enterprises. Application of excessive amounts may contaminate water supplies and

1-56670-199-6/96/$0.00+$.50

Table 1 Relative Value of Animal Wastes Utilized for Different Purposes[a]

	U.S. dollars per metric ton		
Kinds of waste	Fertilizer	Feed	Methane
Beef cattle	25.06	118.14	13.73
Dairy cattle	17.00	118.14	12.74
Swine	18.61	136.57	17.17
Caged layer	36.45	155.14	17.93
Broiler litter	26.54	159.57	16.29

[a] Adapted from Fontenot et al., 1983.

may even decrease crop yields (Mathers and Stewart, 1971). Nevertheless, under some situations, recycling of animal wastes for use as fertilizer may be economically feasible, especially if the wastes do not need to be transported for long distances.

In this chapter, the primary emphasis will be placed on recycling wastes as feed for animals. As shown in Table 1, the value of the wastes is much higher if wastes are used as sources of nutrients for animals than for other uses (Fontenot, 1991). However, even in recycling by feeding to animals, some waste will be produced that could then be recycled indirectly as plant nutrients, especially if the waste is fed to grazing animals.

Over 70 years ago, the value of animal wastes was recognized as a source of vitamins when it was observed that feces of normal rats contained "fat soluble A, and water soluble B" (McCollum, 1922). Animal wastes can be processed and used as animal feed. Because of the high-fiber and nonprotein nitrogen content of the wastes, ruminants are best suited for utilization of the wastes. Reviews on feeding animal wastes include those prepared by Bhattacharya and Taylor (1975), Smith and Wheeler (1979), Fontenot et al. (1983), and Fontenot (1991). Reviews concerning health aspects of feeding animal wastes were published by Fontenot and Webb (1975) and McCaskey and Anthony (1979).

NUTRITIONAL VALUE OF ANIMAL WASTES

The relative nutritive value of animal wastes for ruminants in descending order is excreta of young poultry, deep litter of young poultry, hog feces, excreta of laying hens, manure solids of hogs and laying hens, and excrement of cattle (Koriath, 1975). Ranking of feeding value of wastes based on *in vitro* cell wall digestibility was in general agreement with that ranking, except swine wastes ranked lower than caged laying hen wastes (Smith, 1973). Cell walls from nonruminant feces are generally more digestible than those from ruminant wastes.

The nutritional value of wastes produced by animals in confinement is given in Table 2. Miron et al. (1990) reported higher values for digestibility of organic matter in poultry litter than the total digestible nutrient (TDN) values reported by Bhattacharya and Taylor (1975). Poultry wastes are higher in crude protein and available energy than wastes from cattle. All of the wastes are good sources of phosphorus (P). Calcium is especially high in caged layer waste. Macro- and

Table 2 Nutritional Value[a] of Animal Wastes

Item	Broiler[b]	Dehydrated caged layer waste[b]	Steer waste[b]	Cow waste[b]	Swine waste[c]
Crude protein (%)	31.3	28	20.3	12.7	23.5
True protein (%)	16.7	11.3		12.5	15.6
Digestible protein (%)	23.3	14.4	4.7	3.2	
Crude fiber (%)	16.8	12.7			14.8
Ether extract (%)	3.3	2.0		2.5	8.0
NFE (%)	29.5	28.7		29.5	38.3
Dig. energy (kcal/g)	2440[d]	1884[d]			2134[e]
Metab. energy (kcal/g)	2181[d]				2088[e]
TDN (%)[d]	59.8	52.3	48	16.1	
Ash (%)	15.0	28	11.5	16.1	15.3
Calcium (%)	2.4	8.8	0.87		2.72
Phosphorus (%)	1.8	2.5	1.60		2.13
Magnesium (%)	0.44	0.67	0.40		0.93
Sodium (%)	0.54	0.94			
Potassium (%)	1.78	2.33	0.50		1.34
Iron (ppm)	451	2000	1340		63
Copper (ppm)	98	150	31		63
Manganese (ppm)	225	406	147		
Zinc (ppm)	235	463	242		530

[a] Dry matter basis.

[b] Adapted from Bhattacharya and Taylor, 1975.

[c] Adapted from Kornegay et al., 1977.

[d] For ruminants.

[e] For swine.

micromineral concentrations usually reflect levels in the diet. The amount of bedding used in broiler houses affects the composition of litter, especially crude protein and fiber. In Alabama, crude protein of broiler litter averaged 25%, DM basis (Ruffin and McCaskey, 1990), compared to 30% in Virginia (Fontenot et al., 1971), apparently because of a difference in quantity of bedding, which was reflected in differences in fiber levels.

Performance of ruminants fed different levels of animal wastes was summarized by Smith and Wheeler (1979). Feeding diets containing an average of 24% poultry litter (DM basis) to cattle resulted in a 5% depression in rate of gain and 10% increase in feed DM per kilogram gain, probably reflecting the lower energy value of litter than of the portion of the diet that was replaced. Daily gains were 1.07 and 1.10 kg, respectively, for control and experimental cattle fed caged layer waste. Respective feed to gain ratios were 7.25 and 6.49. Brosh et al. (1993) reported similar liveweight gains for beef cows fed 15, 30, or 45% poultry litter, DM basis. Dry matter intake increased with percentage litter in the diet. The inclusion of 12% cattle waste (DM basis) did not substantially affect daily gain in cattle (1.16 kg for controls vs. 1.14 kg for waste fed) (Smith and Wheeler, 1979). The average feed DM per unit of gain was slightly higher for the cattle fed cattle waste, indicating that cattle wastes are lower in energy value than the traditional components of the diet. Milk production was similar for dairy cows fed 0 or 12% caged layer waste (DM basis). Arave et al. (1990) reported that feeding lactating dairy cows up to 17% poultry excreta, DM basis, had no effect on DM intake, fat-corrected milk yield, or milk fat. Treating cattle waste with sodium hydroxide, calcium hypochlorite, or sodium chlorite resulted in increased DM digestibility of the waste (Smith et al., 1971; Lucas et al., 1975).

Poultry do not utilize fiber and nonprotein nitrogen efficiently. The metabolizable energy (ME) content of dried poultry wastes for layers was reported to be only 6% of that of corn (Rinehart et al., 1973). Digestibility of energy of swine feces by swine was 46.7% (Kornegay et al., 1977).

PROCESSING ANIMAL WASTES

Processing is necessary for destruction of pathogens, improvement of storage and handling characteristics, and maintenance or enhancement of palatability (CAST, 1978). Processes that have been used include dehydration, pelleting, ensiling alone or with other ingredients, deep stacking, and composting.

Dehydration

Heat drying has been used to process poultry, cattle, and swine wastes (Fontenot and Ross, 1980). The resultant product is rather dusty but has good keeping qualities. Heat drying may result in loss of nitrogen (N), at least with poultry wastes. A major problem with this process is the high cost of fossil fuel.

Ensiling

Ensiling animal wastes alone or in combination with other ingredients has been successful. Good silage was made by mixing corn forage and broiler litter (Harmon et al., 1975a). For good ensiling, the level of waste should probably not exceed 30% of the DM. Feeding this kind of silage has resulted in efficient N utilization in sheep (Harmon et al., 1975b) and performance in finishing cattle similar to that of cattle fed corn silage plus conventional protein supplement (McClure and Fontenot, 1985). Poultry litter can be ensiled alone, but the moisture level should be about 40% for good ensiling (Caswell et al., 1978). Satisfactory ensiling was achieved with mixtures of broiler litter and rumen contents (Chaudhry, 1990). The final pH was 5.6.

Cattle waste was ensiled successfully with grass hay (Anthony, 1971). The ensiled mixture of 57% cattle waste and 43% grass hay was termed "wastelage." Satisfactory performance was obtained in cattle fed wastelage. Cattle wastes have been successfully ensiled with crop residues also. Good ensiling with a final pH below 5 was obtained with 70:30 to 30:70 mixtures (wet basis) of cattle wastes and rye straw (Cornman et al., 1981). The low pH was reached after 1 week of ensiling. Mixtures of 50:50 cattle waste and rye straw ensiled similarly with DM levels of 30, 40, or 50% (Aines et al., 1982).

Satisfactory ensiling was reported with mixtures of 30:70 to 70:30 swine waste and grass hay (wet basis) (Berger et al., 1981a). The smell of the silage was similar to that of good-quality hay-crop silage, with no swine fecal odor remaining. Digestibility values obtained with sheep fed the silages with 60:40 and 40:60 swine waste and hay indicated that the ensiled waste was digested to a similar extent as the grass hay (Berger et al., 1981b). Caged layer waste has been successfully ensiled with hay (Saylor and Long, 1974), sugarcane bagasse (Samuels et al., 1980), corn stover (Moriba et al., 1982), corn forage (Goering and Smith, 1977), and sorghum forage (Richter and Kalmbacher, 1980).

Deep Stacking

A maximum temperature of 50°C was reached 4 to 8 d after deep stacking broiler litter to a depth of 1.2 m (Hovatter et al., 1979). The maximum temperature appears to be related to moisture level. The highest temperatures recorded for litter deep stacked at 15% moisture was 57.5°C, compared with 61.2 and 60.2°C for litter deep stacked at 25 and 35% moisture, respectively (Chaudhry, 1990). Moisture levels above 35% resulted in lower maximum temperatures (Abdelmawla, 1990). Highest temperatures at 46 cm from the top of 1.2-m stacks were 65.9, 63.9, and 62.8°C for litter stacked at 35, 40, and 45% moisture, respectively. Nitrogen utilization was similar for ruminants fed either ensiled or deep-stacked poultry litter (Abdelmawla et al., 1988). Performance was similar for finishing cattle fed ensiled broiler litter-corn forage or ensiled corn forage and deep-stacked broiler litter (McClure and Fontenot, 1985).

Excessive heating may occur during deep stacking, resulting in a dark, charred-appearing litter. Excessive heating may reduce DM digestibility (Ruffin and McCaskey, 1990). Evidence was obtained that the N in "charred" litter from a stack exposed to the weather was less soluble and lower in rumen degradability than normal deep-stacked litter (Kwak, 1990). When charred litter was fed to sheep, apparent digestibilities of DM, acid detergent fiber, and crude protein were lower than for sheep fed normal deep-stacked litter from the same stack (Bargahit and Fontenot, 1990). However, no difference was obtained in N retention. Covering the stack with polyethylene resulted in a lower temperature in the stack (Rankins et al., 1993).

Composting

This process involves initial stacking of the wastes and mixing to enhance aerobic fermentation. Composting is described as the "rapid but partial decomposition of most solid organic matter by the use of aerobic microorganisms under controlled conditions" (Anonymous, 1970). This process has been used to process animal wastes for land application, but animal wastes may be composted for use as animal feed (CAST, 1978). Considerable loss of N may occur during composting. Composting litter by stacking, mixing after 2 d, then at weekly intervals for 6 weeks resulted in a 15% decrease in crude protein (Abdelmawla et al., 1988). Apparent digestibilities of DM and crude protein, and N utilization by sheep fed composted litter, were similar to those for sheep fed deep-stacked litter.

QUALITY OF ANIMAL PRODUCTS

An important aspect is the effect of feeding animal wastes on quality of edible animal products. In an early experiment, Fontenot et al. (1966) found that feeding broiler litter to finishing beef steers did not adversely affect carcass quality or taste of the meat. In a subsequent experiment with cattle fed mixtures with 25 or 40% broiler litter with different litter materials, carcass grades and dressing percentages were similar as for those fed a control diet.

Carcass quality was evaluated in an experiment in which cattle were fed diets containing 0, 25, and 50% broiler litter (Fontenot et al., 1971). Carcasses were graded USDA low choice for the cattle fed the different levels of litter. In taste (organoleptic) tests, no differences were detected by a taste panel. Cross et al. (1978) reported no difference in carcass grade, marbling, ribeye muscle area, back fat, or taste panel evaluation from feeding diets with 0 to 50% broiler litter silage, DM basis. Feeding cattle waste to cattle had no effect on carcass quality or taste of the meat (Anthony, 1966).

Feeding up to 4.1 kg dried caged layer waste had no adverse effect on milk composition or flavor (Bull and Reid, 1971). Overall flavor quality scores were similar for milk from cows fed 0, 10, 20, or 30% dried poultry waste (Silva et al., 1976). The flavor of milk from dairy cows fed a diet containing dried layer waste was similar to that of cows fed diets not containing waste (Thomas et al., 1972). Arave et al. (1990) reported no adverse effect of feeding up to 17% processed poultry excreta to dairy cows on milk flavor.

Egg weight and shell thickness were not adversely affected by including up to 40% dried poultry waste in the diet of hens (Flegal and Zindel, 1971). A taste panel did not detect differences in the taste of eggs from hens fed diets containing 0, 10, 20, and 30% dried poultry waste.

SAFETY OF FEEDING ANIMAL WASTES

Toxicity

Copper toxicosis has been documented in sheep fed broiler litter with high copper (Cu) levels (Fontenot and Webb, 1975). The litter, which was fed at levels of 25 and 50% of the diet, contained 195 ppm Cu, resulting from feeding high levels of Cu to chicks. The first fatality occurred after 137 d on test. At the end of 254 d, 64% of the ewes fed the higher level of litter and 55% of those fed 25% litter had died. Suttle et al. (1978) reported elevated Cu levels in livers of lambs fed dried battery or broiler waste, but no signs of toxicity were observed. Feeding molybdenum (Mo) and sulfate may help in preventing the Cu toxicity problem with sheep. Copper accumulation in the liver of ewes fed poultry litter was decreased by about 50% from feeding 25 ppm Mo and 5 g sulfate per kilogram of diet (Olson et al., 1984).

The problem would not be as severe in cattle since they are not as sensitive to high dietary Cu. Beef females were fed a diet high in Cu, resulting from feeding high levels of broiler litter with high Cu levels during the winter period for 7 years with no deleterious effects (Webb et al., 1980). The cows were not fed litter for the remainder of the year (7 to 8 months). Rankins et al. (1993) reported increased liver Cu in cattle fed high-Cu litter for 84 d, but they did not report any clinical signs of Cu toxicity. Practicing veterinarians have reported a limited number of cases of Cu toxicity in cattle fed poultry litter (Pugh et al., 1994), but most veterinarians have not observed such toxicities.

Shlosberg et al. (1992) reported cardiomyopathy in cattle fed litter from broilers fed the coccidiostat maduramycin. They indicated that the coccidistat is frequently found in higher concentrations in litter than in the original feed, whereas monensin concentration in litter may be up to 25% of the concentration in the original feed.

Pathogenic Bacteria and Parasites

Heat processing destroys potential pathogens (Fontenot and Webb, 1975). Proper ensiling of animal wastes also appears to be effective in destroying pathogens (McCaskey and Anthony, 1979). A pH of 4 to 4.5 and a temperature of over 25°C are important for destruction of *Salmonella.* Ensiling feedlot cattle manure and grass hay was effective in eliminating parasites. Apparently, because of the ammonia and minerals in poultry wastes, it is rather difficult to reach a pH of less than 5 without additional materials such as whole plant corn forage. However, ensiling of broiler litter with added water has been shown to destroy coliforms even when the pH did not go below 5.4 (Caswell et al., 1978).

The potential risk of clostridia in ensiled waste-containing rations was suggested by an alleged botulism outbreak in cattle fed poultry wastes in Israel (Egyed et al., 1978). The organism *Clostridium botulinum* (type D) appeared to be endemic in Israel as outbreaks have been reported in animals fed other types of feeds (Tagari, 1978). A major outbreak of type C botulism was reported in cattle fed ensiled poultry litter in Northern Ireland (McLoughlin et al., 1988). The outbreak occurred 24 h after the introduction of purchased ensiled litter. Decomposed poultry carcasses were observed in the purchased ensiled litter. Neill et al. (1989) obtained data indicating there was uneven distribution of botulism toxin in the litter. They stated that it is essential that the silage be made carefully and be free from poultry carcasses.

In Quebec, 28 of 41 cattle died that had been fed a diet of poultry litter, crushed corn, and straw (Bienvenu and Morin, 1990). Botulism was suggested as the cause of the problem. The litter fed when the problem occurred contained numerous chicken carcasses, and the presence of *C. botulinum,* type C was determined in a feed sample. Outbreaks of a paralytic disease, suggested to be botulism, occurred on three farms in Australia (Trueman et al., 1992). Poultry litter was included at 4% in the diet on one of the farms, and chicken carcasses were found in the litter.

Clegg et al. (1985) and Hogg et al. (1990) reported that botulism was diagnosed in the United Kingdom in cattle grazing on pasture that had been fertilized with poultry litter. They suggested that the source of the toxin was poultry carcasses containing types C and D *C. botulinum.* Other cases of botulism have been reported in cattle grazing pastures fertilized with poultry litter containing poultry carcasses infected with *C. botulinum* (Appleyard and Mollison, 1985; Smart et al., 1987; McIllroy et al., 1987). Ruffin and McCaskey (1990) suggested that broiler litter used for composting of dead birds not be used as a source of feed for beef cattle.

No botulism problem has been reported in animals fed waste-containing diets in the United States. The survival of *C. sporagens* as a model for *C. botulinum* was studied by inoculating in a bovine waste-blended diet and in corn forage. After ensiling for 60 d, the number of organisms declined in both silages (McCaskey and Anthony, 1979).

Residues in Animal Products

Mycotoxins pose no greater problem in poultry litter than in common feedstuffs (Lovett, 1972). No evidence of pesticide accumulation in wastes or in animal tissue from animals fed wastes has been reported (Fontenot et al., 1983). Heavy metals have not been found to be sufficiently high in cattle waste and poultry litter to present a problem (Westing et al., 1980). Liver Cu is increased by feeding waste with high Cu levels.

Medicinal drug residues were present in broiler litter in variable amounts if the drugs had been included in the broiler diet (Webb and Fontenot, 1975). However, residues of the three drugs that were in litter, chlortetracycline, nicarbazin, and amprolium, did not accumulate in the tissue of finishing beef cattle after a 5-d withdrawal. Thus it appears that with a modest withdrawal period there is no serious tissue residue problem from feeding broiler litter. Litter should not be fed to cows producing milk or hens producing eggs for human consumption since insufficient data are available on these aspects.

REGULATION OF FEEDING ANIMAL WASTE IN THE UNITED STATES

The Food and Drug Administration (FDA) published a policy (21 CFR 500.4) in the September 2, 1967 *Federal Register* not sanctioning the use of poultry litter as animal feed (Kirk, 1967). Broad interpretation subsequently extended this policy to include all animal wastes used as ingredients in animal feeds. The FDA took this position because the amount of information then available was not believed adequate to conclude that animal wastes were safe when used as feed ingredients. The FDA (1980) revoked 21 CFR 500.4 on the use of poultry litter as an animal feed ingredient on December 30, 1980 (45 FR 86272) and left the regulation of feeding animal wastes to the individual states. In some states, regulation is through the AAFCO (1990) model regulation for processed animal wastes.

The salient points of the AAFCO regulation are (1) waste must be processed so it will be free of pathogenic organisms; (2) if it can be documented by records that animals producing the waste have not been fed drugs, no withdrawal period is required, and the waste can be fed to any class of animals; and (3) if it cannot be documented by records that the animals producing the waste were not fed drugs, a 15-d withdrawal is required prior to slaughtering animals or prior to using milk or eggs for human consumption.

FEEDING VALUE OF ANIMAL WASTES

Although nonruminants can utilize certain wastes, ruminants are best suited for utilization of the high fiber and frequently high nonprotein N in animal wastes. Smith and Wheeler (1979) estimated the monetary value of different kinds of animal wastes. These values were compared to the value of wastes for alternate uses (Fontenot et al., 1983). Values given in Table 1 show that wastes have considerable monetary value and are much more valuable as sources of feed than for fertilizer or methane generation. Zimet et al. (1988) estimated the value of broiler litter as cattle feed ingredient using computer simulation. These values were much higher than the values estimated by Smith and Wheeler (1979).

Although any of the collectible animal wastes can be used as animal feed, poultry litter is the primary waste that is used extensively. Poultry litter is the most nutritious, and the low moisture content facilitates handling, processing, and storage. One of the problems has been the conception that poultry litter had to be fed in close proximity to the poultry houses from which it was collected. However, when the values of the wastes given in Table 1 are considered, obviously the litter can be transported for considerable distances. In some states, it is being sold and transported to other locations, where it is fed to cattle.

A program was initiated jointly by the U.S. Environmental Protection Agency (EPA), Virginia, Maryland, Pennsylvania, and Washington, D.C. to reduce contamination of the Chesapeake Bay (Shuyler, 1994). One of the programs in Virginia under this effort is the subsidizing of structures to store poultry litter. These structures are designed to store the litter from cleaning all the poultry houses on that farm. The structures are well suited for deep stacking litter, since they are very well ventilated. They vary in capacity, up to 1000 tons. Litter from these structures is sold after deep stacking. This method provides a continuous supply of litter that has been processed and is ready for feeding. The litter is transported for distances varying up to over 300 km. Even after transporting for 300 km, the price of litter, including transportation (about $35/metric ton), is only about 35% of its value.

Practical Feeding of Poultry Litter

Poultry litter can be fed to any class of beef cattle, with withdrawal periods imposed where applicable. Diets have been developed to use poultry litter (Gerken, 1977; Ruffin and McCaskey, 1990).

The beef cattle producer needs to know the amount of litter available in order to plan a feeding program. About 900 g of dry broiler litter are produced per bird during a production cycle (Van Dyne and Gilbertson, 1978). Most of the poultry litter fed is used by livestock producers. However, the feed industry could make use of substantial amounts of poultry litter. Some processing would be required to reduce the moisture to a level suitable for storage and to destroy the pathogens. Litter could probably be blended with other ingredients and pelleted. The pelleting process, including cooling the pellets, would probably reduce moisture to a level low enough for safe storage.

Availability of sufficient quantities of uniform litter is essential for a feed manufacturer to start blending feeds containing animal wastes. In 1992, 6,388,990,000 broilers were produced in the United States (USDA, 1993). Thus, using the value of 900 kg of litter DM per 1000 broilers produced, 5.75 million metric tons of litter DM are produced per year. As the trend for constructing storage structures continues, a constant supply of uniform litter should be available.

RECYCLING TO THE SOIL BY FEEDING BROILER LITTER

In order to assess the feasibility of recycling broiler litter to the soil by feeding, an experiment was conducted with lambs fed different levels of deep-stacked broiler litter to quantify excretion of N and minerals. This was followed by an experiment in which cattle are fed broiler litter on pasture, which is in progress.

Excretion of Nitrogen and Minerals by Sheep

A metabolism trial was conducted with 18 wether lambs fed three diets containing 0, 33.3, or 66.7% deep-stacked broiler litter, DM basis (6 lambs per diet). The basal diet was composed of wheat straw, corn grain, soybean meal, and minerals. The litter was substituted for proportional amounts of the straw, corn, and soybean meal, and limestone was deleted from these diets. The trial consisted of a 10-d preliminary period followed by 10 d for total collection of feces and urine.

Apparent digestibility of DM, organic matter, and crude protein showed a linear decrease *($p < 0.01$)*, and hemicellulose increased *($p < 0.01$)* linearly with increased level of litter (Table 3). Excretion of N and minerals (Ca, P, Mg, K, and Na), expressed as grams per day, increased linearly *($p < 0.01$)* with an increased level of broiler litter fed, reflecting increases in dietary levels (Table 4). Expressed as percent of intake, N excretion increased *($p < 0.05$)* and Ca excretion decreased *($p < 0.01$)* as the level of litter in the diet increased. For the other minerals, no significant effect *($p > 0.05$)* of the level of litter on excretion, expressed as percent of intake, was recorded. Excretion of all minerals for the lambs fed poultry litter amounted to 79% or more of the intake. Thus, if litter were fed to cattle on pasture, about 80% or more of the minerals would be applied to the soil via excretion.

Table 3 Apparent Digestibility of Diets by Sheep Fed Different Levels of Broiler Litter

	Level of broiler litter[a]			
Component	**0**	**33.3**	**66.7**	**SE**
Dry matter[b]	69.2	61.6	55.4	0.78
Organic matter[b]	70.6	56.0	60.8	0.75
Crude protein[b]	68.9	59.4	53.8	0.89
Neutral detergent fiber	52.5	51.6	51.4	1.83
Acid detergent fiber[c]	44.9	34.1	35.6	1.17
Cellulose	51.2	50.3	54.0	1.53
Hemicellulose[b]	63.4	66.3	73.0	1.81

[a] Percent, DM basis.

[b] Linear effect ($p < 0.01$).

[c] Quadratic effect ($p < 0.01$).

Nutrient Management of Broiler Litter for Cattle on Pasture

As shown by the results presented previously, ruminants retain only a small percentage of dietary N and mineral elements. Thus feeding a waste such as broiler litter should result in almost comparable applications of plant nutrients

Table 4 Nutrient Excretion by Lambs Fed Different Levels of Broiler Litter

		Level of broiler litter, %[a]		
Nutrient	**Item**	**0**	**33.3**	**66.7**
Nitrogen	Intake, g/d[b]	15.3	19.6	23.5
	Excretion			
	g/d[b]	10.0	14.2	18.4
	% of intake[c]	65.4	72.6	78.0
Calcium	Intake, g/d[b]	3.6	10.1	16.3
	Excretion			
	g/d[b]	3.5	9.5	14.6
	% of intake[b]	98.0	94.6	89.9
Phosphorus	Intake, g/d[b]	2.2	5.9	9.5
	Excretion			
	g/d[b]	1.5	5.2	7.9
	% of intake	69.4	87.1	83.0
Magnesium	Intake, g/d[b]	1.1	2.4	3.6
	Excretion			
	g/d[b]	1.0	2.2	3.4
	% of intake	90.8	95.5	94.5
Potassium	Intake, g/d[b]	9.3	14.0	18.7
	Excretion			
	g/d[b]	6.8	11.2	16.4
	% of intake	74.1	79.5	87.9
Sodium	Intake, g/d[b]	4.3	6.4	8.4
	Excretion			
	g/d[b]	3.9	5.4	7.4
	% of intake	90.3	83.7	87.8

[a] Percent of DM.

[b] Linear effect ($p < 0.01$).

[c] Linear effect ($p < 0.05$).

as if applied directly to the soil. An experiment was initiated to determine the relative efficiency of recycling nutrients from broiler litter by feeding or soil application.

The experiment, in progress, consists of stocker cattle grazing tall fescue pastures and fed fescue hay as needed. Endophyte-free KY-31 fescue was established on 14 ha. The treatments are as follows: (1) no supplementary feeding of broiler litter or soil application of fertilizer or litter; (2) feeding broiler litter; (3) soil application of broiler litter; and (4) soil application of N, P, K, and Ca (inorganic fertilizer). The amount of litter applied to the soil (treatment 3) is to be equal to the amount fed for treatment 2. The litter is fed to the cattle of treatment 2 mixed with ground corn grain. The other cattle are fed the same amount of corn as those fed the corn-litter mixture. Inorganic fertilizer is to be applied to supply plant nutrients (N, P, K, and Ca) equal to levels applied as litter.

There are three pasture replications of each treatment, with four steers per paddock (total of 48 steers per year). Fescue on 55% of the area is grazed during the spring and summer. About 45% of the area is used for making hay and stockpiling. The steers graze stockpiled fescue starting November-December, and hay is fed during periods of heavy snow accumulation or after the stockpiled forage supply is exhausted. Steers continue grazing until the following October. Data include performance of cattle; yield and chemical composition of forages; forage species distribution; and soil levels of N, P, K, Ca, Mg, S, Cu, and Zn.

The first steers were started on May 3 and removed from test on October 24, 1994. The second group of steers were started on December 1, 1994. Since no fertilizer or broiler litter have been applied thus far, the only difference among the treatments is that litter has been fed to cattle on that treatment.

By April 20, 1995, N, P, K, and Cu tended to be higher for the forage on pastures in which steers were fed broiler litter (Table 5). The rate of gain during the 1994 grazing season and from December 1994 to April 1995 were not different among the cattle grazing the different pastures (Table 6). However, in April 1995, daily gain tended to be higher for steers fed litter. Blood serum mineral values for 1994 and up to April 1995 were in the normal range for the minerals determined (Table 7). For the 1994 grazing season, blood serum levels were not affected by feeding broiler litter. In the cattle on trial since November 1994, serum P was higher ($p < 0.05$) in January and April 1995 in cattle fed litter. The plan is to continue this experiment for a minimum of 5 years to evaluate the effect of nutrient management of animal waste by feeding or applying the waste to the soil directly.

CONCLUSIONS

Animal wastes can be used as feedstuffs for animals if processed properly to eliminate pathogens. Performance for animals fed animal wastes is similar to that of control animals if the nutrient levels are equalized. With good management and appropriate withdrawal, feeding wastes does not result in harmful residues in animal products. Feeding wastes will reduce environmental pollution, since up to 60% of the waste DM is digested. If animals are fed balanced diets, excretion by

Table 5 Nitrogen and Mineral Levels in Pastures Grazed by Steers

	Treatment			
		Broiler litter		
Item	None	Fed	Soil application	Inorganic fertilizer
May 3, 1994				
Nitrogen (%)	2.88	2.82	2.92	2.88
Calcium (%)	0.49	0.44	0.47	0.51
Phosphorus (%)	0.48	0.46	0.49	0.47
Magnesium (%)	0.31	0.29	0.31	0.33
Potassium (ppm)	2.09	2.48	2.44	2.02
Copper (ppm)	7.65	7.28	7.28	7.94
Zinc (ppm)	30.6	30.9	33.5	36.5
October 18, 1994				
Nitrogen (%)	1.15	1.82	1.69	1.72
Calcium (%)	0.45	0.46	0.43	0.47
Phosphorus (%)	0.48	0.46	0.44	0.45
Magnesium (%)	0.31	0.33	0.35	0.33
Potassium (%)	1.87	1.96	2.10	1.73
Copper (ppm)	7.12	8.84	7.42	7.08
Zinc (ppm)	27.1	29.7	27.5	26.6
January 25, 1995				
Nitrogen (%)	1.35	1.39	1.50	1.22
Calcium (%)	0.51	0.45	0.45	0.50
Phosphorus (%)	0.34	0.28	0.30	0.32
Magnesium (%)	0.23	0.20	0.20	0.21
Potassium (%)	0.79	0.75	0.77	0.67
Copper (ppm)	6.08	5.83	6.72	6.91
Zinc (ppm)	23.6	22.4	23.8	27.8
April 20, 1995				
Nitrogen (%)	2.59[a]	3.47[b]	2.93[a,b]	2.84[a,b]
Calcium (%)	0.39	0.36	0.37	0.42
Phosphorus (%)	0.34	0.37	0.34	0.34
Magnesium (%)	0.30	0.29	0.29	0.29
Potassium (%)	2.44	2.96	2.76	2.48
Copper (ppm)	4.34	10.12	7.68	8.11
Zinc (ppm)	45.6	44.3	49.4	47.7

[a,b] Values with unlike superscripts differ ($p < 0.05$).

animals fed the waste should be no greater than if the animals were fed conventional diets. The higher value of wastes as feedstuffs than fertilizer justifies transportation of the wastes outside of areas where the wastes are produced. Potentially, feeding wastes such as poultry litter to animals on pasture can be used as a method to apply the plant nutrients to the soil. The experiment in progress will supply definitive data on this aspect.

ACKNOWLEDGMENTS

Appreciation is expressed to the John Lee Pratt Animal Nutrition Program for supporting some of the research; to Nancy B. Frank, Rachael K. Crabtree, and Lyle L. Harlow for technical assistance; and to Catherine S. Johnson for typing the manuscript.

Table 6 Performance of Steers on Fescue Pasture or Fed Fescue Hay

	Treatment			
		Broiler litter		
Item	None (kg)	Fed (kg)	Soil application (kg)	Inorganic fertilizer (kg)
1994				
Initial wt, 5/3	263	264	264	256
Wt, 10/18	439	440	432	419
Daily gain, 168 d	1.05	1.05	1.00	0.97
1994–1995				
Initial wt, 12/1/94	226	229	228	232
Wt, 4/20/95	299	309	291	300
Daily gain, 140 d	0.52	0.57	0.45	0.48

Table 7 Serum Mineral Levels of Steers Grazed on Pastures

	Treatment			
		Broiler litter		
Item	None	Fed	Soil application	Inorganic fertilizer
May 3, 1994				
Calcium (mg/dl)	9.33	9.31	9.18	9.57
Phosphorus (mg/dl)	6.35	6.91	6.96	6.42
Magnesium (mg/dl)	2.59	2.54	2.50	2.57
Potassium (mg/dl)	32.5	33.0	30.9	33.6
Copper (μg/dl)	64.4	63.4	71.3	70.6
Zinc (μg/dl)	79.6	82.0	75.3	82.4
October 18, 1994				
Calcium (mg/dl)	9.49	9.40	9.58	9.62
Phosphorus (mg/dl)	7.07	7.85	7.51	7.25
Magnesium (mg/dl)	2.51	2.56	2.44	2.51
Potassium (mg/dl)	27.1	26.1	28.4	28.5
Copper (μg/dl)	64.3	66.9	67.7	64.8
Zinc (μg/dl)	104.9	108.6	107.4	105.0
January 25, 1995				
Calcium (mg/dl)	9.90	9.49	9.43	9.76
Phosphorus (mg/dl)	7.03[a]	8.96[b]	7.42[a]	7.84[a]
Magnesium (mg/dl)	2.51	2.63	2.54	2.39
Potassium (mg/dl)	30.5	33.8	32.6	30.5
Copper (μg/dl)	70.7	72.8	61.5	74.9
Zinc (μg/dl)	91.83	99.58	91.54	98.20
April 20, 1995				
Calcium (mg/dl)	9.59[a]	8.77[b]	9.48[a]	9.29[a]
Phosphorus (mg/dl)	8.83[a]	10.32[b]	9.05[a]	8.96[a]
Magnesium (mg/dl)	2.30	2.47	2.25	2.40
Potassium (mg/dl)	26.0	27.7	25.9	25.6
Copper (μg/dl)	68.6	66.2	56.7	68.7
Zinc (μg/dl)	87.1	84.9	85.8	85.5

[a,b] Values with unlike superscripts differ ($p < 0.05$).

REFERENCES

AAFCO, *Recycled Animal Waste Products,* Association of American Feed Control Officials, College Station, TX, 1990, 188.

Abdelmawla, S. M., A Study on the performance of weaned male buffalo calves fed rations containing poultry wastes, Ph.D. Dissertation, Virginia Polytechnic Institute and State University, Blacksburg, 1990.

Abdelmawla, S. M., Fontenot, J. P., and El-Ashry, M. A., Composited, deepstacked and ensiled broiler litter in sheep diets: chemical composition and nutritive value study, VPI & SU, *Anim. Sci. Res. Rep.,* 7, 127. 1988.

Aines, G. E., Lamm, W. D., Webb, K. E., Jr., and Fontenot, J. P., Effects of diet dry matter level and sodium hydroxide treatment on the ensiling characteristics and utilization by lambs of cattle waste-rye straw mixtures, *J. Anim. Sci.,* 54, 504, 1982.

Anonymous, *Municipal Refuse Disposal,* 3rd ed., Institute of Solid Wastes, American Public Works Association, Chicago, IL, 1970.

Anthony, W. B., Utilization of animal waste as feed for ruminants, *Proceedings of the National Symposium on Animal Waste Management,* ASAE, St. Joseph, MI, 1966, 109.

Anthony, W. B., Animal waste value nutrient recovery and utilization, *J. Anim. Sci.,* 32, 799, 1971.

Appleyard, W. T. and Mollison, A., Suspected bovine botulism associated with broiler litter waste, *Vet. Rec.,* 116, 522, 1985.

Arave, C. W., Dobson, D. C., Walters, J. W., Gilbert, B. J., Jr., and Arambel, M. J., Effect of added processed poultry waste on dairy heifers' preference for concentrates, *J. Dairy Sci.,* 71, 3021, 1990.

Bargahit, G. A. and Fontenot, J. P., Unpublished data, 1990.

Berger, J. C. A., Fontenot, J. P., Kornegay, E. T., and Webb, K. E., Jr., Feeding swine waste. I. Fermentation characteristics of swine waste ensiled with ground hay or ground corn grain, *J. Anim. Sci.,* 52, 1388, 1981a.

Berger, J. C. A., Fontenot, J. P., Kornegay, E. T., and Webb, K. E., Jr., Feeding swine waste. II. Nitrogen utilization, digestibility and palatability by sheep of ensiled swine waste and orchardgrass hay or corn grain, *J. Anim. Sci.,* 52, 1404, 1981b.

Bhattacharya, A. N. and Taylor, J. C., Recycling animal waste as a feedstuff: a review, *J. Anim. Sci.,* 41, 438, 1975.

Bienvenu, J. and Morin, M., Poultry litter associated botulism (type C) in cattle, *Can. Vet. J.,* 31, 711, 1990.

Brosh, A., Holzer, Z., Aharoni, Y., and Levy, D., Intake, rumen volume, retention time and digestibility of diets based on poultry litter and wheat straw in beef cows before and after calving, *J. Agric. Sci. Camb.,* 121, 103, 1993.

Bull, L. S. and Reid, J. T., Nutritive value of chicken manure for cattle, *Proceedings of the International Symposium on Livestock Wastes,* ASAE, St. Joseph, MI, 1971, 297.

Calvert, C. C., Use of animal excreta for microbial and insect protein synthesis, *J. Anim. Sci.,* 48, 178, 1979.

CAST, Feeding animal waste, *Counc. Agric. Sci. Technol. Rep.,* 75, 48, 1978.

Caswell, L. F., Fontenot, J. P., and Webb, K. E., Jr., Fermentation and utilization of broiler litter ensiled at different moisture levels, *J. Anim. Sci.,* 46, 547, 1978.

Chaudhry, S. M., Processing and nutritional value of poultry litter and slaughterhouse by-product, Ph.D. Dissertation, Virginia Polytechnic Institute and State University, Blacksburg, 1990.

Clegg, F. G., Jones, T. O., Smart, J. L., and McMurty, M. J., Bovine botulism associated with broiler litter waste, *Vet. Rec.,* 117, 22, 1985.

Cornman, A. W., Lamm, W. D., Webb, K. E., Jr., and Fontenot, J. P., Ensiling cattle waste with rye straw as a diet supplement for ruminants, *J. Anim. Sci.,* 52, 1233, 1981.

Cross, D. L., Skelley, G. C., Thompson, C. S., and Jenny, B. F., Efficacy of broiler litter silage for beef steers, *J. Anim. Sci.,* 47, 544, 1978.

Egyed, M. N., Schlosberg, C., Klopfer, U., Nokel, T. A., and Mayer, E., Mass outbreaks of botulism in ruminants associated with ingestion of feed containing poultry waste. I. Clinical and laboratory investigations, *Refuah Vet.,* 35, 93, 1978.

Flegal, C. J. and Zindel, H. C., Dehydrated poultry waste (DPW) as a feedstuff in poultry rations, *Proceedings of the International Symposium on Livestock Wastes,* ASAE, St. Joseph, MI, 1971, 305.

FDA, Recycled animal waste, *Fed. Reg.,* 45, 86272, 1980.

Fontenot, J. P., Recycling animal wastes by feeding to enhance environmental quality, *Prof. Anim. Sci.,* 7(4), 1, 1991.

Fontenot, J. P., Bhattacharya, A. N., Drake, C. L., and McClure, W. H., Value of broiler litter as feed for ruminants, *Proceedings of the National Symposium on Management of Farm Animal Waste,* ASAE, St. Joseph, MI, 1966, 105.

Fontenot, J. P. and Ross, I. J., Animal waste utilization, in *Livestock Waste: A Renewable Resource,* Proceedings of the 4th International Symposium on Livestock Wastes, ASAE, St. Joseph, MI, 1980, 4.

Fontenot, J. P., Smith, L. W., and Sutton, A. L., Alternative utilization of animal wastes, *J. Anim. Sci.,* 57 (Suppl. 2), 221, 1983.

Fontenot, J. P. and Webb, K. E., Jr., Health aspects of recycling animal wastes by feeding, *J. Anim. Sci.,* 40, 1267, 1975.

Fontenot, J. P., Webb, K. E., Jr., Harmon, B. W., Tucker, R. E., and Moore, W. E. C., Studies of processing, nutritional value and palatability of broiler litter for ruminants, *Proceedings of the International Symposium on Livestock Wastes,* ASAE, St. Joseph, MI, 1971, 301.

Freeman, O. and Bennett, I. I., Jr., Control of Agriculture-Related Pollution. A Report to the President Submitted by the Secretary of Agriculture and the Director of the Office of Science and Technology, Washington, D.C., 1969.

Gerken, H. J., Feeding Broiler Litter to Beef Cattle and Sheep, Coop. Ext. Serv. Publ. 754, Virginia Polytechnic Institute and State University, Blacksburg, 1977.

Goering, H. K. and Smith, L. W., Composition of the corn plant ensiled with excreta or nitrogen supplements and its effects on growing wethers, *J. Anim. Sci.,* 44, 452, 1977.

Harmon, B. W., Fontenot, J. P., and Webb, K. E., Jr., Ensiled broiler litter and corn forage. I. Fermentation characteristics, *J. Anim. Sci.,* 40, 144, 1975a.

Harmon, B. W., Fontenot, J. P., and Webb, K. E., Jr., Ensiled broiler litter and corn forage. II. Digestibility, nitrogen utilization and palatability by sheep, *J. Anim. Sci.,* 40, 156, 1975b.

Hogg, R. A., White, V. J., and Smith, G. R., Suspected botulism in cattle associated with poultry litter, *Vet. Rec.,* 126, 476, 1990.

Hovatter, M. D., Sheehan, W., Dana, G. R., Fontenot, J. P., Webb, K. E., Jr., and Lamm, W. D., Different levels of ensiled and deep stacked broiler litter for growing cattle, *VPI SU Res. Div. Rep.,* 175, 77, 1979.

Kirk, J. K., Statements of general policy or interpretation, *Fed. Reg.,* 22, September 2, 1967.

Koriath, von H., Moglichkeiten zur verwertung vow derexkementen. Leipzig. Universitat. Sektion Tierproduction und Veterinarmedizin. Wissenschaftlclhe Tagung, Moglichkeiten und Probleme der Nuzung von Abprodukten in der Tiererahrungein Beitrag zur Umweltgesaltung in der sozilaisischen Landwirschaft, Leipzig, Karl-Marx Universitat, Teil, 2–218, 1975.

Kornegay, E. T., Holland, M. R., Webb, K. E., Jr., Bovard, K. P., and Hedges, J. D., Nutrient characterization of swine fecal waste and utilization of these nutrients by swine, *J. Anim. Sci.,* 44, 608, 1977.

Kwak, W., Solubility, degradability and utilization by ruminants of broiler litter processed by ensiling, deepstacking and composting, Ph.D. Dissertation, Virginia Polytechnic Institute and State University, Blacksburg, 1990.

Lovett, J., Toxigenic fungi from poultry feed and litter, *Poultry Sci.,* 51, 309, 1972.

Lucas, D. M., Fontenot, J. P., and Webb, K. E., Jr., Digestibility of untreated and sodium hydroxide treated steer fecal waste, *VPI SU Res. Div. Rep.,* 163, 15, 1975.

Mathers, A. C. and Stewart, B. A., Crop production and soil analyses as affected by applications of cattle feedlot waste, in Livestock waste management and pollution abatement, *Proceedings of the International Symposium on Livestock Wastes,* ASAE, St. Joseph, MI, 1971, 229.

McCaskey, T. A. and Anthony, W. B., Human and animal health aspects of feeding livestock excreta, *J. Anim. Sci.,* 48, 163, 1979.

McClure, W. H. and Fontenot, J. P., Feeding broiler litter deep stacked or ensiled with corn forage to finishing cattle, in *Agricultural Waste Utilization and Management,* ASAE, St. Joseph, MI, 1985, 154.

McCollum, E. V., *The Newer Knowledge of Nutrition,* 2nd ed., Macmillan, New York, 1922, 77.

McIllroy, S. G., McCracken, R. M., and Huey, J. A., Botulism in cattle grazing pasture dressed with poultry litter, *Irish Vet. J.,* 41, 245, 1987.

McLoughlin, M. F., McIlroy, S. G., and Neill, S. D., A major outbreak of botulism in cattle being fed ensiled poultry litter, *Vet. Rec.,* 122, 579, 1988.

Miron, J., Solomon, R., Yosef, E., and Ben-Ghedalia, D., Carbohydrate digestibility and nitrogen metabolism in sheep fed untreated or sulphur dioxide-treated wheat straw and poultry litter, *J. Agr. Sci. (Camb.),* 114, 115, 1990.

Moriba, J. N., Fontenot, J. P., and Webb, K. E., Jr., Digestibility and palatability of caged layer waste and corn stover ensiled alone and with molasses, *VPI SU Anim. Sci. Res. Rep.,* 2, 160, 1982.

Neill, S. D., McLoughlin, M. F., and McIllroy, S. G., Type C botulism in cattle being fed ensiled poultry litter, *Vet. Rec.,* 124, 558, 1989.

Olson, K. J., Fontenot, J. P., and Failla, M. L., Influence of molybdenum and sulfate supplementation and withdrawal of diet containing high copper broiler litter on tissue copper levels in ewes, *J. Anim. Sci.,* 59, 210, 1984.

Pugh, D., Rankins, G. D. L., Powe, T., and D'Andrea, G., Feeding broiler litter to beef cattle, *Vet. Med.,* July, 1994, 661.

Rankins, D. L., Jr., Eason, J. T., McCaskey, T. A., Stephenson, A. H., and Floyd, J. G., Jr., Nutritional and toxicological evaluation of three deep-stacking methods for the processing of broiler litter as a foodstuff for beef cattle, *Anim. Prod.,* 56, 321, 1993.

Richter, M. F. and Kalmbacher, R. S., Nutrient metabolism and quality of corn and sorghum silages made with caged layer manure, *Proc. Soil Crop Sci. Soc. Fl.,* 39, 125, 1980.

Rinehart, K. E., Snetsinger, D. C., Ragland, W. W., and Zimmerman, R. A., Feeding value of dehydrated poultry waste, *Poultry Sci.,* 52, 2078 (Abstr.), 1973.

Ruffin, B. G. and McCaskey, T. A., Broiler litter can serve as feed ingredient for beef cattle, *Feedstuffs,* 62(15), 13, 1990.

Samuels, W. A., Fontenot, J. P., Lamm, W. D., and Webb, K. E., Jr., Fermentation characteristics of caged layer waste ensiled with sugarcane bagasse, *VPI SU Res. Div. Rep.,* 156, 189, 1980.

Saylor, W. W. and Long, T. A., Laboratory evaluation of ensiled poultry waste, *J. Anim. Sci.,* 39, 139 (Abstr.), 1974.

Shuyler, L. R., Impact of regulations and laws promulgated in response to the Chesapeake Bay program, *Proceedings of the 1994 National Poultry Waste Management Symposium,* 1994, 25.

Shlosberg, A., Harmelin, A., Perl, S., Pano, G., Davidson, M., Orgad, U., Kali, U., Bor, A., Van Ham, M., Hoida, G., Yakobson, B., Avidar, Y., Israeli, B. A., and Bogin, E., Cardiomyopathy in cattle induced by residues of the coccidiostat maduramicin in poultry litter given as a feedstuff. *Vet. Res. Commun.,* 16, 45, 1992.

Silva, L. A., Van Horn, H. H., Olaloku, E. A., Wilcox, C. J., and Harris, B., Jr., Complete rations for dairy cattle. VII. Dried poultry waste for lactating cows, *J. Dairy Sci.,* 59, 2071, 1976.

Smart, J. L., Jones, T. O., Clegg, F. G., and McMurtry, M. J., Poultry waste associated type C botulism in cattle, *Epidemiol. Inf.,* 98, 73, 1987.

Smith, L. W., Nutritive evaluation of animal manure, in *Symposium: Processing Agricultural and Municipal Wastes,* Inglett, G. E., Ed., AVI Publishing Co., Westport, CT, 1973, 55.

Smith, L. W., Calvert, C. C., Frobish, L. T., Dinius, D. A., and Miller, R. W., Animal Waste Reuse-Nutritive Value and Potential Problems From Feed Additives, USDA-ARS 44-224, 1971.

Smith, R. J., Hein, M. E., and Greines, T. H., Experimental methane production from animal excreta in pilot-scale and farm-size units, *J. Anim. Sci.,* 48, 202, 1979.

Smith, L. W. and Wheeler, W. E., Nutritional and economic value of animal excreta, *J. Anim. Sci.,* 48, 144, 1979.

Suttle, N. F., Munro, C. S., and Field, A. C., The accumulation of copper in the liver of lambs on diets containing dried poultry waste, *Anim. Prod.,* 16, 39, 1978.

Tagari, H., Personal communication, 1978.

Thomas, J. W., Yu, Y., Tinnimitt, P., and Zindel, H. C., Dehydrated poultry waste as a feed for milking cows and growing sheep, *J. Dairy Sci.,* 55, 1261, 1972.

Trueman, K. F., Bock, R. E., Thomas, R. J., Taylor, J. D., Green, P. A., Roeger, H. M., and Ketterer, P. J., Suspected botulism in three intensively managed Australian cattle herds, *Vet. Rec.,* 130, 398, 1992.

USDA, *Agricultural Statistics,* U.S. Government Printing Office, Washington, DC, 1993.

Van Dyne, D. C. and Gilbertson, G. B., Estimating U.S. Livestock and Poultry Manure and Nutrient Production, U.S. Department of Agriculture, ESCS-12, 1978.

Webb, K. E., Jr. and Fontenot, J. P., Medicinal drug residues in broiler litter and tissue from cattle fed litter, *J. Anim. Sci.,* 41, 1212, 1975.

Webb, K. E., Jr., Fontenot, J. P., and McClure, W. H., Performance and liver copper levels of beef cows fed broiler litter, *VPI SU Res. Div. Rep.,* 156, 130, 1980.

Westing, T. W., Fontenot, J. P., McClure, W. H., Kelly, R. F., and Webb, K. E., Jr., Mineral element profiles of animal wastes and edible tissues from cattle fed animal wastes, *Proceedings of the 4th International Symposium on Livestock Wastes,* ASAE, St. Joseph, MI, 1980, 81.

Zimet, D. J., Ouart, M. D., and Prichard, D. L., Research note: use of computer simulation to determine the value of broiler litter as a cattle feed ingredient, *Poultry Sci.,* 67, 1352, 1988.

CHAPTER 15

Livestock Methane: Current Emissions and Mitigation Potential

Donald E. Johnson, Gerald M. Ward, and Jon J. Ramsey

INTRODUCTION

Methane emissions by livestock facilitate enteric microbial function, but result in a loss of feed energy to the animal. Measurements of emissions, particularly from ruminants, over the last century by animal nutritionists have sought to characterize and lead to methods to minimize methane emissions. Data describing these emissions have taken on a new importance since the recognition of the significance of methane as a greenhouse gas (Figure 1) and its role in atmospheric chemistry, including likely effects on ozone concentrations (Wuebbles and Tamaresis, 1993).

VARIATION IN METHANE LOSS BY RUMINANTS

Cattle and sheep eructate a relatively constant fraction of the energy of their diet throughout life, except during the early suckling phase before rumen development. There are a few notable exceptions, but for the majority of the circumstances of growth, pregnancy, lactation, or maintenance of the worlds' sheep and cattle, the loss is 6 ± 0.5% of gross diet energy consumed. Thus the variation in daily loss per animal is related mostly to the total amount of diet consumed to serve the maintenance and production functions of the animal. This conclusion was reached in spite of the wide range of percentage methane losses found in the literature.

Reported losses from cattle with fully developed gastrointestinal tracts range from 2 (Carmean, 1991) to nearly 12% (Whitelaw et al., 1984) of the gross energy of the diet. Both extremes occurred with cattle fed very high concentrate diets, the low loss from a corn-based diet fed ad libitum, and the high percentage loss from barley fed in amounts restricted to approximately 50% of ad libitum intake. Four of the principle interacting factors causing variation are digestibility of dietary carbohydrate, type of carbohydrate, amount of diet consumed per day (percent of ad libitum), and the propensity of the sometimes vacillating rumen microbial species to interact in the conversion of carbohydrate to acetic or propionic acid. Fermentation dynamics dictate that for each mole of carbohydrate

1-56670-199-6/96/$0.00+$.50

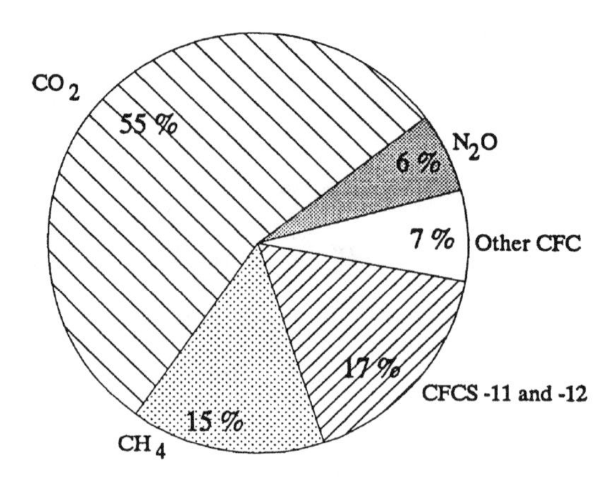

Figure 1 Contribution of each greenhouse gas to global warming in the 1980s (IPCC, 1990).

converted to acetic acid, 4 mol of hydrogen will be available for methanogens to reduce carbon dioxide to 1 mol of methane. On the other hand, conversion of carbohydrate to propionic acid requires 2 mols of hydrogen, a recovery of $1/2$ mol of methane. Interactions of many of these factors are encompassed in regression relationships predicting methane from substrate content or use. For instance, considerable variation in beef cattle emissions can be accounted for in the expression:

$$CH_4, \text{ Mcal/d} = 0.54 + 0.39 \text{ SOL} + 0.08 \text{ STAR} + 0.68 \text{ CW}, \; r^2 = 0.82 \qquad (1)$$

where SOL = soluble, STAR = starch, and CW = cell wall, each as g carbohydrate fermented in the rumen per animal per day (Torrent et al., 1994). The low starch coefficient, for example, represents microbial propensity toward propionate production from starch in contrast to either soluble sugars or cell wall.

Most of the high percentage methane losses can be ascribed to experimentally limiting diet intake, which seldom occurs in the livestock industry. The interaction of three of the factors mentioned previously when diets are fed ad libitum commonly results in a 6% loss. Low-quality diets are not very digestible, but are consumed in small quantities, pass slowly through the gut, and the major digestible component is the cell wall. These conditions tend to support microbial populations producing large proportions of acetic acid. High-quality diets, on the other hand, are readily digested, consumed at high levels, passed rapidly through the gut, and have more starch and solubles, which favor propionic acid producing microbial populations. Acetic acid stoichiometrically leaves more hydrogen, and propionic acid less, to be converted to methane. Exceptions to the typical 6% loss at ad libitum intakes do occur and some will be described. Further detail can be found in another review (Johnson et al., 1993), and undoubtedly, some are yet to be discovered.

Physical and Chemical Processing of Forages

Fine grinding and/or pelleting of forage diets can reduce methane losses by 20 to 40% (Blaxter, 1989). The decrease undoubtedly relates to the faster rate of passage and lowered cell wall carbohydrate digestibility that occurs when these small particle feedstuffs are fed ad libitum. Faster passage rates per se may reduce methane, conceivably through changes in microbial species or site of digestion. Okine et al. (1989) found 29% reductions in methane when weights were placed in the rumen, which accelerated passage rate, but did not change digestibility. Chemical treatments such as sodium hydroxide (Robb et al., 1979) or ammoniation (Birkelo et al., 1986) of straws have been shown to increase methane loss in proportion to increased digestibility. Recent research with minimal basal supplementation (Moss et al., 1994), however, found a constant 6% of gross energy loss from treated and untreated straws.

High Grain Diets

The very high, 90% or greater, concentrate levels typical of the United States beef feedlot industry result in lower percentage methane emissions. Relatively few measurements are available where animals consumed diets at levels found in feedlots. Nevertheless, experiments with corn-based diets, including the classic experiments of Blaxter and Wainman (1964) and more recently Aboomar (1989) and Carmean (1991), reported individual animal losses as low as 1.8%. When extrapolated to feedlot intake levels, the results are interpreted to suggest approximately 3% mean methane losses. There are several factors that contribute to methane suppression when animals are fed these diets. Starch substrate, low pH, rapid passage, partial small intestine digestion of carbohydrate, high propionate production, and depressed protozoal populations likely contribute to these results. Some feed grains may respond differently. Measurements available with high levels of barley (Whitelaw et al., 1984) indicate higher methane losses than when feeding comparable levels of corn cited previously.

Ionophores

Ionophores effect a 5 to 6% improvement in feed efficiency (Goodrich et al., 1984) when added to ruminant diets. One commonly held mode of action is to shift the fermentation pattern to result in a higher propionate production and reduced methane loss. Many *in vitro* trials (e.g., Fuller and Johnson, 1981) and several *in vivo* chamber calorimetry trials (e.g., Thornton and Owens, 1981; Wedegaertner and Johnson, 1983) have shown 20 to 25% reductions in methane losses resulting from ionophore additions. However, studies over time of dosing of ionophores at common industry use levels to cattle fed grain (Aboomar, 1989) or forage diets (Saa et al., 1993) show an adaptation response with methane emissions returning to normal in about 2 weeks. This likely accounts for the minor, or lack of, methane suppression reported by Garrett (1982).

Lipid Additions

Several factors likely contribute to lipid-depressing effects on methane. Small reductions in methane can result from uptake of hydrogen by unsaturated fatty acid additions (Czerkawski, 1972). Lipids are largely nonfermentable; thus they decrease percentage methane by a dilution effect. Depression of protozoa and cellulolytics can also be caused by fat addition, which in turn is methane suppressive. Van der Honing et al. (1981) found 10 to 15% decreases in methane loss from dairy cows fed 5% tallow or soybean oil, to cite one of many examples.

Other Competing Hydrogen Sinks

Inorganic sulfur, nitrates, and oxygen swallowed or diffused from the blood may serve to dispose of excess hydrogen in the rumen preferential to the reduction of carbon dioxide to methane. Major reductions in methane resulted from experimental nitrate, sulfide, or cysteine additions (Takahashi and Young, 1995). The cysteine depression, however, does not appear to be through its effect on oxidation reduction potential (Takahashi et al., 1994). Also, Demeyer et al. (1989) found no methane depression from sodium sulfate additions to *in vitro* incubations with washed cell suspensions of rumen microorganisms. Toxic effects of nitrate or sulfide make major additions to ruminant diets impractical as well.

BACKGROUND: SOURCES, WARMING POTENTIAL, AND STABILIZATION

Methane levels in the atmosphere have more than doubled to approximately 1800 ppb over the last 200 years (IPCC, 1990). Although the rate of increase slowed (Khalil et al., 1993) and for unknown reasons did not increase this last year in the Northern Hemisphere (Crutzen, 1994), expectations are for continued increases in the absence of mitigation strategies. The increased concentrations, in addition to intense infrared energy absorption by increased methane, some 20 to 30 times that of carbon dioxide per mole (Rodhe, 1990), causes concerns about its global warming effects. Trends prior to last year suggest that methane increases caused 15% of the radiative forcing in the 1980s (IPCC, 1990) and will cause 18% of warming effects over the next 50+ years (EPA, 1994a).

The atmospheric increases are fueled by the entry of 500 to 550 million metric tons (teragram, Tg) into the atmosphere annually (Khalil and Schearer, 1993). Relative to other greenhouse gases, methane half-life is short, 10 to 12 years, requiring a decreased entry rate of only about 10% to stabilize present concentrations. About one third comes from natural sources (i.e., swamps); another third is oil, gas, and refuse related; and the balance is from agricultural sources (Table 1). Recent estimates rank methane produced in the gastrointestinal tracts of animals as the largest agricultural source. The other agricultural sources of methane are rice biomass burning and manure disposal.

Table 1 Global Sources and Sinks of Atmospheric Methane[a]

Sources	Tg/year	(range)[b]	Sinks	Tg/year
Natural sources				
Wetlands	110	(100–200)	Hydroxyl (OH)	450
Oceans, hydrates	35	(10–80)	Soils	30
Termites	20	(10–100)	Cl and O	40
Burning and other	5	(2–15)		
Agricultural				
Rice	65	(25–150)		
Livestock	80	(50–110)		
Manure	10	(10–35)		
Biomass burning	30	(10–50)		
Energy and waste				
Gas and oil industries	70	(25–85)		
Coal mines	40	(20–43)		
Charcoal/wood	10	(5–30)		
Wastewater	25	(15–35)		
Landfills	40	(15–70)		
Total, all	540			510

[a] Compiled from Cicerone and Oremland, 1988; Crutzen et al., 1986; NATO Workshop Proceedings, 1993; and EPA, 1994a.

[b] Tg = Teragram = 10^{12}g = million metric ton, range = range of estimates.

Livestock Methane Emission Estimates

Estimates made in the early 1980s of global methane emissions from domestic livestock ranged from 90 Tg* (Sheppard et al., 1982) to 120 Tg*/year (Khalil and Rasmussen, 1983). A thorough analysis by Crutzen et al. (1986) found 74 Tg/year. His analysis extrapolated to livestock census changes indicated a 4.5-fold increase over the last century. Recently published estimates (Gibbs and Johnson, 1994) that used 1990 census data conducted by a method independent from Crutzens' for the major contributors, cattle and buffalo, largely confirmed their estimates (Table 2). Little change has been expected since 1990 because of static cattle numbers. The world's 1.3 billion cattle with their large body size, feed intake, and extensive fermentation produce 73% of the livestock methane. Water buffalo contribute another 10%, sheep and goats 12%, camels and swine about 1% each, and horses and donkeys about 2%.

There have been many questions about the contribution of wild ruminants. It appears that the vast herds of bison on the American continent in previous centuries, for instance, had largely disappeared prior to the parallel rise in domestic livestock and atmospheric methane during the last century. Estimates of present-day wild, large animals (Crutzen et al., 1986; Lerner and Mathews, 1988) suggest 4 Tg/year, only about 5% of that from domestic livestock. They also estimated human methane emissions at 0.1 Tg/year. Biomass and methane emission rates of insects have been recently compiled (Hackstein and Stumm, 1994); however, unknown reoxidation at or below the soil surface where many reside prohibit accurate estimates of emissions to the atmosphere of this potentially large source (>100 Tg).

* Tg = teragram = 10^{12} g = million metric ton.

Table 2 Global Estimates of Gastrointestinal Methane Emissions by Livestock During Recent Years, 1983 and 1990[a]

	Crutzen 1983 estimate		Gibbs and Johnson 1990 estimate		
Species	No. head × 10^3	Methane (Tg/year)	No. head × 10^3	Methane (Tg/year)	%
Cattle	1225	54.3	1279	58.1	73.4
Buffalo	124	6.2	141	7.7	9.7
Sheep	1137	6.9	1191	7.0	8.8
Goats	476	2.4	557	2.8	3.5
Camels	17	1.0	19	0.9	1.1
Pigs	774	0.9	857	1.0	1.3
Equine	117	1.7	119	1.7	2.1
Total		73.4		79.2	

[a] Data from Crutzen et al., 1986 and Gibbs and Johnson, 1994.

In the United States, approximately 6.1 Tg of methane are produced from fermentation in the gastrointestinal tracts of livestock. Cattle producing beef contribute 4.1 Tg, while dairy cattle contribute 1.8 Tg (Table 3). Mature cows account for 56 and 71% in the beef and dairy herds. The feedlot phase contributes less than 7% of cattle methane due to the short time and low fractional loss from high-grain diets. The balance is produced by growing replacement and stocker calves, yearlings, and breeding bulls.

Methane Emissions from Livestock Manure Disposal

Because the animal digestive process is not complete and because of endogenous and microbial residues in fecal material, there is always considerable nonlignin organic matter in manure that can be converted to methane if decomposed under warm, moist, and anaerobic conditions such as an anaerobic lagoon, a landfill, or a methane generator specifically designed to capture gas as a fuel. The types of

Table 3 Livestock Gastrointestinal Methane Emissions in the United States in 1994[a]

Species/class	Average no. head × 10^6	Methane % of GE	Methane Tg/year
Beef cattle			
Cows	36.3	6.2	2.46
Feedlot	10.5	3.5	.36
Other	43.5	6.5	1.55
Dairy cattle			
Cows	9.6	5.8	1.17
Replacements	8.6	6.5	0.46
Sheep and goats	11.9	6.0	0.09
Swine	58.1	0.6	0.09
Horses	5.2	2.5	.09
			Total: 6.27

[a] Beef and dairy extrapolated from Johnson et al., 1993; other species extrapolated from Crutzen et al.,1986.

Table 4 Estimates of Methane Emissions from Livestock Manure Disposed in the United States and Globally[a]

	Tg/year	
Species	**United States**	**Global**
Dairy	0.71	2.89
Beef	0.19	3.16
Swine	1.11	5.29
Sheep and goat	—	0.71
Poultry	0.23	1.28
Other	0.24	0.51
Total	2.48	13.84

[a] Data from the EPA, 1994a, 1994b.

manure disposal systems employed, the fraction of potential methane emissions expected, and their use by each major livestock species in each state in the United States and country of the world has been partially characterized (Safley et al., 1992). This information can be used to estimate methane emissions from manure, although some modifications are suggested to reflect the effect of drying or aeration and result in <1% of potential emissions for range livestock manure deposition (Lodman, et al., 1993; Williams, 1993). Making adjustments for these factors results in an estimated 14 Tg of methane entering the atmosphere annually from manure disposal (Table 4; EPA, 1994b). Also, the amount of reoxidation at the surface of anaerobic systems could be considerable (Cicerone and Oremland, 1988) and would revise the estimate downward to 10 Tg. Nonruminant animals are significant contributors of methane from manure disposal. Swine are the single most important species in the United States, producing 49% of all manure methane, in comparison to 32 and 8% for dairy and beef cattle (EPA, 1994b).

POTENTIAL FOR MITIGATION OF LIVESTOCK METHANE EMISSIONS

Many specific diet manipulations, as discussed previously, can and do moderately reduce methane emissions from livestock. The generally recommended strategy, however, is one that has already made large differences in reducing methane per unit of product in many countries. It is an indirect effect through increased productivity per animal and the related improved feed efficiency. Increased meat or milk per animal per day means a dilution of the methane produced and feed used to meet the maintenance requirement of the animal, an approximately fixed amount regardless of the amount of product. Higher productivity thus contributes strongly toward lower feed input per product. Also, methane emissions are mainly a constant proportion of feed consumed, and any decrease in feed required per unit of meat or milk production will decrease the methane per unit of product. In product-saturated markets such as dairy and beef in developed countries, feed and methane savings are nearly wholly realized from increased

Table 5 Projected Methane Emissions from Cattle in the Ukraine Producing Constant Total Milk and Meat, but at Varying Rates[a]

	Productivity level		
Item	**Baseline**	**Double**	**Triple**
Milk (kg/cow/year)	2200	4300	6300
Gain (g/d)	300	600	900
Feed use (T $\times$ 10^9)	66	47	44
Feed for maintenance (%)	73	62	60
Methane (Tg/year)	1.32	0.94	0.88
Milk/methane	35	56	69
Gain/methane	2.9	5.1	6.5

[a] Data from Martinez et al., 1995. Milk cows decreased and beef cows increased to hold milk and meat/year constant.

efficiencies. In other markets, the savings will be partial, as some increased animal product output is likely to follow rising income. Nevertheless, these strategies still make sense because they should at least blunt methane increases and are likely to be cost effective, thus increasing the likelihood they will be adopted.

Enhancing ruminant productivity generally requires simultaneous improvements in nutrition, genetics, and management, no matter what the base scenario is. Single factor changes such as protein or mineral deficiencies can make important strides, but these soon run into "the next weakest link" of diet energy, genetic potential, ketosis prevention, etc. Potential for progress is large when starting with very low productivity. Methane emissions per kilogram of milk decreased from 138 to 34 g in a simulation of a cow producing 2 vs. 8 kg of milk per day (Ward et al., 1993). This response required improvement in diet quality, i.e., digestibility increasing by about 25%, although ancillary crop production changes (discussed later) may not be insignificant. A study of recent milk and meat production in Ukraine (Martinez et al., 1995) found an average dairy cow of under 2100 kg/year and liveweight gains of 300 g/d. Under these circumstances, a replacement heifer must be fed for over 3 years before beginning to produce milk, with a comparable long time of feeding bulls before slaughter. The fraction of feed used for maintenance of the collective herd is about 73%. Simulations of double and triple productivity (Table 5) while holding milk and meat production constant indicate reductions of feed and methane of 29 and 34%. Much of the decreased feed requirements and methane emissions arise from decreased numbers of slow-growing and developing animals. Total stable herd numbers declined by 49%, but they were fed better, grew faster, and produced more milk per animal. Feed use for maintenance of the overall herd fell to about 60%. For individual producing cows, milk per kilogram of methane increased from 35 to 69 g, when comparing the baseline to the triple productivity. The genetic milk production potential is higher than the present productivity since the country average was 30% higher in 1990. With nearly all cows being successfully bred by artificial insemination (AI), they have the technical capability to change genotypes to reach the higher goals. Simultaneous changes in feed quality, protein and mineral supplementation, and management are perhaps more problematic, but are certainly attainable as repeatedly demonstrated by the better producers in Ukraine and in many other countries.

Although not as dramatic, productivity enhancement can reduce methane emissions from cattle in U.S. herds as well. The impact of the use of bovine somatotropin, as summarized by Bauman (1990), recapitulates a few years of productivity and efficiency improvement through simultaneous nutritional, genetic, and management changes. A study of the environmental impacts of bovine somatotropin adoption (Johnson et al., 1992) indicates that this method of increasing milk production by about 13% will result in a decrease in methane emissions from the U.S. dairy cattle herd by about 9%. Bovine somatotropin treatment, like genetic selection, results in fewer cows, more feed per cow, and less feed and methane overall — the principle of dilution of maintenance at work. Another principle, that of the law of diminishing returns, will undoubtedly limit the eventual progress potential of this route; however, the limit, unknown presently, has not been reached, at least in the average U.S. herd.

Strategies Proposed by U.S. EPA Teams

In response to the Clean Air Act Amendments of 1990, a series of methane mitigation strategies have or are being proposed for a wide range of anthropogenic methane sources including livestock internationally (EPA, 1993a; IPCC, 1995) and in the United States (EPA, 1993b). Techniques targeting livestock systems internationally (Table 6) include feed processing, strategic supplementation, production-enhancing agents, and methods to improve genetics and reproduction. These are all technologies that are currently available, have low to medium capital needs, and are estimated to reduce ruminant livestock methane per unit of product by 25 to 75%. Existing technology for manure methane recovery systems is also outlined to reduce livestock manure emissions by 25 to 80%. At constant product levels, total livestock and manure emissions would decrease 23 to 70 Tg/year, enough to stabilize atmospheric methane simply from mitigation of emissions from this source alone.

Closer examination of the technologies leads to some skepticism. Many, for example, alkali or ammonia treatment of low-quality forages or molasses-urea blocks, have been available for more than 50 years and have not been widely accepted for a variety of reasons, largely lack of cost effectiveness. Additionally, the most widely successful technologies that have been used to increase productivity around the world, such as improved forage quality and increased grain supplementation, are not even discussed.

Economically justified reductions in methane emissions in the year 2010 from the U.S. livestock industry (EPA, 1993b) have been estimated at 12 to 25% of enteric and manure methane. Combined reductions from these two expected 2010 emissions thus reach about 2 Tg/year, some 6% of U.S. anthropogenic emissions. The enterically related reductions would largely result from four productivity enhancement practices by beef and dairy producers (Table 7). Developing marketing and pricing structure, which move weaned beef calves directly into the feedlot without a growing/backgrounding phase, and management practices to increase animal calf crop will each contribute 25% of the enteric reductions. Changing dairy milk pricing to a solids-nonfat base and continued improved management

Table 6 Methane Mitigation Strategies for Ruminant Livestock and Animal Manures[a]

Strategy	Technology availability	Capital requirements	Methane reduction (per unit of product)
Feed processing			
Alkali/ammonia-straw	Currently	Low	≥10%
Chopping straws	Currently	Low	≥10%
Rice straw wrap	Proposed	Low/medium	≥10%
Strategic supplementation			
Molasses urea blocks	Current	Low	≤40%
Blocks plus escape protein	Current	Low/medium	≤60%
Mineral, protein supplements	Current	Low	5–10%
Defaunation	Future	Low	≤25%
Productivity enhancers			
Bovine somatotropin	Current	Low	10%
Anabolic implants	Partial	Low	5–10%
Genetic/reproductive/other[b]			

[a] Data from EPA, 1993a.

[b] Several currently available techniques are suggested, but without estimates of methane reduction.

and genetics-driven milk production efficiencies will reduce methane another 15 and 12%.

While the above efficiency enhancement methods are well researched, rapid industry adoption is not guaranteed. Consumer attitudes toward new technologies, such as bovine somatotropin or lowered marbling requirements, for choice beef have slowed producer adoption rates. Also, many beef cow producers are part-time or hobby farmers and have been less inclined to invest the effort and capital required to use the best technology available.

Future Potential Technologies

The current research literature contains several promising leads. Ionophores reduce feed requirements for maintenance and growth by 5 to 6%, thus reducing methane by this amount in addition to some reduced methane per unit of feed for a short time upon introduction to the diet. Increased U.S. and global use will

Table 7 Economically Feasible Methane Reduction Options for the United States to 2010[a]

Strategy	Reduction (% of total cattle CH_4)[b]
Increase dairy productivity by genetics, bST, and management	17
Price milk on solids-nonfat	15
Increase beef cow calf crop	25
Ionophore use by cow-calf sector	10
Refine marketing to decrease waste fat	5
Place weaned calves directly on feed	25

[a] Data from EPA, 1993b.

[b] Fraction of 1 to 1.2 Tg total reductions in emissions.

provide proportional depressions. Additionally, if ionophores or antibiotics can be found that have more persistent action in decreasing methane per unit of feed, they would be doubly effective. Another antibiotic or inhibitor that could prove useful is a defaunating agent (protozoal inhibitor). Protozoa metabolize carbohydrate to end products with a high proportion of hydrogen, so much so that methanogens either attach to the outside (Stumm et al., 1982) or exist inside the protozoa (Finlay et al., 1994) to convert the hydrogen to methane. Defaunation has been shown to reduce methane emissions by steers consuming barley diets by one half (Whitelaw et al., 1984). Others have also found defaunation to depress methane loss by ruminants, but not on all diets (Itabashi et al., 1984). Apparently, the depression depends on the diet and thus the bacteria that replace the protozoa. Also, simple, effective defaunating agents are not presently available.

An intriguing long-term possibility is to encourage the use of hydrogen by acetogens to produce acetic acid in the rumen (Demeyer et al., 1989). In the gut of some termites and rodents (Breznak and Kane, 1990), these microbes convert excess hydrogen to acetic acid, which then can be absorbed and utilized by the host. Acetogens also are found in the rumen (Leedle and Greening, 1988); however, in this environment they cannot compete with methanogens for the hydrogen. Fostering their ability to use hydrogen as well or better than methanogens could prove advantageous to the ruminant and the environment. Interestingly, inhibition of methane by cysteine (Takahashi and Young, 1995) and by mucins (Demeyer et al., 1989) increased acetate levels, which could suggest stimulation of acetogens.

NEW FRONTIERS NEEDING CONSIDERATION

As described earlier, application of more advanced technology has been the principle means of reducing methane emissions per kilogram of milk and beef. However, the tradeoff is that more advanced technology nearly always requires fossil fuel inputs that increase the greenhouse gas budget. An example can be computed from a model of the fossil fuel energy requirements for U.S. beef production (Combs et al., 1983). Methane emissions are calculated from feed inputs and fossil fuel energy converted to CO_2 using the factors of Rodhe (1990). Two scenarios are considered: a baseline of actual production practices in the United States and a second assuming that no grain or corn silage is fed, that is, without a feedlot phase. The results of the model exercise (Table 8) indicate lower CH_4 per kilogram of beef for the baseline scenario, which includes feedlot finishing, while the CO_2 from fossil fuels is higher. The sum of global warming equivalents from methane and fossil fuels is higher for beef produced from forage-only diets. The principle reason is the slower growth and longer period required to reach slaughter weight. It also should be noted that the latter option produces a lower-priced type of beef, only 2% reaching the choice grade vs. 40% currently, which would have a dramatic impact upon producers' profits under the current pricing system.

Table 8 Model Output of Methane and Fossil Fuel CO_2 from U.S. Beef Produced Under the Baseline System and Without Grain Feeding[a]

	Baseline	No grain–no silage
kg CH_4/kg beef	0.45	0.59
CO_2 equivalent of CH_4	9.0	11.80
Fossil fuel inputs as CO_2	3.0	2.28
Total CO_2 equivalent	12.0	14.08

[a] Data from Combs et al., 1983.

Two other studies by the authors also have evaluated combined methane emissions and CO_2 from fossil fuel inputs for cattle production. A model of the U.S. dairy industry assuming complete adoption of bovine somatotropin (bST) and a constant milk supply indicated that methane emissions would be reduced by 9% and fossil fuel requirements by 6% (Johnson et al., 1992). The second analysis was of improved feed quality to increase milk production of Indian cattle as described earlier. Methane (grams per kilogram of milk) decreased considerably, but fossil fuel requirements were increased for the higher milk production scenario, mostly for irrigation and fertilizer. The CO_2 equivalents per kilogram of milk were still reduced by about one third, but not as markedly as methane/milk. On the other hand, replacement of draft bullocks with tractors was estimated to result in more than a twofold increase in CO_2 equivalents (Ward et al., 1993).

These preliminary analyses indicate the need to consider not only CH_4 emissions or livestock production systems, but the fossil fuel contributions to greenhouse gases (GHG). The United States leads in high technology applications in livestock production, such as grain feeding, which require significant fossil fuel subsidies. Increased animal productivity generally can only be accomplished with major fossil fuel inputs. Exceptions such as bST are not common. The trends in the United States are being followed in other countries and to a lesser extent in the less-developed countries.

CONCLUSION

The synergism of anaerobic microbial digestion in the gut of ruminants allows the production of meat, milk, and draft power from fibrous plant resources and results in eructated methane of approximately 6% of the animal's diet energy. Globally, this enterically produced methane from all livestock, primarily cattle, approximates 80 Tg/year, with another 10 to 14 Tg of emissions from manure.

Methane emissions are approximately constant per unit of diet; thus, mitigation or amelioration of methane emissions from livestock is most effectively approached by strategies to reduce feed input per unit of product output. This approach also encourages adoption because it is usually also economically advantageous. Nutritional, genetic, and management strategies to improve feed efficiency take one of two routes: (1) increasing rate of product (milk, meat) output per animal increases efficiency by decreasing the maintenance subsidy, and (2) decreasing the energy density of the product (i.e., fat content of meat and milk) reduces the feed input requirements. Implementation of these strategies with present technology has been estimated to reduce methane per product by 12 to 25% in the United States and likely larger fractional improvements for lower production rate systems. Other resource inputs required to improve the rate of productivity and their environmental impact, however, have not been well characterized.

REFERENCES

Aboomar, J. M., Methane losses by steers fed ionophores singly or alternatively, Ph.D. Dissertation, Colorado State University, Fort Collins, 1989.

Bauman, D. E., Bovine somatotropin: review of an emerging animal technology, Office of Technology Assessment, Washington, D.C., 1990.

Birkelo, C. P., Johnson, D. E., and Ward, G. M., Net energy value of ammoniated wheat straw, *J. Anim. Sci.,* 63, 2044, 1986.

Blaxter, K. L., *Energy Metabolism in Animals and Man,* Cambridge University Press, London, 1989.

Blaxter, K. L. and Wainman, F. W., The utilization of the energy of different rations by sheep and cattle for maintenance and fattening, *J. Agric. Sci. Camb.,* 113, 1964.

Breznak, J. A. and Kane, M. D., Microbial H_2-CO_2 acetogenesis in animal guts: Nature and nutritional significance, *FEMS Microbiol. Rev.,* 87, 309, 1990.

Carmean, B. R., Persistence of monensin effects on nutrient flux in steers, M.S. Thesis, Colorado State University, Fort Collins, 1991.

Cicerone, R. J. and Oremland, R. R. S., Biogeochemical aspects of atmospheric methane, *Global Biogeochem. Cycles,* 2, 299, 1988.

Combs, J. J., Ward, G. M., Miller, W. C., and Ely, L. O., Cattle as competitors for biomass energy, *Energy Agric.,* 1, 251, 1983.

Crutzen, P. J., The role of ruminants as a source for methane: Relevance to atmospheric chemistry, *Proc. Soc. Nutr. Phys.,* 3, 175, 1994.

Crutzen, P. J., Aselmann, I., and Seiler, W., Methane production by domestic animals, wild ruminants, other herbivorous fauna, and humans, *Tellus,* 38B, 271, 1986.

Czerkawski, J. W., Fate of metabolic hydrogen in the rumen, *Proc. Nutr. Soc.,* 31, 141, 1972.

Demeyer, D., DeGrave, K., Duran, M., and Stevani, J., Acetate: A hydrogen sink in hindgut fermentation as opposed to rumen fermentation, *ACTA Vet. Scand.,* 86, 68, 1989.

EPA (U.S. Environmental Protection Agency), Options for Reducing Methane Emissions Internationally, EPA-430-R-93-006B, U.S. EPA, Washington, D.C., 1993a.

EPA (U.S. Environmental Protection Agency), Opportunities to Reduce Anthropogenic Methane Emissions in the United States, EPA-430-R-93-012, U.S. EPA, Washington, D.C., 1993b.

EPA (U.S. Environmental Protection Agency), International Anthropogenic Methane Emissions: Estimates for 1990, EPA-230-R-93-010, U.S. EPA, Washington, D.C., January 1994a.

EPA (U.S. Environmental Protection Agency), Inventory of U.S. Greenhouse Gas Emissions and Sinks: 1990–1993, EPA230-R-94-014, U.S. EPA, Washington, D.C., September 1994b.

Finlay, D. J., Esteban, G., Clarke, K. J., Williams, A. G., Embley, T. M., and Hirt, R. P., Some rumen ciliates have endosimbiotic methanogenesis, *FEMS Microbiol. Lett.*, 117, 157, 1994.

Fuller, J. R. and Johnson, D. E., Monensin and lasalocid effects on fermentation *in vitro, J. Anim. Sci.,* 53, 1574, 1981.

Garrett, W. N., Influence of monensin on the efficiency of energy utilization by cattle, in *Energy Metabolism of Farm Animals,* Eckern, A., Ed., European Association for Animal Production, 1982, 29.

Gibbs, M. and Johnson, D. E., Methane emission from the digestive processes of livestock, in *International Anthropogenic Methane Emissions: Estimates for 1990,* EPA 230-R-93-010, 2-1, January 1994.

Goodrich, R. D., Garrett, J. E., Gast, D. R., Kirick, M. A., Larson, D. A., and Meiske, J. C., Influence of monensin on the performance of cattle, *J. Anim. Sci.,* 58, 1484, 1984.

Hackstein, J. H. P. and Stumm, C. K., Methane production in terrestrial arthropods, *Proc. Natl. Acad. Sci. U.S.A.,* 92, 5441, 1994.

IPCC (Intergovernmental Panel on Climate Change), *Climate Change: The IPCC Scientific Assessment,* Cambridge University Press, London, 1990.

IPCC (International Panel on Climate Change), *Mitigation Options in Agriculture,* Cole, V., Ed., Cambridge University Press, New York, 1995.

Itabashi, H., Kobayashi, T., and Matsumoto, M., The effects of rumen ciliate protozoa on energy metabolism in some constituents in rumen fluid and blood plasma of goats, *Jpn. J. Zootech. Sci.,* 55, 248, 1984.

Johnson, D. E., Hill, T. M., Ward, G. M., Johnson, K. A., Branine, M. E., Carmean, B. R., and Lodman, D. W., Ruminants and other animals, in *Atmospheric Methane: Sources, Sinks and Role in Global Change,* Khalil, M. A. K., Ed., Springer-Verlag, New York, 1993, 199.

Johnson, D. E., Ward, G. M., and Torrent, J., The environmental impact of bovine somatotropin use in dairy cattle, *J. Environ. Qual.,* 21, 157, 1992.

Khalil, M. A. K. and Schearer, M. J., Sources of methane: An overview, in *Atmospheric Methane: Sources Sinks and Role in Global Change,* Khalil, M. A. K., Ed., Springer-Verlag, New York, 1993, 180.

Khalil, M. A. K. and Rasmussen, R. A., Sources, sinks and seasonal cycles in atmospheric methane, *J. Geophys. Res.,* 88, 5131, 1983.

Khalil, M. A. K., Rasmussen, R. A., and Moraes, F., Atmospheric methane at Cape Meares: Analysis of a high resolution database and its environmental implications, *J. Geophys. Res.,* 98, 14753, 1993.

Leedle, J. A. Z. and Greening, R. C., Postprandial changes in methanogenic and acetogenic bacteria in the rumen of steers fed high or low forage diets once daily, *Appl. Environ. Microbiol.,* 54, 502, 1988.

Lerner, J. and Mathews, E., Methane emissions from animals: A global high resolution database, *Global Biogeochem. Cycles,* 2, 139, 1988.

Lodman, D. W., Branine, M. E., Carmean, B. R., Zimmerman, P., Ward, G. M., and Johnson, D. E., Estimates of methane emissions from manure of U.S. cattle, *Chemosphere,* 26, 189, 1993.

Martinez, A., Johnson, D. E., and Rust, J., Ukraine Cattle Herd: Methane Emissions and Mitigation Potential, Report to EPA by Winrock International, Morrilton, AK, 1995.

Moss, A. R., Givens, D. I., and Garnsworth, P. C., The effect of alkali treatment of cereal straws on digestibility and methane production of sheep, *Anim. Feed Sci. Technol.,* 49, 245–259, 1994.

NATO Workshop Proceedings, *Atmospheric Methane: Sources, Sinks and Role in Global Methane,* Khalil, M. A. K., Ed., Springer Verlag, New York, 1993.

Okine, E. K., Matheson, G. W., and Hardin, R. T., Effects of changes in frequency of reticular contractions on fluid and particle rates of passage in cattle, *J. Anim. Sci.,* 67, 3388, 1989.

Robb, J., Evans, T. J., and Fisher, C., A study of the nutritional energetics of sodium hydroxide treated straw pellets in rations fed to growing lambs, in *Energy Metabolism, European Association for Animal Production,* 1979, 26.

Rodhe, H., A comparison of the contribution of various gases to the greenhouse effect, *Science,* 248, 1217, 1990.

Saa, C. F., Hill, T. M., and Johnson, D. E., Persistence of methane suppression by ionophore and a glycopeptide in steers fed a brome hay diet, *Colorado State Univ. Beef Progr. Rep.,* 123, 1993.

Safley, L. M., Jr., Casada, M. E., Woodbury, J. W., and Roos, K. F., Global Methane Emissions from Livestock and Poultry Manure, EPA 400-1-91-048, U.S. EPA, Washington, D.C., February 1992.

Sheppard, J. C., Westberg, H., Hopper, J. F., Ganesan, K., and Zimmerman, P., Inventory of global methane sources and their production rates, *J. Geophys. Res.,* 87, 1305, 1982.

Stumm, C. K., Ngzijzen, H. J., and Vogels, G. D., Association of methanogenic bacteria with ovine ciliates, *Br. J. Nutr.,* 47, 95, 1982.

Takahashi, J., Beneke, R. G., Aoki, M., Fukushima, M., Nakano, M., and Young, B. A., Effects of L-cysteine and archaebacterial supplementation on ruminal methanogenesis in sheep, *Proc. Soc. Nutr. Phys.,* 3, 180, 1984.

Takahashi, J. and Young, B. A., The regulation of energy metabolism in sheep by nitrate and L-cysteine, in *Energy Metabolism of Farm Animals,* Vermorel, M., Ed., European Association for Animal Production, 1995, 387.

Thornton, J. H. and Owens, F. N., Monensin supplementation and *in vivo* methane production by steers, *J. Anim. Sci.,* 52, 628, 1981.

Torrent, J., Johnson, D. E., and Reverter, A., Predicting methane production using rates of passage in digestion, Unpublished data, Colorado State Univ., 1994.

van der Honing, Y., Wieman, B. J., Steg, A., and van Donselaar, B., The effect of fat supplementation of concentrates on digestion and utilization of energy by productive dairy cows, *Neth. J. Agric. Sci.,* 29, 79, 1981.

Ward, G. M., Doxtader, K. G., Miller, W. C., and Johnson, D. E., Effects of intensification of agricultural practices on emissions of greenhouse gases, *Chemosphere,* 26, 87, 1993.

Wedegaertner, T. C. and Johnson, D. E., Monensin effects on digestibility, methanogenesis and heat increment of a cracked corn-silage diet fed to steers, *J. Anim. Sci.,* 57, 168, 1983.

Whitelaw, F. G., Eadie, J. M., Bruce, L. A., and Shand, W. J., Methane formation in faunated and ciliate-free cattle and its relationship with rumen volatile fatty acid production, *Br. J. Nutr.,* 52, 261, 1984.

Williams, D. J., Methane emissions from the manure of free range cows, *Chemosphere,* 26, 179, 1993.

Wuebbles, D. J. and Tamaresis, J. S., The role of methane in the global environment, in *Atmospheric Methane: Sources, Sinks and Role in Global Change,* Khalil, M. A. K., Ed., Springer Verlag, New York, 1993, 469.

CHAPTER 16

Use of Endophyte-Infected Fescue Pasture and Avoidance of Herbicide Pollution

R. L. Rice, G. G. Schurig, W. S. Swecker, D. E. Eversole, C. D. Thatcher, and D. J. Blodgett

OVERVIEW OF PROBLEM

Impact of Endophyte-Infected Fescue Pastures on Nutrient Management and Environmental Consequences

Tall fescue *(Festuca arundinacea* Schreb.) is a perennial grass that covers an estimated 15 million ha in the United States (Bacon et al., 1986). Tall fescue is often infected by the endophytic fungus *Acremonium coenophialum* (Morgan-Jones and Gams, 1982). *A. coenophialum* synthesizes ergopeptide alkaloids such as ergovaline within the fescue plant. The fungus imparts to fescue many desirable attributes of the plant. Desirable attributes of endophyte-infected (EI) tall fescue are its ease of establishment, wide adaptation, long grazing season, tolerance to environmental stress, and pest resistance. Evaluation of digestible dry matter, crude protein, and mineral levels of EI fescue forages indicated that livestock grazed on tall fescue should exhibit good performance (Ball et al., 1991).

However, despite the excellent forage quality of tall fescue, livestock grazed on EI fescue pastures often exhibit poor body weight gains, reproductive dysfunction, and decreased production that results in economic losses for producers in the horse and cattle industries. Consumption of EI fescue grass and hay is associated with summer slump syndrome, fescue foot, and fat necrosis in cattle. In 1990, the economic loss in the beef cattle industry alone was estimated to be greater than $800 million annually (Hoveland, 1990).

Effective measures to control the endophyte within the fescue plant or to alleviate adverse health effects in cattle to allow them to benefit from the excellent forage quality of fescue do not exist. Application of fungicides to EI fescue pastures does not destroy the fungus. A definitive treatment does not exist for cattle with summer slump syndrome, and treatment for fescue foot and fat necrosis is based on clinical signs.

Pasture renovation of EI fescue pastures and reseeding with endophyte-free (EF) fescue is often recommended to producers. However, EF fescue grass is not as insect and drought resistant as the EI grass nor as tolerant to overgrazing by

1-56670-199-6/96/$0.00+$.50

livestock. Astute grazing management of EF pastures is required to maintain stands and productivity. Establishment and maintenance of EF fescue pastures may be difficult in areas where fescue is marginally adapted. Additionally, EI fescue sod will eventually dominate EF fescue grass or other grasses to reestablish its presence within a few short years if not completely destroyed during pasture renovation.

Pasture renovation of EI fescue pastures also is environmentally unsound, since large amounts of nonselective herbicides are required to destroy the EI sod. The destroyed fescue contributes to environmental pollution when it decomposes, since excess nitrogen and phosphorus are released into surrounding watersheds. Application of herbicides and replanting pastures with another grass also expends valuable fuel reserves. Pasture renovation with EF fescue is also expensive, costing up to $375 per ha, and pastures cannot be utilized for up to 1 year (White, 1990). Cost of pasture renovation of all 15 million ha of EI fescue pastures is estimated to be $5.6 billion and would require 47.3 million L of herbicide (White, 1990). Many EI fescue pastures are too hilly for tillage and, if tilled, would result in severe soil erosion.

Since renovation of EI pastures is monetarily and environmentally unsound, methods to ameliorate the deleterious effects need to be investigated. Preventative methods would be preferable to new therapeutic drugs, which may cause residue problems. Development of a vaccine to protect cattle against fescue toxicosis may be an environmentally safe and economical solution. A vaccine against fescue toxicosis may save producers millions of dollars and avoid environmental damage associated with pasture renovation, while allowing cattle to benefit from the nutritional value of EI fescue forage.

Toxins Associated with Endophyte-Infected Fescue Pastures and Their Effects

The causative agent(s) of fescue toxicosis has (have) not been definitively proven. Ergopeptide alkaloids synthesized by *A. coenophialum* present in fescue plant leaves and seeds are associated with fescue toxicosis. Synthesis of ergot alkaloids in EI fescue grass does not require special environmental or host conditions. Several ergot alkaloids identified in *Acremonium*-infected fescue are chanoclavine, ergosine, ergosinine (Porter et al., 1979), ergotamine, ergovaline, ergocrystine, ergocornine (Yates et al., 1985), and lysergic acid amide (Petrosky and Powell, 1991). Ergovaline (EV) is thought to be the alkaloid responsible for fescue toxicosis, since it constitutes 84 to 97% of the total concentration of ergopeptide alkaloids in EI fescue samples (Lyons et al., 1986).

Clinical signs of summer slump syndrome in cattle are poor weight gains, reduced calf weaning weights, heat intolerance, increased respirations and rectal temperatures, decreased conception rates, rough hair coats, failure to shed winter coats, decreased serum prolactin concentrations, reduced milk production, and nervousness (Hemken et al, 1984; Stuedemann and Hoveland, 1988). Serum cholesterol and alkaline phosphatase (ALP) concentrations are decreased in cattle grazed on EI fescue pastures for long time periods (Lipham et al., 1989; Bond et al., 1984; Stuedemann et al., 1985; Garner and Cornell, 1978).

Serum prolactin concentrations are often decreased in cattle grazed on EI fescue pastures (Thompson et al., 1987) and may interfere with the ability of cattle to mount humoral and cell-mediated immune (CMI) responses. Prolactin, a polypeptide pituitary hormone produced in the anterior pituitary and usually associated with lactation, also is important in the normal maintenance of the immune system. Experimentally, decreases in serum prolactin concentrations reduce antibody responses to sheep red blood cells (SRBC) vaccination (Spangelo et al., 1987), which is reversed by administration of exogenous prolactin (Nagy et al., 1983). Prolactin also is important for the induction of interleukin-2 (IL-2) receptors on lymphocytes (Mukherjee et al., 1990), which are pivotal to T-cell-mediated immune responses. Impaired CMI responses in rodents induced by decreased prolactin concentrations are restored by exogenous prolactin administration (Nagy and Berczi, 1981).

Mice fed EI fescue seed diets exhibit similar clinical responses as observed in cattle. Similar clinical responses observed in mice include depressed feed intake, weight gains, serum prolactin concentrations, reduced milk production, and decreased conception rates and litter sizes (Zavos et al., 1987, 1988). Therefore, mice may serve as an animal model for fescue toxicosis.

Historical Perspective in Vaccination Against Plant Toxins

Nutritional management of livestock is often complicated by plant or fungal toxins in forages and feeds intended for livestock consumption. This contamination may interfere with the nutritional benefits to animals by decreasing feed consumption, producing toxicosis, and rendering the feed unsuitable for livestock consumption. Producers may use chemical means to reduce contamination that may negatively impact the environment. Immunization of animals to protect against plant toxins may allow livestock to benefit nutritionally, while avoiding adverse health effects of toxins in livestock and chemical treatment of affected forages in the environment.

Previous parenteral vaccines that induce specific immunoglobulin G (IgG) antibodies against plant or fungal toxins resulted in various degrees of animal protection. Degree of protection against genistein (Cox, 1985), lantana toxin (Stewart et al., 1988), and lupinosis (Payne et al., 1993) is related to levels of specific serum IgG titers. Jonas and Erasmunson (1979) report that mice vaccinated against the toxic effects of sporidesmin are protected. However, protection in mice against sporidesmin is related to the carrier and not to serum antibody levels (IgG and IgM).

Parenteral vaccines against plant or fungal toxins are not always protective and in some cases exacerbate toxin-induced toxicosis. For instance, vaccination against zearalenone (MacDonald et al., 1990; Smith et al., 1992) and senecionine (Culvenor, 1978) worsen clinical signs of toxicosis instead of conferring protection to animals. Ewes vaccinated against sporidesmin develop worse clinical signs than nonimmunized ewes, despite the protection demonstrated in mice immunized against sporidesmin (Fairclough et al., 1984).

To address the problem of EI fescue pastures and avoid the use of herbicides, research to develop a vaccine against fescue toxicosis was initiated. Because hypoprolactemia in cattle with fescue toxicosis may impair immune responses, we first determined whether cattle could mount adequate humoral immune responses against vaccination when grazed on fescue pastures. Then we assessed carrier proteins and their ability to induce prolonged titers against fescue toxicosis. Dietary EV is not immunogenic because of its small molecular weight and will not elicit an anamnestic immune response; thus a vaccine that prolongs titers will be necessary. After evaluating the magnitude and duration of titers induced, we evaluated the ability of antibodies to protect against fescue toxicosis in mice.

HUMORAL IMMUNE RESPONSES OF CATTLE MAINTAINED ON FESCUE PASTURES

Hypoprolactemia in Fescue Toxicosis May Affect the Immune System

Serum prolactin concentrations are often decreased in cattle grazed on EI fescue pastures (Thompson et al., 1987) and may interfere with the ability of cattle to mount an immune response to vaccination. Ergot alkaloids are potent inhibitors of prolactin secretion via dopamine D2 receptors. Field reports suggest that cattle grazed on EI fescue pastures may not respond to vaccination as well as those on EF pastures (Sprowls, 1987). Research in rats supports this anecdotal evidence, since rats exhibit a lower titer to SRBC vaccination when fed EI fescue seed (Dew et al., 1990). However, decreases in humoral immune responses against SRBC vaccination in rats fed EI fescue diet are not in agreement with humoral immune responses in cattle fed EI fescue diet (Dew et al., 1990).

The ability of steer calves to mount humoral immune responses against causative agents of bovine respiratory complex (BRC) when fed EI fescue diet was previously investigated (Dew, 1989). Steer calves vaccinated against infectious bovine rhinotracheitis (IBR), parainfluenza type 3 (PI3), and bovine virus diarrhea (BVD) viruses when fed EI fescue diet had similar humoral immune responses as calves fed EF fescue diet. In fact, antiviral titers of steers fed EI fescue diet tended to be higher in general. However, calves fed EI fescue diet exhibited only mild signs of fescue toxicosis, as indicated by body temperatures, weight gains, and serum ALP concentrations. Ergovaline content of treatment diets and serum prolactin concentrations also were not measured and correlated to antiviral titers.

Cattle with decreased serum prolactin concentrations maintained on EI fescue pastures may also have impaired CMI responses that predispose cattle to infection. Cattle shipped from EI fescue pastures to feedlots experience a higher incidence of respiratory disease during the first few weeks (Sprowls, 1987). Hypoprolactemia impairs CMI responses by decreasing both γ-interferon (IFN-γ) production by T lymphocytes and lymphocyte proliferation responses to mitogens (Bernton et al., 1988). Mitogen-induced lymphocyte proliferation is suppressed in rodents fed an EI fescue seed diet (Dew, 1989).

Additionally, nonspecific immune mechanisms may be negatively affected by fescue toxicosis. Phagocytosis and free radical production by alveolar macrophages and neutrophils are nonspecific immune mechanisms important in the defense of the lungs. Superoxide anion secretion by neutrophils, which is important in the killing of phagocytosed pathogens, is dependent on prolactin (Fu et al., 1992). In cattle with fescue toxicosis, Saker et al (1995) observed a significant decrease in hydrogen peroxide release, major histocompatibility complex (MHC) II expression, and phagocytic activity of monocytes.

Since a reduction of basal levels of prolactin concentrations may attenuate selected immune responses, it is important to determine if cattle can adequately mount a humoral immune response when grazed on EI fescue pastures to optimize vaccination programs. If cattle grazed on EI fescue pastures had impaired humoral immune responses, cattle could be strategically vaccinated before going on EI fescue pastures. In this trial, cattle grazed on EI and EF fescue pastures were vaccinated with the T-cell-dependent antigens, SRBC and Concanavalin A (Con A). The magnitude of humoral immune responses induced were correlated to serum prolactin concentrations.

Materials and Methods

Twenty-four Angus steers were blocked by weight and allocated into two groups of 12 steers each. One group grazed on established EI fescue pastures (EI group) that were 82% endophyte infected. The other group grazed on EF pastures (EF group) that were 2% endophyte infected. The steers of both groups were further allocated into groups of six to graze on two EI and two EF fescue pastures. Steers grazed on the pastures for 1 month prior to the beginning of the study in mid-June. Ergovaline concentrations were quantified by high-performance liquid chromatography (HPLC) analysis in the pastures according to published methods (Hill et al., 1993) at the initiation and termination of the study. All steers were injected intramuscularly with 1 ml of two different antigen preparations. One antigen preparation contained 800 µg/ml of lysozyme without adjuvant. The second preparation consisted of 250 µg/ml of Con A with 10% SRBC in Freund's incomplete adjuvant (FIA). Steers were revaccinated 19 d after the first injection with the same antigen preparations. Whole blood was collected on days 0, 19, and 33 postvaccination for titer determinations. Indices measured were serum prolactin concentrations, weight gains, rectal temperatures, hemagglutination titers to SRBC, and IgG titers against Con A and lysozyme. Rectal temperatures and body weights were measured on days that whole blood was collected. Serum IgG titers against Con A and lysozyme were determined by enzyme linked immunosorbent assay (ELISA). The mean $\log_2$

Table 1 Body Temperatures, Serum Prolactin Concentrations, and Average Daily Gains of Cattle Grazed on EI and EF Fescue Pastures

	Rectal temperature (°C)		Prolactin level (ng/ml)		Average daily weight gain (kg/steer/day)[a]	
Day of Study	EI	EF	EI	EF	EI	EF
0	40.8	40.0	5.5	113.2	0.27	1.05
19	40.5	39.8	8.6	102.8	0.27	0.27
33	40.5	39.7	9.1	121.9	0.77	0.73

[a] Weights for gain calculation are not cumulative and measurements made on days 0, 19, and 33 on pasture correspond to acclimation period, d 0 to 19, and days 19 to 33, respectively.

titers and other indices between the steers grazed on EI and EF fescue pastures were analyzed by the two-tailed Student's t-test. Probability values < 0.05 were considered significant.

Results and Discussion

Overall EV concentration in the EI fescue pastures was 280 ppb. Ergovaline was not detected in the EF fescue pastures. Rectal temperatures were elevated ($p < 0.05$) and serum prolactin concentrations were lower ($p < 0.05$) in the EI group throughout the 33-d trial period (Table 1). On day 0, after 1 month of acclimation, average daily weight gains (ADG) (Table 1) were decreased ($p < 0.05$) in cattle grazed on EI fescue pastures. Thereafter, ADG were not different ($p > 0.05$) between the two groups; however, body weight differences were maintained. Elevated rectal temperatures, lower serum prolactin concentrations, and decreased ADG measured in cattle maintained on EI fescue pasture indicated overt fescue toxicosis.

The primary immune response to Con A tended to be higher ($p = 0.09$) in the EI group. The primary immune response to SRBC was higher ($p < 0.05$) for the EI group. The secondary immune responses to Con A and SRBC were higher ($p < 0.05$) for the EI group (Table 2). Titers to lysozyme were not induced in either group, indicating an adjuvant was needed to elicit an immune response at the dose used. The titer results in this study are in agreement with the results of antiviral titers in calves housed in temperature-controlled rooms fed EI fescue diet (Dew, 1989).

The significantly higher humoral immune responses to vaccination in the EI group indicated that despite hypoprolactemia, adequate titer responses were mounted. Perhaps the ability of cattle with hypoprolactemia to mount humoral immune responses when grazed on EI fescue is related to concentrations of growth hormone and balance of T helper (TH) cells.

Both prolactin and growth hormone are involved in the maintenance of humoral immunocompetence. Although growth hormone was not measured in this study, increases of basal growth hormone concentrations in cattle grazed on

Table 2 Humoral Immune Responses of Cattle Vaccinated with Con A and SRBC Grazed on EI and EF Fescue Pastures

	Primary immune response (Log_2 titer)		Secondary immune response (Log_2 titer)	
Treatment	**Con A**	**SRBC**	**Con A**	**SRBC**
EI fescue pasture	5.8	5.6	9.2	6.0
EF fescue pasture	3.3	3.1	7.4	3.2

EI fescue pastures are reported (Thompson et al., 1987). Growth hormone is similar in structure to prolactin (Hiestand and Mekler, 1986) and may be able to bind prolactin receptors to exert prolactin-like actions on prolactin target tissues (Bernton et al., 1988). Prolactin receptors are located on lymphocytes (Russell et al., 1984). Growth hormone restores antibody responses to SRBC vaccination (Spangelo et al., 1987) and restores selected CMI responses (Berczi et al., 1983) in hypophysectomized rats. Thus increased growth hormone may aid the humoral immune responses in cattle grazed on EI fescue pastures.

In this study, not only did cattle with fescue toxicosis mount humoral immune responses, but significantly higher humoral immune responses despite hypoprolactemia. Fescue toxicosis may significantly increase humoral immune responses by altering the balance of TH cell activities. T-cell-dependent antigens require both T cells and B cells for generation of a humoral immune response. T cells may be divided into helper T cells and T suppressor cells on the basis of their function. Helper T cells can be further divided into TH1 or TH2 subsets based on the lymphokines produced (Mosmann and Coffman, 1989). Lymphokines secreted by TH1 cells are IFN-γ, IL-2, and IL-3. Functionally, TH1 cells are generally thought to activate cell-mediated immunity. Lymphokines secreted by TH2 cells are IL-3, IL-4, and IL-5. The TH2 cells mediate hapten-carrier helper activity and augment IgG1 and IgE production (Lise and Audibert, 1989) to activate humoral immunity.

Relative numbers of TH cells are similar in cattle grazed on EI fescue pastures and those grazed on EF fescue pastures (Dew, 1989). However, TH2 cells may dominate functionally since secretion of selected cytokines by TH1 cells may be inhibited in fescue toxicosis. Hypoprolactemia decreases IFN-γ secretion by TH cells in mice (Bernton et al., 1988) to decrease CMI. Gamma interferon is important in preventing viral replication within cells, enhancing the killing ability of macrophages and augmenting expression of Class II proteins. Decreases in IFN-γ production are associated with normal or increased humoral immune responses, despite low T cell immune responsiveness in elderly people (Candore et al., 1993). Additionally, the TH2-derived cytokine, IL-10, may inhibit TH1 cell function to decrease CMI and increase humoral immune responses (Powrie et al., 1993).

Decreases in IFN-γ production by TH1 cells and increases in IL-4 production by TH2 cells often increase specific IgG production. Production of IL-4, IL-5, and antigen-specific IgG_1 are increased by TH2 cells in mice with the inability to secrete IFN-γ (Graham et al., 1993). Transgenic mice expressing IL-4 exhibit increased serum levels of IgG_1 and antigen-specific antibody immune responses (Burstein et al., 1991). Additionally, IL-4 synergizes with other mediators such as prostaglandin E2 to significantly increase IgG_1 production up to 26-fold (Roper et al., 1990).

Elevated body temperature also may augment immune responses. Cattle grazed on EI fescue pastures often have elevated body temperatures, as seen in this study. A plausible mechanism for increased body temperature and augmented humoral immune responses in cattle with fescue toxicosis may be through IL-1 production. Interleukin-1 is an endogenous pyrogen that elevates body temperature, which has increased humoral immune responses against SRBC vaccination *in vivo* (Reed et al., 1989).

Conclusion

Cattle with overt fescue toxicosis mounted significantly higher humoral immune responses against two antigens, despite hypoprolactemia. Increased basal concentrations of growth hormone and a predominance of TH2 cells with IL-4 production induced by fescue toxicosis could explain the augmentation of humoral immune responses in cattle maintained on EI fescue pastures. Suppression of TH1 cell functions by a decrease in IFN-γ production may permit TH2 cells to synthesize higher levels of IgG to generate higher titers. Decreased serum prolactin concentrations may also negatively affect nonspecific immune responses that may predispose cattle to respiratory infections. Therefore, increased morbidity from BRC in cattle fed EI fescue may be a result of the inability of nonspecific defense mechanisms in the lungs to kill invading pathogens and not the failure of cattle to respond to immunization.

PREPARATION AND EVALUATION OF A VACCINE AND EVALUATION OF PROTECTION AGAINST FESCUE TOXICOSIS IN MICE

To investigate the development of an oral or parenteral vaccine to protect cattle against fescue toxicosis, the ergopeptide alkaloid ergotamine (EG) was conjugated to protein carriers via the Mannich reaction. Protein carriers assessed were cholera toxin subunit B (CTB), Con A, and bovine serum albumin (BSA). The magnitude and duration of humoral and mucosal immune responses induced by parenteral and oral administration of these protein-EG conjugates were evaluated in mice. Duration and magnitude of specific immune responses against protein-EG conjugates given orally or parenterally were important to determine the frequency of vaccine administration and possibly the best route of vaccine administration. Additionally, parenterally and orally administered protein-EG conjugates and passively immunized anti-EV antibodies were assessed for the ability to alleviate signs of fescue toxicosis in the mouse model.

Vaccine Considerations

In the development of a vaccine for fescue toxicosis, EG was the model antigen used in this study because of its availability and similarities to EV. Ergotamine is inexpensive and available commercially. Ergovaline is not commercially available and its synthesis is complicated with many chemical steps (Smith, 1993). Ergotamine and EV are both ergopeptine alkaloids with similar structures (Shelby and Kelley, 1991), mechanism of actions (Marple et al., 1988; Kerley et al., 1994), and physiological effects (McCollough et al., 1994). In evaluation by ELISA, monoclonal anti-EV antibodies crossreact strongly with EG.

Ergotamine must be conjugated to a protein carrier for immunogenicity because of its low molecular weight. An indole nitrogen in the structures of EG and EV enables conjugation to protein carriers that contain lysine residues via the Mannich reaction. Conjugation of ergot peptide alkaloids to carrier proteins via the indole nitrogen directs antibody specificity to the tricyclic peptide moiety of the alkaloid (Berde and Schild, 1978).

The Mannich reaction is often used to conjugate small neurotransmitter ligands and indolealkylamines with carrier proteins using formaldehyde as a chemical linker as described by Ranadive and Sehon (1967). The Mannich reaction was previously used to prepare protein conjugates for induction of polyclonal and monoclonal antibodies (Flurkey et al., 1985; Castro et al., 1973; Taunton-Rigby et al., 1973; Miwa et al., 1977). Previously, the Mannich reaction was used to conjugate EG (Shelby and Kelley, 1990) and EV (Kelley and Shelby, 1990) to BSA for the production of monoclonal and polyclonal antibodies in mice.

Both the carrier protein used for protein-EG conjugates and the route of administration of conjugate vaccines may modify the magnitude and duration of the immune responses against EG. In our research, CTB was used as a protein carrier for oral and parenteral vaccinations, and Con A was used in parenteral vaccinations. The success of the Mannich reaction to conjugate CTB and Con A to EG was determined by comparing these conjugates to BSA-EG conjugates prepared by the Mannich reaction as previously described (Shelby and Kelley, 1990).

Cholera toxin (CT) contains the same number of lysine groups on a molar basis as BSA (Azcona-Olivera et al., 1992). This supports use of CTB as a protein carrier for conjugation to EG via the Mannich reaction. Cholera toxin is a protein consisting of a 28,000-Da A subunit and a 57,500-Da B subunit that is produced by *Vibrio cholera*. The CTB functions both as a carrier protein, because of its large molecular size, and as an adjuvant. The B subunit binds GM1 ganglioside receptors, which are located on virtually all membrane surfaces of nucleated cells. Cholera toxin subunit B binds lymphoid cells to activate immunocompetent cells in the mucosal-associated lymphoid tissues (Tamura et al., 1989) that augment immune responses. The A subunit is responsible for the toxicity of CT (Elson, 1989). Extraction of the subunit B from the holotoxin removes the toxic effects of the subunit A to produce a protein that retains its adjuvant/carrier properties.

The main advantages of CTB as an protein carrier/adjuvant are that low concentrations of CTB and hapten are needed for immunization and the immunity produced is long lasting (Russell and Wu, 1991). Specific IgA antibodies and IgG antibodies that are long lived are induced when CTB is coadministered orally or intranasally (Czerkinsky et al., 1989) and parenterally (Hirabayashi et al., 1990) with unrelated antigens. A prolonged anti-EG titer is important, since dietary EV present in EI fescue forage will not induce an anamnestic immune response because of its low molecular weight.

The T-cell mitogen, Con A, also was examined as a potential carrier for the hapten EG. The use of carrier proteins that are mitogens may significantly increase the immunogenicity of conjugate vaccines and may also potentially avoid epitopic suppression. Concanavalin A, a glycoprotein extract from the jack bean, binds to mannose residues on the surface of T lymphocytes and induces the cells to proliferate (Liener et al., 1986). Concanavalin A has not previously been used as a carrier protein for small haptens in vaccination. It was expected that conjugation of Con A to EG via the Mannich reaction would result in anti-EG antibody production because of its large molecular size and mitogenic effects on T cells.

In the development of a vaccine against fescue toxicosis, the route of vaccine administration may be an important determining factor of protection. Parenteral vaccination induces systemic IgM and IgG isotypes, but rarely serum or secretory IgA (sIgA) isotypes. Oral vaccination can produce high levels of mucosal antigen-specific sIgA. Secretory IgA is the major immunoglobulin class found in exocrine secretions, since it functions in the host defense of mucosal surfaces. Daily output of sIgA (50 to 100 mg/kg of body weight per day) (Holmgren et al., 1992) exceeds all other immunoglobulins combined. Lymphoid tissues associated with mucosal membranes contain more immunocytes, including B and T lymphocytes and plasma cells, than any other tissue in the body. Moreover, intestines are the richest lymphoid tissue present in the body (Mestecky, 1987). The IgA responses originating at one mucosal site disseminate to other mucosal-associated lymphoid tissues by migration of sensitized B and T cells, hence the basis for oral and intranasal immunization strategies. Neutralization of EV at the site of absorption (i.e., mucosal surfaces) with specific antibodies induced by oral vaccination may be beneficial in the protection against fescue toxicosis.

Preparation of Protein-Ergotamine Conjugates and Vaccination of Mice to Determine Magnitude and Duration of Anti-Ergotamine Titers

Ergotamine (tartrate salt: Sigma Chemical Co.) was conjugated to BSA, Con A, and CTB for vaccination, and to ovalbumin for use as a solid phase antigen in the indirect ELISA. Proteins were linked to EG by modifications of the Mannich reaction as previously described (Kelley and Shelby, 1990). Conjugates were evaluated on 250-µm silica gel thin-layer chromatographic (TLC) plates developed in chloroform/methanol (9:1, v/v) solvent system and observed under a long wave lamp. Protein content was determined by bicinchoninic acid protein assay (BCA). The relative alkaloid content of each conjugate was determined by titration in the ELISA.

Three groups each consisting of six Balb-c female mice, 6 weeks of age, were vaccinated intraperitoneally with BSA-EG, Con A-EG, and CTB-EG conjugates emulsified in Freund's complete adjuvant (FCA) and revaccinated 14 d later with the protein-EG conjugates emulsified in FIA. A fourth group of six mice also was vaccinated intraperitoneally with CTB-EG conjugate without adjuvant. A fifth group of mice was orally vaccinated with CTB-EG conjugate along with unconjugated CT and revaccinated 10 d later. Orally vaccinated mice were revaccinated intraperitoneally 93 d post-oral vaccination with CTB-EG conjugate to induce a mucosal anamnestic immune response. Mice were bled retroorbitally under anesthesia. Fecal pellets from the orally vaccinated group were collected, and sIgA and IgG were quantally determined by measuring the optical density (OD) at a wavelength of 490 nm as described previously (DeVos and Dick, 1991). Anti-EG serum IgG titers were determined in the indirect ELISA using ovalbumin-EG conjugate as the antigen coating.

Results and Discussion

As previously reported (Shelby and Kelley, 1990), the protein-EG conjugates on TLC plates exhibited intense fluorescence at the origin, whereas free EG fluoresced and migrated. Unconjugated protein controls remained at the origin and did not fluoresce. Ergotamine content of CTB and BSA conjugates was similar, and the EG content of ovalbumin and Con A conjugates was similar. The CTB and BSA conjugates bound more EG than the Con A or ovalbumin conjugates. The average number of EG residues per molecule of carrier protein was previously determined as 7.7 for BSA and 1.3 for ovalbumin (Shelby and Kelley, 1990).

The magnitude and duration of anti-EG IgG antibodies (Figure 1) induced by parenteral administration were greatest for the BSA-EG conjugate. Titers increased linearly postvaccination to peak around day 28, after which titers slowly declined by day 100 postvaccination. The Con A-EG and CTB-EG conjugates administered with adjuvant induced IgG antibodies with similar kinetics to each other. In contrast to BSA-EG conjugate, titers induced by Con A-EG and CTB-EG conjugates did not increase rapidly after vaccination. Titers were increasing at day 36 postvaccination and peaked between days 36 and 84 postvaccination. Titers were similar to prevaccination levels by day 105 postvaccination. The duration of the antibody response elicited by the CTB-EG conjugate with adjuvant was not prolonged, but was similar to that induced by a fivefold larger dose of Con A-EG. The CTB-EG conjugate without adjuvant induced low anti-EG IgG titers. Thus CTB-EG conjugates did require an adjuvant to achieve adequate titers when administered parenterally. The CTB, as a protein carrier for EG, did not prolong titers in this study as reported for other

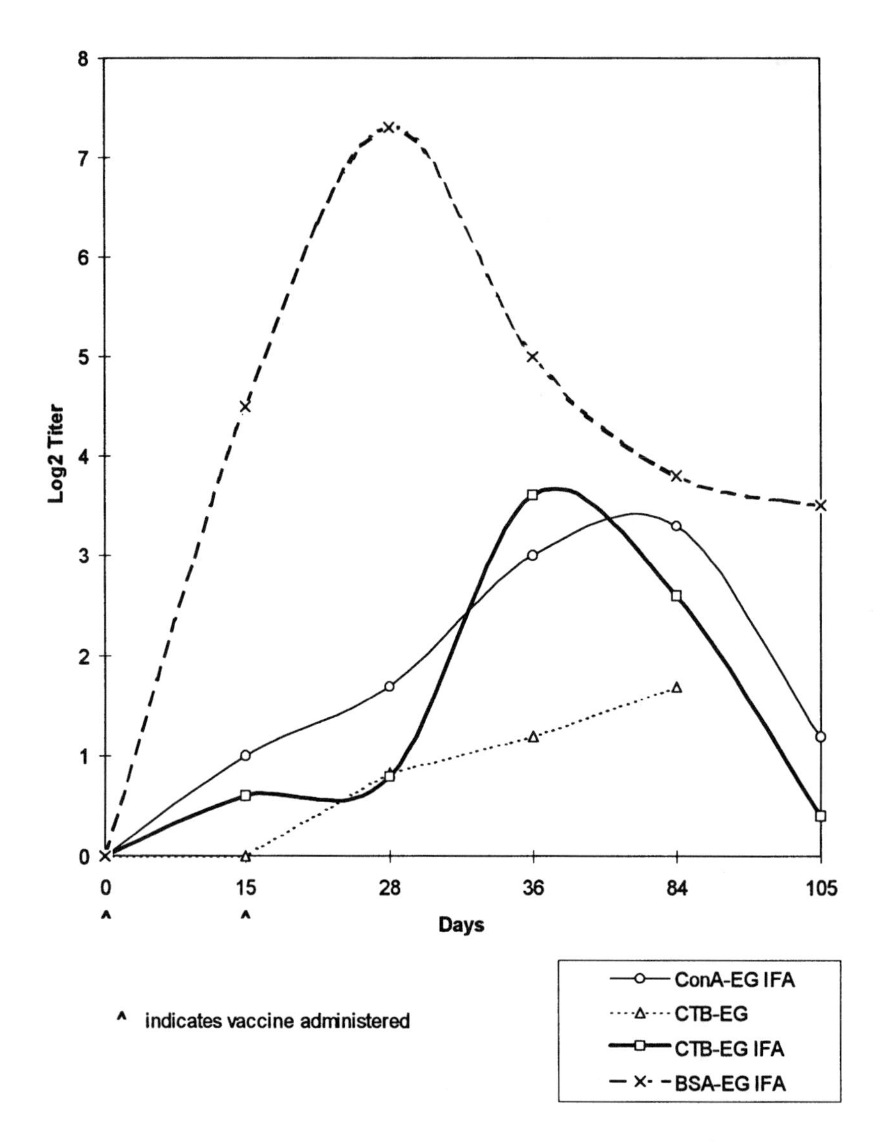

Figure 1 Magnitude and duration of anti-EG IgG titers induced by protein-EG conjugates in mice.

antigens. Perhaps the conjugation procedure negatively affected the ability of CTB to bind GM1 receptors, thus inhibiting activity of CTB.

The CTB-EG conjugate given orally induced both mucosal and systemic anti-EG antibodies. Determination of specific sIgA in oral vaccination protocols is important because sIgA can provide protection in the absence of demonstrable serum IgG antibodies. Thus protection is correlated best with local sIgA secretion rather than circulating antibody. A systemic IgG response was observed from days 9 to 21 postvaccination and decreased by day 30. The systemic response was

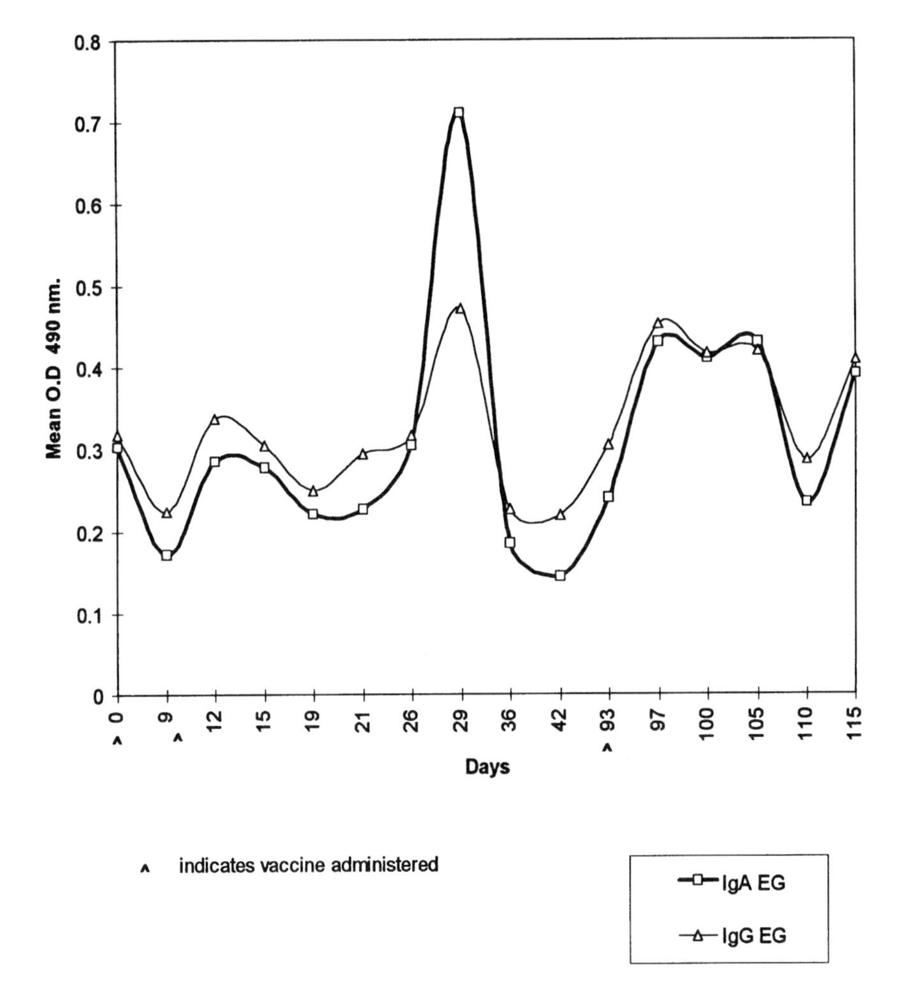

Figure 2 Mucosal anti-EG IgA and IgG levels induced by CTB-EG oral immunization in mice.

lower than the other treatments shown in Figure 1. The systemic IgG response preceded the appearance of IgA and IgG anti-EG antibodies at the mucosal level. The mucosal response (Figure 2) peaked at day 29 postvaccination, and the duration was short lived. Kinetics of mucosal IgA and IgG were similar with IgA attaining higher peak levels.

Intraperitoneal administration of CTB-EG conjugate 93 d after oral vaccination induced an immediate mucosal anamnestic immune response for both IgG and IgA isotypes (Figure 2). The magnitude of IgA and IgG anamnestic mucosal immune responses was decreased after intraperitoneal administration of CTB-EG conjugate, but the duration was extended as compared to only oral vaccination. These results are in agreement with previous research demonstrating that parenteral immunization with an antigen, to which an individual was previously immunized by the oral or intranasal routes, induces sIgA responses at mucosal surfaces (Mestecky, 1987).

Besides inducing sIgA, an orally administered vaccine against fescue toxicosis would be advantageous to producers. Oral vaccination would eliminate the possibility of vaccine reactions in meat and would decrease time, labor, and stress on animals. The vaccine could be added to feed at necessary intervals to induce sIgA responses in mucosal surfaces of the gastrointestinal tract to neutralize EV prior to systemic absorption to afford protection. Repeated oral vaccination would also prolong antibody production that will be necessary in cattle exposed to EI fescue forage throughout the grazing season.

However, oral vaccination to induce mucosal immune responses is more complicated in cattle because of the complex nature of their gastrointestinal tract. The limitations of inducing a mucosal immune response by oral vaccination in cattle may be overcome by the use of microspheres or aided by intranasal vaccination. Microspheres as the vaccine vehicle for an oral vaccine against respiratory infection caused by *Pasteurella haemolytica* decreased morbidity and mortality in calves (Bowersock et al., 1994). Novel immunization protocols developed against fescue toxicosis could involve primary mucosal (oral or intranasal) immunization to induce anti-EG IgA effector and memory cell populations followed by subcutaneous immunization. Subcutaneous administration of microsphere-encapsulated vaccine may prolong mucosal and systemic titers for months to decrease the frequency of vaccination. Additionally, systemic anti-EG IgG antibodies induced by parenteral administration could neutralize any EV systemically absorbed.

In conclusion, the Mannich reaction was used to conjugate EG to Con A and CTB protein carriers. Both protein carrier–EG conjugates, when injected with adjuvant, induced systemic IgG anti-EG antibodies with similar kinetics. The CTB-EG conjugate did not prolong antibody titers, although smaller doses were needed to elicit titers. Oral vaccination with the CTB-EG conjugate induced both mucosal IgA and IgG isotypes, in addition to a systemic IgG response. Intraperitoneal administration of CTB-EG conjugate to orally vaccinated mice elicited an anamnestic mucosal immune response. Our research suggested that an anti-EG antibody response of adequate magnitude and duration will require either frequent vaccine administration or an enhanced vaccine delivery system, such as microspheres, to protect against fescue toxicosis.

IMMUNIZATION AND EVALUATION OF PROTECTION AGAINST FESCUE TOXICOSIS IN MICE

Previous research supports vaccination as a potential solution to fescue toxicosis. The ability of a murine monoclonal antibody to protect cattle against decreases in prolactin concentrations observed in fescue toxicosis is reported. Serum prolactin concentrations increase by 7 ng/ml in cattle passively immunized with a murine monoclonal antibody specific to the lysergic base moiety of ergot alkaloids when grazed on EI fescue pastures (Hill et al., 1994). The increases in prolactin concentrations after passive immunization suggest that systemic ergot alkaloid toxins have a higher affinity for the specific antibody than for the dopamine D2 receptor, which further supports the feasibility of vaccination as a potential solution to fescue toxicosis.

Research conducted to explore the possibility of vaccine development against fescue toxicosis in cattle induced specific antibodies against lysergol, the base ring structure of ergot alkaloids. Vaccination of cattle with protein-lysergol conjugates induced specific IgG antibodies against lysergol that were short lived (Hill et al., 1994). However, the ability of specific antibodies against lysergol to protect cattle *in vivo* against fescue toxicosis was not assessed.

In our research, both parenteral and oral routes of vaccination were explored in the protection against fescue toxicosis in mice. In parenteral vaccination protocols, plant toxins absorbed from the mucosal surfaces of the gastrointestinal tract may enter systemic circulation, where interaction with target receptors may occur. If specific IgG antibody responses are insufficient to neutralize all absorbed toxins or if the plant toxins have a higher affinity for nonantibody receptors, eventually enough toxins may escape the IgG immune response to produce toxicosis. Therefore, neutralization of EV at the site of absorption (i.e., mucosal surfaces) with specific antibodies induced by oral vaccination may be beneficial.

Direct immunization of mucosal surfaces by oral vaccination to protect against mycotoxicosis in animals has not previously been investigated. Oral vaccination is increasingly used in immunization protocols. Oral vaccines are currently used in humans to protect against polio and cholera. Oral vaccines are being investigated for protection against tetanus, *Salmonella typhi,* and *Streptococcus mutans* in humans and *P. haemolytica* and *Brucella abortus* in cattle.

As discussed previously, anti-EG and anti-EV IgG antibodies have been induced *in vivo*. However, the ability of monoclonal or polyclonal IgG antibodies specific for EV or EG to confer a protective effect *in vivo* against fescue toxicosis in the murine model has not been investigated. The purpose of our research was to generate a systemic and mucosal immune response against EG and to evaluate the protection afforded in mice against fescue toxicosis. Passive immunization of mice with monoclonal anti-EV antibodies of the IgG isotype was to show the importance of serum IgG in the protection against fescue toxicosis. Weight gains, concentrations of serum ALP, cholesterol, and prolactin were the indices used to evaluate the ability of anti-EG/anti-EV antibodies to confer protection in mice.

Diet Preparation and Immunization of Mice

An identical genotype of EI and EF Kentucky 31 fescue seed was obtained from International Seed Inc. and stored at –20°C. Concentrations of EV in the seed were determined by HPLC analysis as previously described (Hill et al., 1993). Certified ground rodent chow was mixed with equal parts of ground EI and EF fescue seed by weight. The seed-chow diets were analyzed for nutrient content. Dietary intake was limited to 5 to 6 g of ground food per mouse daily. Water was provided ad libitum.

Fifty Balb-C male mice, 6 weeks of age, were blocked by weight and randomly allocated into five groups of 10 mice each. Treatment groups were as follows: (1) mice passively immunized with affinity purified monoclonal anti-EV IgG antibodies fed EI fescue seed diet (Mab group), (2) mice parenterally immunized with BSA-EG conjugate fed EI fescue seed diet (BSA-EG group), (3) mice orally immunized with CTB-EG conjugate and free CT fed EI fescue seed diet (CTB-EG group), (4) nonimmunized mice fed EI fescue seed diet (EI group), and (5) nonimmunized mice fed EF fescue seed diet (EF group).

Mice were individually housed in clear polystyrene shoebox cages and acclimatized to ground rodent chow. In the BSA-EG group, mice were injected IP with BSA-EG conjugate in FCA and revaccinated 14 d later with BSA-EG conjugate in FIA. Anti-EG IgG titers were determined on days 14 and 34 postvaccination. In the Mab group, mice were passively immunized intraperitoneally with anti-EV antibodies on days 0 and 7 of the trial period. In the CTB-EG group, mice were orally vaccinated twice with CTB-EG conjugate and cholera toxin at 10-d intervals. Titers were determined prior to the start of the study period.

After 13 d on treatment, mice were anesthetized with halothane, and blood was collected via the retroorbital sinus. The sera obtained was analyzed for prolactin, ALP, cholesterol concentrations, and anti-EG/anti-EV IgG antibodies. Feces were collected from all groups prior to the start of the study and at the completion of the study. Fecal supernatants (DeVos and Dick, 1991) were assayed for anti-EG sIgA and IgG by ELISA. Mice were weighed on days 0 and 12 of the trial period.

Statistical Analysis: Differences in weight gain, ALP, cholesterol, and serum prolactin concentrations among the treatment groups were determined to be significant at probability values < 0.05 by analysis of variance. Bonferroni test on pairwise comparisons between the means was used to control type I errors. Spearman rank correlation was used for correlation analysis.

Results and Discussion

The EV content of the dietary treatments was 1500 ppb for the EI fescue diet. No EV was detected in the EF fescue diet. The respective diets were similar upon nutritional analysis (data not shown).

Parenteral immunization resulted in elevated systemic anti-EG IgG. Anti-EG IgG titers (Log_2 titers = 8) were elevated in the BSA-EG group throughout the feeding trial. The anti-EV titers (Log_2 titers = 3.1) of the Mab group were lower than the anti-EG IgG titers of the BSA-EG group. Oral immunization with CTB-EG conjugate resulted in low systemic anti-EG IgG titers (Log_2 titers = 1.0).

Oral and parenteral immunization resulted in elevated levels of mucosal sIgA and IgG anti-EG antibodies. The sIgA concentrations in the feces of the CTB-EG group were increased from days 17 to 23 and decreased to background levels by the end of the study. However, sIgA concentrations in the feces of the BSA-EG group were increased throughout the 13-d trial period. Secretory IgA induction by intraperitoneal immunization of BSA-EG conjugate was greater

Table 3 Effects of Vaccination on Serum ALP, Prolactin, and Cholesterol Concentrations in Mice Fed EI and EF Fescue Seed

Treatment group	ALP (U/l)	Cholesterol (mg/dl)	Prolactin (ng/ml)
BSA-EG (intraperitoneally)	86.0	90.6	8.5
CTB-EG (orally)	106.7	95.0	7.4
EI	104.4	91.5	6.5
MAB	96.3	89.9	8.2
EF	114.9	102.6	7.3

than that induced by oral administration of CTB-EG conjugate. Parenteral immunization usually does not stimulate sIgA immune response at mucosal surfaces because of the compartmentalization of systemic and secretory immune systems. Induction of sIgA by intraperitoneal immunization of BSA-EG conjugate possibly is a result of systemic absorption of the conjugate followed by biliary excretion into the intestines where a local mucosal immune response was elicited. Additionally, direct stimulation of mesenteric-associated lymph nodes by intraperitoneal administration of the conjugate may have resulted in the production of specific anti-EG committed B cells. Migration of the sensitized B cells to Peyer's patches may have resulted in sIgA production in mice immunized with BSA-EG conjugate.

Although prolactin concentrations (Table 3) were not significantly different among groups, immunized groups tended to have higher prolactin levels. Prolactin levels were positively correlated ($p < 0.05$) with titers. Anesthesia stimulates prolactin release in rodents (Chi and Shin, 1978). Mice of the age used in this study are resistant to bromocryptine-induced hypoprolactinemia (McMurray et al., 1991). Perhaps mice of this age were not the best model to demonstrate the effects of fescue toxicosis on serum prolactin concentrations measured in our study.

Concentrations of ALP (Table 3) tended to be lower in the EI group than in the EF group. However, ALP concentrations were lower ($p < 0.05$) in the BSA-EG and Mab groups than in the EF group. Also, ALP concentrations in the BSA-EG group were lower ($p < 0.05$) than the EI group. The Mab group tended to have lower ($p = 0.06$) ALP concentrations than the EI group. The ALP levels were negatively correlated ($p < 0.001$) to specific serum IgG titers.

Cholesterol concentrations (Table 3) tended to be lower in the EI, CTB-EG, BSA-EG, and Mab groups as compared to the EF group. However, cholesterol levels were not correlated to titers.

Although lower serum ALP and cholesterol values may not have overt clinical significance, these results demonstrated a worsening of indices associated with antibody production against fescue toxicosis. Decreases were not caused by an antigen depot effect of the BSA-EG conjugate vaccine, because the effect of monoclonal IgG antibodies was similar. Even a low IgG titer decreased ALP and tended to decrease cholesterol concentrations as demonstrated in the Mab group. Decreases in ALP and cholesterol concentrations were likely an effect of the specific IgG antibodies, since the effect was not observed in the orally immunized CTB group.

Serum ALP is usually of hepatic origin because this isoenzyme has a long half-life. Ergotamine may decrease hepatic ALP levels by inhibition of cyclic adenosine monophosphate (cAMP). Ergotamine levels are fairly high in the liver, lung, and kidney after intravenous administration (Berde and Schild, 1978). In addition to the affinity of EG for the liver possibility for biliary excretion, IgG and IgA immune complexes are often cleared from circulation by the liver. Thus ergot alkaloid-immunoglobulin complexes could increase ergot alkaloid concentrations in the liver to further decrease ALP activity. As suggested previously in the case of sporidesmin, the specific antibody binding EV could alter elimination kinetics to allow higher levels of the alkaloid to accumulate in target organs.

Weight gain (Figure 3) was increased ($p < 0.05$) in the BSA-EG group vs. the EI group. Weight gain tended to increase in the Mab and CTB-EG groups as compared to the EI group. Weight gain was positively correlated ($p < 0.05$) to titers. Systemic IgG titers against EV were low and may not have been sufficient to neutralize all the dietary EV absorbed. A higher oral dose of the CTB-EG conjugate may have induced higher specific sIgA concentrations in mice to increase weight gains. The contribution of sIgA to the protection against adverse effects on weight gains observed in the BSA-EG group or the exacerbation of clinical indices cannot be determined from this study, since sIgA also was induced in the BSA-EG group.

Conclusion

The EG conjugated to a protein carrier induced specific antibodies that protected against adverse effects on weight gain in the murine model of fescue toxicosis. Decreased serum ALP and cholesterol concentrations, and weight gain in the EI group vs. the EF treatment group further validates the use of the murine model to study fescue toxicosis. Titers required to provide protection against adverse effects on weight gains negatively influenced serum ALP concentrations, the significance of which is unknown. Immunization did not affect serum prolactin concentrations as previously demonstrated in cattle (Hill et al., 1994), although prolactin concentrations were positively correlated with titers. Further research in cattle is indicated to determine if cattle can mount high titers to be protective against fescue toxicosis and whether the protective effects of vaccination against adverse effects on weight gains will translate into economic savings for producers. Vaccination of mucosal surfaces (oral or intranasal) are routes infrequently used in prevention of toxicosis and, in the case of fescue toxicosis, further research is needed.

OVERALL CONCLUSIONS

The Mannich reaction was used to conjugate the model hapten EG to Con A and CTB protein carriers. Both protein carrier-EG conjugates induced systemic

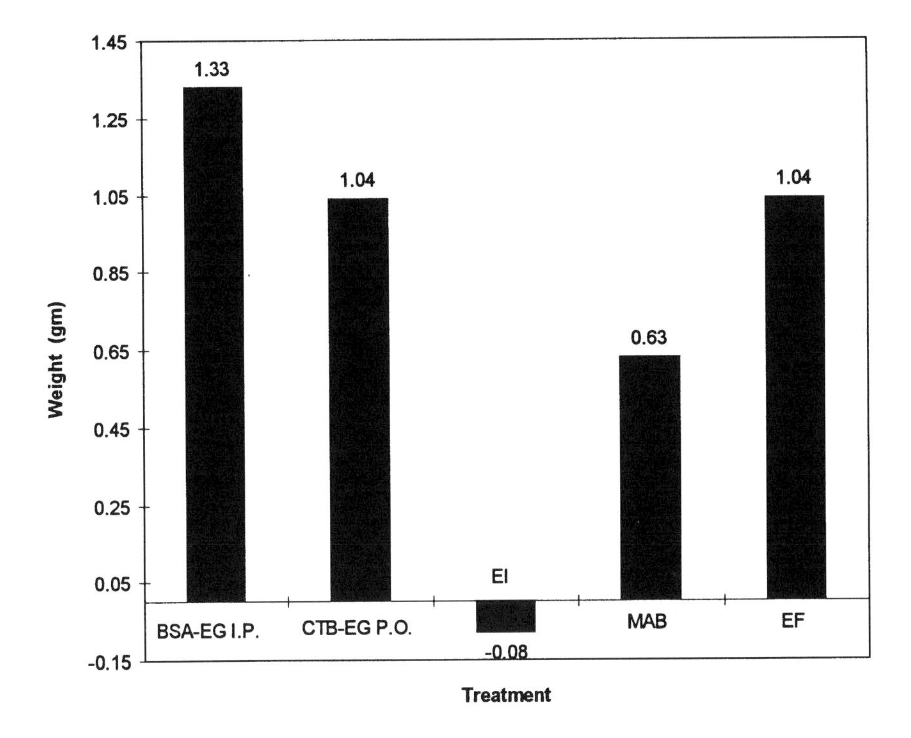

Figure 3 Average weight gain during a 13-day immunization study in mice.

IgG anti-EG antibodies in mice that had similar kinetics. The CTB-EG conjugate did not prolong antibody titers, although smaller doses were needed to elicit titers. Oral vaccination with the CTB-EG conjugate induced both mucosal IgA and IgG isotypes, in addition to a systemic IgG response. Parenteral revaccination of the CTB-EG conjugate in the orally vaccinated mice elicited an anamnestic mucosal immune response. Thus, novel vaccination protocols against fescue toxicosis may be developed based on the induction of mucosal immune responses.

Parenteral immunization with BSA-EG conjugate induced specific anti-EG antibodies that protect against adverse effects on weight gain in the murine model of fescue toxicosis. Our data supports the use of EG as a hapten in the development of a vaccine and validates the murine model in fescue toxicosis. Although vaccination protected against adverse effects on weight gains in mice, it exacerbated indices associated with fescue toxicosis. Titers required to protect against adverse effects on weight gains negatively influenced serum ALP concentrations, the significance of which is unknown. Serum prolactin concentrations were positively correlated with titers, but immunization did not significantly increase prolactin concentrations as previously demonstrated in cattle.

Our research in mice suggested that to achieve an anti-EG antibody response of adequate magnitude and duration will require either frequent vaccine administration or an enhanced vaccine delivery system, such as microspheres. To protect against fescue toxicosis, cattle could be strategically vaccinated during the summer months when the rate of weight gains in growing animals are most often affected. Cattle grazed on EI fescue pastures mounted significantly higher humoral immune responses, despite decreases in serum prolactin concentrations. Immunization against fescue toxicosis needs to be evaluated for its ability to reverse the adverse effects on CMI responses, so cattle losses incurred because of health problems can be minimized. Also, effects of antibody production on ALP and cholesterol concentrations need to be investigated to ensure these indices are not further compromised.

A protective vaccine against the adverse effects on weight gains in cattle grazed on EI fescue pastures could significantly decrease environmental pollution associated with pasture renovation. Decreased weight gains in cattle with fescue toxicosis causes the greatest economic loss for producers. A successful vaccine against fescue toxicosis would provide producers with an alternative to pasture renovation that would decrease herbicide usage and nutrient and soil runoff into surrounding watersheds. Our study indicates the feasibility of vaccine development by demonstrating protection against adverse effects on weight gains in mice. Further research in cattle is indicated to determine if cattle can mount high titers to protect against fescue toxicosis and whether the positive effects of vaccination on weight gains will translate into economic savings for producers and a decrease in pasture renovation of EI fescue pastures.

ACKNOWLEDGMENTS

The authors would like to thank the John Lee Pratt Animal Nutrition Program for financial support of this research. Also, we gratefully acknowledge Dr. J. P. Fontenot and Dr. V. G. Allen for use of the cattle grazed on fescue pastures. The authors would also like to thank Dr. M. R. Akers and Pat Boyle for their work in RIA determination of serum prolactin concentrations and Mary Nickle for her expert animal care.

REFERENCES

Azcona-Olivera, J. I., Abouzied, M. M., Plattner, R. D., Norred, W. P., and Pestka, J. J., Generation of antibodies reactive with fumonisins B1, B2, and B3 by using cholera toxin as the carrier-adjuvant, *Appl. Environ. Microbiol.,* 58, 169, 1992.

Bacon, C. W., Lyons, P. C., Porter, J. K., and Robbins, J. D., Ergot toxicity from endophyte-infected grasses: A review, *Agron. J.,* 78, 106, 1986.

Ball, D. M., Hoveland, C. S., and Lacefield, G. D., Fescue toxicity, *Southern Forages,* Potash and Phosphate Institute and Foundation for Agronomic Research, Atlanta, GA, 1991, chapter 23.

Berczi, I., Nagy, E., Asa, S. L., and Kovacs, K., Pituitary hormones and contact sensitivity in rats, *Allergy,* 38, 325, 1983.

Berde, B. and Schild, H. O., *Ergot Alkaloids and Related Compounds,* Springer-Verlag, New York, 1978, 741.

Berde, B. and Schild, H. O., *Ergot Alkaloids and Related Compounds,* Springer-Verlag, New York, 1978, 753.

Bernton, E. W., Meltzer, M. S., and Holaday, J. W., Suppression of macrophage activation and T-lymphocyte function in hypoprolactinemic mice, *Science*, 239, 401, 1988.

Bond, J., Powell, J. B., Undersander, D J., Moe, P. W., Tyrrell, H. F., and Oltjen, R. R., Forage composition and growth and physiological characteristics of cattle grazing several varieties of tall fescue during summer conditions, *J. Anim. Sci.*, 59, 584, 1984.

Bowersock, T. L., Shalaby, W., Levy, M., Samuels, M. L., Lallone, R., White, R., Borie, D. L., Lehmeyer, J., and Park, K., Evaluation of an orally administered vaccine, using hygrogels containing bacterial exotoxins of *Pasteurella haemolytica* in cattle, *Am. J. Vet. Res.,* 55, 502, 1994.

Burstein, H. J., Tepper, R. I., Leder, P., and Abbas, A. K., Humoral immune functions in IL-4 transgenic mice, *J. Immunol.,* 147, 2950, 1991.

Candore, G., Di-Lorenzo, G., Melluso, M., Cigna, D., Colucci, A. T., Modica, M. A., and Caruso, C., Gamma-interferon, interleukin-4 and interleukin-6 *in vitro* production in old subjects, *Autoimmunity,* 16, 275, 1993.

Castro, A., Grettie, D. P., Bartos, F., and Bartos, D., LSD radioimmunoassay, *Res. Commun. Chem. Pathol. Pharmacol.,* 6, 879, 1973.

Chi, H. J. and Shin, S. H., The effect of exposure to ether on prolactin secretion and the half-life of endogenous prolactin in normal and castrated male rats, *Neuroendocrinology,* 26, 193, 1978.

Cox, R. I. , *Plant Toxicology,* Seawright, A. A., Hegarty, M. P., James, L. F., and Keeler, R. F., Eds., Queensland Poisonous Plants Committee, Yeerongpilly, 1985, 98.

Culvenor, C. C., Prevention of pyrrolizidine alkaloid poisoning-animal adaptation or plant control? in *Effects of Poisonous Plants and Livestock,* Keeler, R. F., Van Kampen, K. R., and James, L. F., Eds., Academic Press, New York, 1978, 189.

Czerkinsky, C., Russell, M. W., Lycke, N., Lindblad, M., and Holmgren, J., Oral administration of a Streptococcal antigen coupled to cholera toxin B subunit evokes strong antibody responses in salivary glands and extramucosal tissues, *Infect. Immun.,* 57, 1072, 1989.

DeVos, T. and Dick, T. A., A rapid method to determine the isotype and specificity of coproantibodies in mice infected with Trichinella or fed cholera toxin, *J. Immunol. Methods,* 141, 285, 1991.

Dew, R. K., Effects of endophyte-infected tall fescue on cellular and humoral aspects of immune function in the rat, mouse, and bovine, Ph.D. Dissertation, University of Kentucky, Lexington, 1989.

Dew, R. K., Boissonneault, G. A., Gay, N., Boling, J. A., Cross, R. J., and Cohen, D. A., The effects of the endophyte *(Acremonium coenophialum)* and associated toxin(s) of tall fescue on serum titer response to immunization and spleen cell flow cytometry analysis and response to mitogens, *Vet. Immunol. Immunopathol.,* 26, 285, 1990.

Elson, C. O., Cholera toxin and its subunits as potential oral adjuvants, *Curr. Top. Microbiol. Immunol.,* 146, 29, 1989.

Fairclough, R. J., Ronaldson, J. W., Jonas, W. W., Mortimer, P. H., and Erasmuson, A. G., Failure of immunization against sporidesmin or a structurally related compound to protect ewes against facial eczema, *N. Z. Vet. J.,* 32, 101, 1984.

Flurkey, K., Bolger, M. B., and Linthicum, D. S., Preparation and characterization of antisera and monoclonal antibodies to serotonergic and dopaminergic ligands, *J. Neuroimmunol.,* 8, 115, 1985.

Fu, Y. K., Arkins, S., Fuh, G., Cunningham, B. C., Wells, J. A., Fong, S., Cronin, M .J., Dantzer, R., and Kelley, K. W., Growth hormone augments superoxide anion secretion of human neutrophils by binding to the prolactin receptor, *J. Clin. Invest.*, 89, 451, 1992.

Garner, G. B. and Cornell, C. R., Fescue foot in cattle, in *Mycotoxic Fungi, Mycotoxins, and Mycotoxicoses,* Wyllie, T. D. and Morehouse, M. G., Eds., Marcel Dekker, New York, 1978, 45.

Graham, M. B., Dalton, D. K., Giltinan, D., Braciale, V. L., Stewart, T. A., and Braciale, T. J., Response to influenza infection in mice with a targeted disruption in the interferon gamma gene, *J. Exp. Med.,* 178, 1725, 1993.

Hemken, R. W., Jackson, J. A., and Boling, J. A., Toxic factors in tall fescue, *J. Anim. Sci.,* 58, 1011, 1984.

Hiestand, P. C. and Mekler, P., Mechanism of action: Ciclosporin- and prolactin-mediated control of immunity, *Prog. Allergy,* 38, 239, 1986.

Hill, N. S., Rottinghaus, G. E., Agee, C. S., and Schultz, L. M., Simplified sample preparation for HPLC analysis of ergovaline in tall fescue, *Crop Sci.,* 33, 331, 1993.

Hill, N. S., Thompson, F. N., Dawe, D. L., and Stuedemann, J. A., Antibody binding of circulating ergot alkaloids in cattle grazing tall fescue, *Am. J. Vet. Res.,* 55, 419, 1994.

Hirabayashi, Y., Kurata, H., Funato, H., Nagamine, T., Aizawa, C., Tamura, S., Shimada, K., and Kurata, T., Comparison of intranasal inoculation of influenza HA vaccine combined with cholera toxin B subunit with oral or parenteral vaccination, *Vaccine,* 8, 243, 1990.

Holmgren, J., Czerkinsky, C., Lycke, N., and Svennerholm, A. M., Mucosal immunity: Implications for vaccine development, *Immunobiology,* 174, 157, 1992.

Hoveland, C. S., Importance and economic significance of the *Acremonium* endophytes to performance of animals and grass plant, in *Proceedings of the International Symposium on Acremonium/Grass Interactions: Program and Abstracts,* Quisenberry, S. S. and Joost, R. E., Eds., Louisiana Agricultural Experiment Station, Baton Rouge, LA, 1990, 7.

Jonas, W. E. and Erasmuson, A. F., The effect of immunizing mice with a derivative of 2-amino-5-chloro-3, 4-dimethoxy benzyl alcohol coupled to some bacteria on sporidesmin induced bilirubinaemia, *N. Z. Vet. J.,* 27, 61, 1979

Kelley, V. C. and Shelby, R. A., Production and characterization of monoclonal antibody to ergovaline, in *Proceedings of the International Symposium on Acremonium/Grass Interactions: Program and Abstracts,* Quisenberry, S. S. and Joost, R. E., Eds., Louisiana Agricultural Experiment Station, Baton Rouge, LA, 1990, 83.

Kerley, M. S., Larson, B. T., Samford, M. D., and Turner, J. E., Involvement of alpha 2-adrenergic receptors in ergovaline response of cattle, in *Proceedings on Tall Fescue Toxicosis Workshop,* SERAIEG-8, Atlanta, GA, October 23–25, 1994, 39.

Liener, I. E., Sharon, N., and Goldstein, I. J., *The Lectins: Properties, Functions and Applications in Biology and Medicine,* Academic Press, Orlando, FL, 1986.

Lipham, L. B., Thompson, F. N., Stuedemann, J. A., and Sartin, J. L., Effects of metoclopramide on steers grazing endophyte-infected fescue, *J. Anim. Sci.,* 67, 1090, 1989.

Lise, D. and Audibert, F., Immunoadjuvants and analogs of immunomodulatory bacterial structures, *Curr. Opin. Immunol.,* 2, 269, 1989.

Lyons, P. C., Plattner, R. D., and Bacon, C. W., Occurrence of peptide and clavine ergot alkaloids in tall fescue grass, *Science,* 232, 487, 1986.

MacDonald, O. A., Thulin, A. J., Weldon, W. C., Pestka, J. J., and Fogwell, R. L., Effects of immunizing gilts against zearalenone on height of vaginal epithelium and urinary excretion of zearalenone, *J. Anim. Sci.,* 68, 3713, 1990.

Marple, D. N., Osburn, T. G., Schmidt, S. P., Rahe, C. H., and Sartin, J .L., Hypothalamic catecholamines in calves consuming fungus-infected, fungus-free or fungus-free tall fescue plus ergotamine tartrate, *J. Anim. Sci.,* 66 (Suppl. 1), 373, 1988.

McCollough, S., Piper, E., Moubarak, A., Johnson, Z., and Flieger, M., Effects of fescue alkaloids on peripheral blood flow and prolactin secretion in calves, in *Proceedings on Tall Fescue Toxicosis Workshop,* SERAIEG-8 Atlanta, GA, October 23–25, 1994, 9.

McMurray, R., Keisler, D., Kanuckel, K., Izui, S., and Walker, S. E., Prolactin influences autoimmune disease activity in the female B/W mouse, *J. Immunol.,* 147, 3780, 1991.

Mestecky, J., The common mucosal immune system and current strategies for induction of immune responses in external secretions, *J. Clin. Immunol.,* 7, 265, 1987.

Miwa, A., Yoshioka, M., Shirahata, A., and Tamura, Z., Preparation of specific antibodies to catecholamines and L-3,4-dihydroxyphenylal anine, Part 1 (Preparation of the conjugates), *Chem. Pharmacol. Bull.,* 25, 1904, 1977.

Morgan-Jones,G. and Gams, W., An endophyte of *Festuca arundinacea* and the anamorph of *Epichloe typhina*, new taxa in one of two sections of *Acremonium, Mycotaxon,* 15, 311, 1982.

Mosmann, T. R. and Coffman, R. L., TH1 and TH2 cells: Different patterns of lymphokine secretion lead to different functional properties, *Annu. Rev. Immunol.,* 7, 145, 1989.

Mukherjee, P., Mastro, A. M., and Hymer, W. C., Prolactin induction of interleukin-2 receptors on rat splenic lymphocytes, *Endocrinology,* 126, 88, 1990.

Nagy, E. and Berczi, I., Prolactin and contact sensitivity, *Allergy,* 36, 429, 1981.

Nagy, E. I., Berczi, I., and Friesen, H. G., Regulation of immunity in rats by lactogenic and growth hormone, *Acta Endocrinol.,* 102, 351, 1983.

Payne, A. L., Than, K. A., Stewart, P. L., and Edgar, J. A., Vaccination against lupinosis, in *Fourth International Symposium on Poisonous Plants,* Fremantle, W. Australia, 1993, 234.

Petrosky, R. J. and Powell, R. G., Preparative separation of complex mixtures by high-speed countercurrent chromatography, in *Naturally Occurring Pest Bioregulators,* Hadlin, P. A., Ed., ACS Symposium Series 441, American Chemical Society, Washington, D.C., 1991, 426.

Porter, J. K., Bacon, C. W., and Robbins, J. D., Ergosine, ergosinine, and chanoclavine I from *Epichloe typhina, J. Agric. Food Chem.,* 27, 595, 1979.

Powrie, F., Menon, S., and Coffman, R. L., Interleukin-4 and interleukin-10 synergize to inhibit cell-mediated immunity *in vivo, Eur. J. Immunol.,* 23, 3043, 1993.

Ranadive, N. S. and Sehon, A. H., Antibodies to serotonin, *Can. J. Biochem.,* 45, 1701, 1967.

Reed, S. G., Pihl, D. L., Conlon, P. J., and Grabstein, K. H., IL-1 as adjuvant. Role of T cells in the augmentation of specific antibody production by recombinant human IL-1 α, *J. Immunol.,* 142, 3129, 1989.

Roper, R. L., Conrad, D. H., Brown, D. M., Warner, G. L., and Phipps, R. P., Prostaglandin E2 promotes IL-4 induced IgE and IgG1 synthesis, *J. Immunol.,* 145, 2644, 1990.

Russell, D. H., Matrisian, L., Libler, R., Larson, D. F., Poulos, B., and Magun, B. E., Prolactin receptors on human lymphocytes and their modulation by cyclosporin, *Biochem. Biophys. Res. Commun.,* 121, 899, 1984.

Russell, M. W. and Wu, H. Y., Distribution, persistence, and recall of serum and salivary antibody responses to preoral immunization with protein antigen I/II of *Streptococcus mutans* coupled to the cholera toxin B subunit, *Infect. Immun.,* 59, 4061, 1991.

Saker, K. E., Allen, V. G., Kalnitsky, J., and Fontenot, J. P., Select Immune Response in Beef Calves Grazed on Endophyte-Infected Tall Fescue, submitted for publication, 1995

Shelby, R. A. and Kelley, V. C., An immunoassay for ergotamine and related alkaloids, *J. Agric. Food Chem.,* 38, 1130, 1990.

Shelby, R. A. and Kelley, V. C., Detection of ergot alkaloids in tall fescue by competitive immunoassay with a monoclonal antibody, *Food Agric. Immunol.,* 3, 169, 1991.

Smith, F. T., Synthesis of ergovaline, in *Proceedings on Tall Fescue Workshop,* SERAIEG-8, October 25–26 Atlanta, GA, 1993, 5.

Smith, J. F., diMenna, M. E., and Towers, N. R., Zearalenone and its effects on sheep, in *Proceedings of the 12th International Congress on Animal Reproduction,* The Hague, The Netherlands, 1992, 1219.

Spangelo, B. L., Hall, N., Ross, P. C., and Goldstein, A. L., Stimulation of *in vivo* antibody production and concanavalin-A-induced mouse spleen cell mitogenesis by prolactin, *Immunopharmacology,* 14, 11, 1987.

Sprowls, R. W., Fescue toxicosis: Potential Health Problems for Stocker and Feeder Cattle, in *Proceedings of the Symposium on Fescue Toxicosis—From Cow-Calf to Slaughter,* Amarillo, TX, April 15, 1987.

Stewart, C., Lamberton, J. A., Fairclough, R. J., and Pass, M. A., Vaccination as a means of preventing lantana poisoning, *Aust. Vet. J.,* 65, 349, 1988

Stuedemann, J. A., Breedlove, D. L., Pond, K. R., Belesky, D. P., Tate, L. P., Thompson, F. N., and Wilkinson, S. R., Effect of endophyte *(Acremonium coenophialum)* infection of tall fescue and paddock exchange on intake and performance of grazing steers, *XVI International Grassland Congress,* Nice, France, 1989, 1243.

Stuedemann, J. A. and Hoveland, C. S., Fescue endophyte: History and impact on animal agriculture, *J. Prod. Agric.,* 1, 39, 1988.

Stuedemann, J. A., Rumsey, T. S., Bond, T., Wilkinson, S. R., Bush, L. P., Williams, D. J., and Caudle, A. B., Association of blood cholesterol with occurrence of fat necrosis in cows and tall fescue summer toxicosis in steers, *Am. J. Vet. Res.,* 46, 1990, 1985.

Tamura, S. I., Samegai, Y., Kurata, H., Kikuta, K., Nagamine, T., Aizawa, C., and Kurata, T., Enhancement of protective antibody responses by cholera toxin B subunit inoculated intranasally with influenza vaccine, *Vaccine,* 7, 257, 1989.

Taunton-Rigby, A., Sher, S. E., and Kelley, P. R., Lysergic acid diethylamide: Radioimmunoassay, *Science,* 181, 165, 1973.

Thompson, F. N., Stuedemann, J. A., Sartin, J. L., Belesky, D. P., and Devine, O. J., Selected hormonal changes with summer fescue toxicosis, *J. Anim. Sci.,* 65, 727, 1987.

White, H., Personal communication, 1990.

Yates, S. G., Plattner, R. D., and Garner, G. B., Detection of ergopeptine alkaloids in endophyte-infected, toxic Ky-31 tall fescue by mass spectrometry/mass spectrometry, *J. Agric. Food Chem.,* 33, 719, 1985.

Zavos, P. M., Varney, D. R., Jackson, J. A., Hemken, R. W., Siegel, M. R., and Bush, L. P., Lactation in mice fed endophyte-infected tall fescue seed, *Theriogenology,* 30, 865, 1988

Zavos, P. M., Varney, D. R., Jackson, J. A., Siegel, M. R., Bush, L. P., and Hemken, R. W., Effect of feeding fungal endophyte *(Acremonium coenophialum)*-infected tall fescue seed on reproductive performance in Cd-1 mice through continuous breeding, *Theriogenology,* 27, 549, 1987.

CHAPTER 17

Optimizing Mineral Levels and Sources for Farm Animals

Jerry W. Spears

INTRODUCTION

Minerals are critical to the normal functioning of essentially all biochemical processes in the body. Optimizing the concentration of minerals in animal diets should generally mean providing sufficient bioavailable concentrations of essential minerals to maximize economically important processes such as growth, efficiency of gain, milk production, reproduction, and health (immune function). In recent years, there has been increased environmental concern about the excretion of minerals by animals. Presently there are few if any restrictions in the United States regarding the concentration of minerals occurring in animal waste; most regulations are directed to land applications of nitrogen. In the future, regulations will likely be placed on the amounts of certain minerals that can be applied via animal waste per unit of land area. Therefore, it is important to define mineral levels and sources that will optimize animal production and mineral losses in urine and feces.

CURRENT DIETARY MINERAL CONCENTRATIONS

Estimates of the range and median mineral concentrations in swine and lactating dairy cow diets in North Carolina are shown in Tables 1 and 2. These data were obtained from recent feed samples analyzed at the North Carolina Feed Testing Laboratory. Only one sample from a given farm was included in the summaries presented. Although mineral concentrations in diets may vary from one state to another, results presented in Tables 1 and 2 are probably typical of those in most areas of the United States. The median level for each mineral is presented instead of the mean because a very high or low value can drastically affect the mean. The median level for each mineral indicates that 50% of the samples analyzed were below and 50% were above the median value. For example, in dairy diets, 50% of the samples were below and 50% were above 1.03% calcium. For all minerals examined, the median value exceeded the National Research Council (NRC) requirements. The median concentration for several minerals (e.g., iron, copper) was several times higher than the requirements.

1-56670-199-6/96/$0.00+$.50

Table 1 Mineral Concentrations in Total Mixed Rations for Lactating Dairy Cows[a]

	Requirement[b]	Range	Median	Median: requirement
Calcium (%)	0.58	0.63–1.61	1.03	1.77
Phosphorus (%)	0.37	0.36–0.66	0.49	1.32
Sodium (%)	0.18	0.17–0.60	0.32	1.78
Magnesium (%)	0.20	0.26–0.49	0.34	1.70
Sulfur (%)	0.20	0.17–0.35	0.24	1.20
Potassium (%)	0.90	0.93–1.47	1.21	1.34
Copper (ppm)	10	7–46	24	2.40
Iron (ppm)	50	218–1603	488	9.76
Manganese (ppm)	40	42–161	101	2.53
Zinc (ppm)	40	43–399	108	2.70

[a] Results are from analysis conducted recently at the North Carolina Feed Testing Laboratory (n = 31).

[b] Based on a 600-kg cow producing 30 kg of milk per day.

Some of the high mineral concentrations relative to requirements can be explained by commonly used feeds being high in certain minerals. However, in many instances, the high mineral concentrations are undoubtedly due to high supplemental levels. There are a number of reasons often stated as to why nutritionists formulate diets for mineral levels in excess of recommended requirements. Many nutritionists believe that NRC recommendations for some minerals do not meet requirements under practical feeding conditions. For some minerals, requirements are poorly defined, and certainly factors that impact mineral requirements need to be better characterized. Supplementation of minerals above requirements is also practiced as a safety margin to prevent any likelihood of deficiencies. Providing a certain level of safety may be sound, but the question becomes at what magnitude relative to requirements should safety levels be provided. Possible genetic differences in mineral requirements within a species can also be argued as a justification for supplementing minerals well in excess of recommended requirements. Little is known regarding genetic differences in mineral requirements, but in some instances, genetic differences may be of importance. For example, copper absorption has been shown to be different among breeds of sheep (Wiener et al., 1978), and recent research suggests that Simmental and Charolais cattle have a higher copper requirement than Angus (Ward and Spears, 1994).

The trace mineral concentrations in energy and protein components of the diet are often not taken into account when formulating concentrations of trace minerals to provide in the diet. Industry levels of trace minerals being supplemented to diets of different animal species were compiled by Nelson (1987). With most trace minerals, the level added met or exceeded NRC requirements. With the exception of phosphorus, only limited research has been conducted examining bioavailability of minerals from energy and protein feedstuffs. However, we should probably not assume that most trace elements present in feedstuffs are totally unavailable. Variation in the mineral content of feedstuffs presents a problem when formulating mineral supplements. Forages differ considerably in mineral concentrations (Underwood, 1981). However, much less variation in mineral content is usually

Table 2 Mineral Concentrations in Sow and Finishing Swine Diets[a]

	Sow				Finishing swine			
	Requirement	Range	Median	Median: requirement	Requirement	Range	Median	Median: requirement
Calcium (%)	0.75	0.62–2.01	1.21	1.61	0.50	0.57–1.38	0.96	1.92
Phosphorus (%)	0.60	0.45–1.17	0.84	1.40	0.40	0.45–0.78	0.62	1.55
Sodium (%)	0.15	0.13–0.45	0.22	1.47	0.10	0.13–0.29	0.19	1.90
Magnesium (%)	0.04	0.12–0.44	0.21	5.25	0.04	0.13–0.21	0.16	4.00
Potassium (%)	0.20	0.43–1.15	0.78	3.90	0.17	0.48–0.93	0.72	4.23
Copper (ppm)	5	12–222	22	4.40	3	9–281	20	6.67
Iron (ppm)	80	162–698	376	4.70	40	131–503	311	7.76
Manganese (ppm)	10	28–203	77	7.70	2	37–160	62	31.00
Zinc (ppm)	50	79–497	167	3.34	50	103–205	149	2.98

[a] Results are from analysis conducted recently at the North Carolina Feed Testing Laboratory (n = 26 for sow and n = 17 for finishing diets).

found in cereal grains and oilseed meals that are commonly used in swine, poultry, and dairy diets.

POSSIBLE CONSEQUENCES OF EXCESSIVE MINERAL SUPPLEMENTATION

A number of problems may arise from excessive supplementation of minerals. In his classic book, *The Mineral Nutrition of Livestock,* Underwood (1981) indicated that (1) mineral supplements often contain an unnecessarily wide margin of safety as an insurance against deficiency and (2) the provision of minerals beyond the needs of an animal is economically wasteful and can be harmful. In addition, providing minerals in excess of an animal's requirement reduces the efficiency of mineral utilization and thus increases mineral excretion in feces and (or) urine. For most essential minerals, an animal attempts to maintain a relatively constant body concentration via homeostatic control mechanisms that alter absorption and (or) excretion. Depending on the mineral in question, if intake of the mineral relative to the animal's requirement is exceeded, the percentage of the mineral absorbed will decrease or the amount excreted (via urine, bile, etc.) will increase.

Numerous interactions occur between minerals. The balance between dietary minerals, in regard to bioavailable concentrations relative to animal requirements, is an important factor affecting mineral utilization and thus animal requirements. For example, if one mineral is supplemented or present in feed ingredients at concentrations well above requirements, this mineral may negatively affect utilization of other minerals and subsequently lead to a deficiency.

Excessive mineral supplementation may also result in toxicosis. High levels of iron in ruminant diets may result in a number of health problems (Pitzen, 1993). Most of the animal's requirement for iron can be derived from feed ingredients, but most feed companies continue to include iron in their mineral supplements. In animal diets where phosphorus is added, the phosphorus supplement alone probably provides enough iron to meet requirements. Commercial dicalcium and defluorinated phosphates contain approximately 10,000 ppm of iron. Relative to ferrous sulfate, iron from dicalcium phosphate and defluorinated phosphate has been shown to be 67 and 48% available, respectively, in chicks (Henry et al., 1992b). Iron from defluorinated phosphate was found to be at least 50% as available as ferrous sulfate in pigs (Kornegay, 1972).

PHOSPHORUS AND CALCIUM

Phosphorus has received more attention from an environmental standpoint than any other mineral. In the Netherlands, the amount of phosphorus from animal waste that can be applied to soil is already regulated (Jongbloed and Lenis, 1992). Much of the phosphorus in cereal grains and other feedstuffs of plant origin is in the form of phytate phosphorus. In ruminant animals, phytase, produced by microorganisms in the rumen, hydrolyzes phytate, releasing phosphorus in an available form. However, phytate phosphorus is of very low bioavailability in nonruminants. The addition of microbial phytase to nonruminant diets offers considerable potential to increase bioavailability of phosphorus and other minerals such as calcium and zinc in feeds (Jongbloed and Lenis, 1992; Lei et al., 1993). Phytase is discussed in detail in Chapter 18.

Avoiding oversupplementation and selecting highly available supplemental phosphorus sources can reduce excretion of phosphorus by animals. Phosphorus from defluorinated rock phosphate and steamed bonemeal is approximately 87 and 82% as available, respectively, as phosphorus in dicalcium phosphate (actually a mixture of mono- and dicalcium phosphate) in nonruminants (NRC, 1988). In cattle, phosphorus from defluorinated phosphate (Miller et al., 1987) and monoammonium phosphate (Jackson et al., 1988) is equal in availability to phosphorus from dicalcium phosphate.

Most calcium sources are of similar bioavailability in nonruminants. Dolomitic limestone is approximately 65% as available as calcitic limestone (Miller, 1980). Early studies in ruminants reviewed by Miller (1980) indicated that calcium from bonemeal and dicalcium phosphate can be 30 to 40% more available than calcium carbonate.

Calcium and phosphorus requirements are higher for maximal bone strength than for maximal growth. Since bone mineralization decreases with age, there has been interest in omitting or reducing supplemental phosphorus in diets during the last part of the finishing period. Withdrawing dicalcium phosphate from the diets of broilers during the last week of feeding did not affect gain or feed efficiency (Chen and Moran, 1994). However, omitting dicalcium phosphate from the diet tended to increase carcass defects and reduce the percentage of carcasses that graded "A" (Chen and Moran, 1994).

Phosphorus supplementation has traditionally been recommended for beef cattle that are grazing forages. Recent studies (reviewed by Spears, 1994) suggest that NRC recommendations may overestimate phosphorus requirements of beef cattle, and in many instances supplemental phosphorus may not be needed. The endogenous loss of phosphorus constitutes the animal's maintenance requirement for phosphorus and is primarily of fecal origin in ruminants fed forage diets. Endogenous fecal losses of phosphorus consist largely of unabsorbed salivary phosphorus. Phosphorus requirements in ruminants have been difficult to define because fecal endogenous losses vary depending on dietary phosphorus intake and other factors that affect salivary phosphorus output (Spears, 1994).

ELECTROLYTES

The major electrolytes, sodium, potassium, and chloride, are all well absorbed. The level of potassium naturally present in diets usually meets or exceeds animal requirements, with the possible exception of lactating dairy cattle and finishing cattle consuming high-concentrate diets. Sodium chloride has generally been supplemented to diets at levels that probably exceed requirements of nonlactating animals by two- to fourfold. Adding minimal quantities of sodium and chloride needed to meet animal requirements and then formulating for a given electrolyte balance (mEq Na+K-Cl) may optimize animal performance and reduce urinary excretion of these minerals.

Numerous animal species have responded positively to modifying dietary electrolyte or cation-anion balance. Body weight gain was maximized in chicks fed diets containing an electrolyte balance of 25 mEq/100 g of diet (Mongin, 1981). In growing and finishing swine, fed in high ambient temperature, increasing electrolyte balance from 2.5 up to 40 mEq/100 g of diet resulted in a linear improvement in gain and feed intake (Haydon et al., 1990). Increasing electrolyte balance in sow diets from 13 to 25 mEq/100 g of diet tended ($p < 0.07$) to improve 21-d pig weights (Dove and Haydon, 1994). Based on recent studies with feedlot cattle, the optimal electrolyte balance in growing and finishing diets is between 15 and 30 mEq/100 g of diet (Ross et al., 1994a, 1994b). Electrolyte balance has been extensively studied in dry and lactating dairy cows, and a number of responses have been observed (Block, 1994).

BIOAVAILABILITY OF INORGANIC TRACE MINERAL SOURCES

Bioavailability studies with inorganic trace elements have frequently used reagent-grade sources, and these may differ in purity, solubility, and availability from feed-grade sources (Nelson, 1987). Various feed-grade sources of a particular metal (oxide, sulfate, etc.) can also differ in purity and other factors that affect bioavailability of the mineral.

Zinc

Recent studies with chicks (Wedekind and Baker, 1990) and young pigs (Wedekind et al., 1994) indicate that zinc in feed-grade zinc oxide (ZnO) is only 44 to 87% as available as zinc sulfate ($ZnSO_4$). In both studies, animals were fed basal levels of zinc below their requirements. Zinc from ZnO was 44 and 61% as available as $ZnSO_4$ in chicks, based on bone zinc concentrations and weight gain, respectively (Wedekind and Baker, 1990). In pigs, relative availabilities of zinc from ZnO were 67 and 87%, based on bone and plasma zinc concentrations, respectively (Wedekind et al., 1994). When feed-grade zinc sources were supplemented at high concentrations (3000 to 5000 ppm), pigs fed $ZnSO_4$ had much higher plasma zinc concentrations than those fed ZnO (Hahn and Baker, 1993).

In contrast to recent studies with nonruminants, ZnO and $ZnSO_4$ appear to be utilized similarly in ruminants. Kegley and Spears (1992) compared feed-grade ZnO and $ZnSO_4$ in lambs and detected no differences in zinc availability based on gain, plasma zinc, plasma alkaline phosphatase activity, or apparent absorption of zinc. Kincaid (1979) dosed calves with high levels of zinc (20 mg of Zn per kilogram of body weight) from different sources and estimated availability using the increase in plasma zinc concentrations. He concluded that zinc availability from ZnO and $ZnSO_4$ was similar.

Zinc oxide sources can be produced by different processes and probably vary more in availability than different $ZnSO_4$ sources. The ZnO evaluated in recent studies (Wedekind and Baker, 1990; Kegley and Spears, 1992; Wedekind et al., 1994) was prepared by the Waelz process, which produces a high-quality product compared to other ZnO sources (Nelson, 1987). Feed-grade ZnO can contain high concentrations of contaminates such as lead and cadmium, as well as significant quantities of zinc metal that is probably of very poor availability. Therefore, some of the commercially available ZnO sources may be less available than those evaluated in recent trials.

Copper

The primary sources of inorganic copper currently available in the feed industry are copper sulfate ($CuSO_4$) and copper oxide (CuO). Bioavailability of copper from feed-grade CuO is approximately 0%, based on a number of recent studies. Feeding CuO at high concentrations, well in excess of requirements, did not increase liver copper in pigs (Cromwell et al., 1989) or chicks (Ledoux et al., 1991; Baker et al., 1991). In these studies, feeding $CuSO_4$ greatly increased liver copper. In calves deficient in copper, feeding CuO did not increase plasma copper or ceruloplasmin activity relative to noncopper-supplemented calves (Kegley and Spears, 1994). Copper oxide also was ineffective in preventing a decline in copper status when steers were fed the copper antagonist molybdenum (Kegley and Spears, 1994). Feed-grade CuO is primarily in the cupric rather than the cuprous form (Cu_2O). Reagent-grade Cu_2O was as available as reagent-grade $CuSO_4$, based on liver copper accumulation in chicks (Baker et al., 1991).

Feed-grade copper carbonate ($CuCO_3$) also appears to be less available than feed-grade $CuSO_4$. Based on liver copper concentrations in chicks fed high levels of copper, $CuCO_3$ was approximately 61% as available as $CuSO_4$ (Ledoux et al., 1991). However, when reagent-grade sources of $CuSO_4$ and basic $CuCO_3$ were compared at low dietary concentrations in chicks, the two sources were of equal bioavailability (Aoyagi and Baker, 1993a). In copper-deficient steers, feed-grade $CuCO_3$ increased plasma copper to the same extent as $CuSO_4$; however, liver concentrations increased less in animals fed $CuCO_3$ (Ward and Spears, 1994). A feed-grade source of tribasic cupric chloride (Cu_2OH_3Cl) was recently found to be similar in bioavailability to reagent-grade $CuSO_4$ in chicks (Ammerman et al., 1995). Aoyagi and Baker (1993b) reported that the relative bioavailability of copper from reagent-grade copper chloride (CuCl) was 145% that in $CuSO_4$.

Manganese

Manganese sulfate ($MnSO_4$) is the most bioavailable of the inorganic sources of manganese. Compared with reagent-grade $MnSO_4$ (set at 100%), bioavailability values for feed-grade manganese oxide (MnO) have ranged from 53 to 81% (Wong-Valle et al., 1989a, 1989b; Henry et al., 1992a). Furthermore, different feed-grade MnO sources can vary in regard to manganese availability. Three MnO sources evaluated in chicks showed bioavailability estimates of 81, 64, and 62% relative to reagent-grade $MnSO_4$ (Wong-Valle et al., 1989b). Manganese from two MnO sources tested in lambs was 70 and 53% as available as manganese in reagent-grade $MnSO_4$ (Henry et al., 1992a). In both studies, the MnO sources with the highest relative bioavailability also had the highest percentage of manganese (64%) in the product. The variation in manganese availability from different feed-grade MnO sources probably relates to the amount of MnO_2 present. The amount of manganese in feed-grade MnO from MnO_2 can vary from a low of 1 to 3% up to a high of 20 to 25% of the total manganese (Nelson, 1987). Reagent-grade MnO_2 has been shown to be only 40% as available as reagent-grade MnO in chicks (Henry et al., 1987).

Selenium

Selenium is generally supplemented to animal diets as sodium selenite (Na_2SeO_3). Podoll et al. (1992) reported that Na_2SeO_3 and sodium selenate ($NaSeO_4$) were of similar bioavailability in dairy cows, sheep, and horses, based on serum selenium concentrations and glutathione peroxidase activity. Availability of selenium from cobalt selenite (Pehrson et al., 1989) and calcium selenite (Henry et al., 1988) also was similar to Na_2SeO_3.

Iron

Ferrous sulfate ($FeSO_4 \times H_2O$) and ferrous carbonate ($FeCO_3$) are the major sources of iron used in the feed industry. A considerable amount of iron oxide (FeO) is added to feed as a coloring agent, but availability of iron from this source is very low (Miller, 1980). Iron from $FeCO_3$ has been reported to be from 0 to 74% as available as iron from $FeSO_4$ (Miller, 1980). Most of the variation observed between studies in bioavailability of iron from $FeCO_3$ is probably related to the amount of ferrous vs. ferric iron present in the product. A good source of feed-grade $FeCO_3$ should contain only 1 to 3% ferric iron and 39 to 40% ferrous iron (Nelson, 1982). However, the amount of iron in the ferric form can increase with storage because of oxidation of the $FeCO_3$ to the ferric form. A good-quality $FeCO_3$ should be tan in color, whereas one high in ferric iron would likely have a reddish color (Nelson, 1982).

BIOAVAILABILITY OF ORGANIC TRACE ELEMENTS

Recently there has been considerable interest in the use of organic trace minerals in animal diets. The type of ligand or ligands used to form metal complexes or chelates varies, but in most organic products the metal is bound to an amino acid(s), hydrolyzed protein, or a polysaccharide (Spears, 1993). Most of the organic minerals marketed are classified as complexes, amino acid chelates, or proteinates. Organic trace minerals have been developed based on the theory that they are more bioavailable or more similar to forms naturally occurring in the body than inorganic sources. If the metal chelate or complex is stable in the digestive tract, the metal should be protected from forming insoluble complexes with other dietary components that inhibit absorption and thus allow for greater absorption. Certain organic trace minerals also appear to be metabolized differently following absorption and may affect some body processes differently from inorganic sources. Use of organic trace minerals has been limited because of their high cost relative to inorganic sources. However, in the future, if regulations are placed on the amount of certain trace minerals that can appear in animal excreta, organic trace minerals may become more widely used if their availability or effectiveness is found to be higher than that of inorganic trace mineral sources.

Zinc

Wedekind et al. (1992) compared the bioavailability of zinc from zinc-methionine complex (Zn-Met) to feed-grade $ZnSO_4$ in chicks fed either a purified, semipurified (soy isolate-dextrose-based diet), or a practical corn-soybean meal-based diet. Using tibia zinc concentrations as a measure of availability, relative bioavailability of zinc from ZnMet was 177 to 206% compared with $ZnSO_4$ in chicks fed the semipurified and corn-soybean meal diets, respectively. In the purified-crystalline amino acid diet that was devoid of phytate and fiber, Zn-Met bioavailability was 116% relative to $ZnSO_4$. They concluded that Zn-Met was considerably more bioavailable than feed-grade $ZnSO_4$ when added to practical chick diets containing phytate and fiber. In contrast, Pimentel et al. (1991) reported that bioavailability of zinc from Zn-Met and reagent-grade ZnO was similar in chicks fed a semipurified diet based on growth and tibia zinc concentrations. However, chicks fed ZnMet had higher zinc concentrations in the pancreas than those fed ZnO, suggesting that the two forms of zinc were metabolized differently. The zinc-lysine (Zn-Lys) complex was similar in bioavailability to $ZnSO_4$ in chicks based on growth rate and tibia zinc concentrations (Aozagi and Baker, 1993c).

In growing and finishing pigs, no differences were observed in bioavailability between $ZnSO_4$ and ZnMet based on growth and bone zinc concentrations (Hill et al., 1986). More recently, Wedekind et al. (1994) reported that ZnMet and $ZnSO_4$ were similar in availability based on coccygeal vertebrae and plasma zinc concentrations. Based on metacarpal zinc concentrations, ZnMet was only 60% as available as $ZnSO_4$ (Wedekind et al., 1994). When zinc was supplemented at pharmacological levels (3000 ppm), pigs fed ZnMet had higher plasma zinc concentrations than pigs fed $ZnSO_4$ (Hahn and Baker, 1993). Availability studies with ZnLys in pigs have been inconsistent. Zinc from ZnLys was only 24 to 38% as available as $ZnSO_4$ in pigs based on bone zinc concentrations (Wedekind et al., 1994). In contrast to these results, Cheng et al. (1995) reported that ZnLys and

$ZnSO_4$ were equal in availability for pigs based on growth and tissue zinc concentrations. Swinkels et al. (1991) reported that a zinc amino acid chelate was of similar bioavailability to $ZnSO_4$ when fed to zinc depleted pigs.

Apparent absorption of zinc from ZnMet and reagent-grade ZnO was similar in lambs fed a zinc-deficient diet or a hay-based diet (Spears, 1989). In lambs fed the zinc-deficient diet, urinary excretion of zinc tended to be lower in lambs fed ZnMet, resulting in higher zinc retention. Spears (1989) concluded that zinc in ZnO and ZnMet was absorbed to a similar extent, but zinc from these two sources appeared to be metabolized differently following absorption. In steers, apparent absorption of zinc from ZnMet and $ZnSO_4$ was similar, but urinary excretion of zinc was lower in steers fed ZnMet following stress (Nockels et al., 1993).

Zinc methionine has improved performance, carcass quality, and immune response measurements in ruminants above those noted in animals fed isozinc levels from ZnO. Quality grades, marbling scores, and percentage kidney, pelvic, and heart fat were higher in finishing steers fed ZnMet compared with steers in the control or ZnO treatments (Greene et al., 1988). The basal diet used in this study contained 82 ppm of zinc, and the addition of 360 mg of zinc per day from either source did not affect performance. Average daily gain and feed efficiency were similar for growing heifers fed a control diet containing 24 ppm of zinc and those supplemented with 25 ppm of zinc from ZnO (Spears, 1989). However, heifers receiving 25 ppm of zinc from ZnMet gained 8.1% faster ($p < 0.07$) and 7.3% more efficiently ($p < 0.08$) than controls over the 126-d study. Lactating dairy cows fed ZnMet tended to have lower somatic cell counts and higher milk yields than cows fed a similar level of zinc from ZnO (Kincaid et al., 1984). In stressed steers, antibody titers against bovine herpesvirus-1 on day 14 following vaccination were 47 and 31% higher in steers supplemented with ZnMet compared with control and ZnO fed cattle, respectively (Spears et al., 1991). Calves experimentally challenged with infectious bovine rhinotracheitis virus tended to recover from the disease more rapidly when fed ZnMet compared with ZnO (Chirase et al., 1991).

The addition of 40 ppm of ZnMet to broiler breeder diets that contained 72 to 83 ppm of zinc increased cellular immune response in their progeny (Kidd et al., 1993). Zinc oxide supplementation at a similar level did not affect cellular immunity. Supplementing a combination of ZnMet and manganese methionine (MnMet) to turkey diets increased humoral and cell-mediated immune response and reduced mortality and leg abnormalities (Ferket and Qureshi, 1992; Ferket et al., 1992). In this study, increasing supplemental zinc from 80 to 140 ppm and manganese from 120 to 180 ppm from the sulfate forms did not affect immunity, mortality, or leg problems.

Zinc proteinate also has been evaluated in ruminants. Zinc retention was greater in lambs fed zinc proteinate compared with lambs fed ZnO (Lardy et al., 1992). In growing and finishing cattle, zinc proteinate tended to improve gain above that seen in steers fed a similar level of zinc from ZnO (Spears and Kegley, 1994). Steers fed zinc proteinate had heavier hot carcass weights and slightly higher dressing percentages than those fed ZnO.

Copper

Copper from copper lysine (Cu-Lys) has been shown to be similar in availability to $CuSO_4$ when compared at low and high dietary concentrations in chicks (Baker et al., 1991; Aoyagi and Baker, 1993b, 1993c; Pott et al., 1994). A copper-methionine (Cu-Met) complex also was equal in bioavailability to $CuSO_4$ in chicks (Aoyagi and Baker, 1993b). Depending on whether copper was supplemented at high or low concentrations, high dietary concentrations of ascorbate and cysteine inhibited copper utilization from $CuSO_4$ to a greater extent than from CuLys or CuMet (Aoyagi and Baker, 1994).

Pharmacological levels of copper are often fed to swine as a growth promotant. Because large amounts of copper are excreted in the feces when high copper is fed, there has been interest in determining if organic forms of copper are more effective in stimulating growth at lower dietary concentrations than $CuSO_4$. Copper lysine has increased gain and feed intake above that observed in pigs fed $CuSO_4$ in some studies (Coffey et al., 1994; Zhou et al., 1994), but not in others (Apgar et al., 1995). A polysaccharide-copper chelate was found to be equal to $CuSO_4$ in growth promotion when fed to swine at high levels (Stansbury et al., 1990).

In ruminants, there has been considerable interest in the use of copper chelates or complexes because of the strong interactions that occur between copper and molybdenum and sulfur. Kincaid et al. (1986) compared copper proteinate and $CuSO_4$ in terms of their ability to increase copper status in calves fed a hay-concentrate diet naturally high in molybdenum. After 84 d, calves receiving copper proteinate had higher plasma (0.87 vs. 0.75 mg/l) and liver (325 vs. 220 ppm) copper concentrations than calves supplemented with a similar level of copper from the sulfate form. Copper sulfate did not increase liver copper above values observed in noncopper-supplemented calves. Ward and Spears (1994) reported that copper proteinate had a greater bioavailability than $CuSO_4$ when dietary molybdenum was high, but the two sources were similar in bioavailability when dietary molybdenum was low. In contrast to these findings, Wittenberg et al. (1990) reported that availability of copper in copper proteinate and $CuSO_4$ was similar, based on liver and plasma copper concentrations when fed to copper-depleted steers receiving diets high in molybdenum.

Relative availability of copper from Cu-Lys and $CuSO_4$ was similar in growing steers, based on plasma copper concentrations and activity of ceruloplasmin (Ward et al., 1993). In this study, the copper sources were compared in the absence and presence of high dietary molybdenum. When fed to copper-deficient calves, CuLys was 108 to 112% as available as $CuSO_4$ based on plasma copper and ceruloplasmin activity, respectively (Kegley and Spears, 1994). These numerical differences were not significant. Following stress induced by feed and water restriction and ACTH administration, copper retention was much higher in steers fed CuLys compared with those given $CuSO_4$ (Nockels et al., 1993). The in creased retention resulted from both increased absorption and decreased urinary excretion of copper.

Manganese

Relative availability of manganese in Mn-Met was 174% of that present in feed-grade MnO based on bone manganese accumulation in chicks (Fly et al., 1989). When a semipurified diet devoid of phytate and fiber was fed, bioavailability of Mn-Met relative to MnO was 130% (Fly et al., 1989). Compared with reagent-grade $MnSO_4$, relative availability of manganese in Mn-Met was 108% based on bone and 132% based on kidney manganese concentrations in chicks (Henry et al., 1989). Relative availability of manganese from Mn-Met was 120% of that present in reagent-grade $MnSO_4$ based on manganese accumulation in bone, kidney, and liver of lambs fed high dietary manganese (Henry et al., 1992a). In beef heifers fed corn silage-based diets, Mn-Met addition improved gain and feed efficiency in one of two experiments compared with control and MnO supplemented diets (Spears, 1993).

Manganese from manganese proteinate was similar in bioavailability to $MnSO_4$ in chicks fed diets either devoid of or containing fiber and phytate (Baker and Halpin, 1987). In a more recent study, manganese proteinate was 120% as available as $MnSO_4$ when tested in chicks between 1 and 21 d of age (Smith et al., 1994). However, in older birds reared under heat distress, availability of manganese in manganese proteinate relative to $MnSO_4$ was 145% (Smith et al., 1994).

Selenium

Selenomethionine (Se-Met) is the predominant form of selenium in most feedstuffs. Bioavailability of SeMet was considerably less than Na_2SeO_3 based on their ability, when fed at graded levels, to prevent exudative diathesis in chicks fed a selenium-deficient diet (Cantor et al., 1975a). However, when prevention of pancreatic fibrosis was used as the criterion to estimate bioavailability, SeMet was four times as effective as Na_2SeO_3 (Cantor et al., 1975b). Chicks fed SeMet had higher selenium concentrations in pancreas and breast muscle, but lower concentrations in kidney, liver, and heart than chicks fed Na_2SeO_3 (Osman and Latshaw, 1976). Retention of selenium from SeMet has generally been higher than from Na_2SeO_3. This is at least partly due to SeMet being incorporated into nonspecific body proteins in place of methionine (Behne et al., 1991).

Selenium absorption in ruminants is much lower than in nonruminants (Wright and Bell, 1966). The lower absorption of selenium is believed to be related to the reduction of SeO_3 to insoluble forms in the rumen. Selenomethionine and a high selenium-yeast product were both approximately twice as available as Na_2SeO_3 in growing cattle (Pehrson et al., 1989).

CONCLUSION

Environmental concerns regarding the excretion of certain minerals by animals will likely increase in the future. Providing an adequate supply of essential minerals is necessary to maximize efficiency of animal production. However, supplementing minerals in excess of animal requirements greatly increases mineral losses in animal waste. Research is needed to further characterize the availability of minerals from commonly used protein and energy feedstuffs. This would allow nutritionists to more accurately formulate diets for minerals. When mineral supplementation is needed, choosing a highly bioavailable form, whether inorganic or organic, will allow for a lower inclusion rate and thus minimize mineral losses in animal excreta.

REFERENCES

Ammerman, C. B., Henry, P. R., Luo, X. G., and Miles, R. D., Bioavailability of copper from tribasic cupric chloride for nonruminants, *J. Anim. Sci.,* 73 (Suppl. 1), 18, 1995.

Aoyagi, S. and Baker, D. H., Bioavailability of copper in analytical-grade inorganic copper sources when fed to provide copper at levels below the chick's requirement, *Poultry Sci.,* 72, 1075, 1993a.

Aoyagi, S. and Baker, D. H., Nutritional evaluation of copper-lysine and zinc-lysine complexes for chicks, *Poultry Sci.,* 72, 165, 1993b.

Aoyagi, S. and Baker, D. H., Nutritional evaluation of a copper-methionine complex for chicks, *Poultry Sci.,* 72, 2309, 1993c.

Aoyagi, S. and Baker, D. H., Copper-amino acid complexes are partially protected against inhibitory effects of L-cysteine and L-ascorbic acid on copper absorption in chicks, *J. Nutr.,* 124, 388, 1994.

Apgar, G. A., Kornegay, E. T., Lindemann, M. D., and Notter, D. R., Evaluation of copper sulfate and a copper lysine complex as growth promotants for weanling swine, *J. Anim. Sci.,* 73, 2640, 1995.

Baker, D. H. and Halpin, K. M., Efficiency of a manganese-protein chelate compared with that of manganese sulfate for chicks, *Poultry Sci.,* 66, 1561, 1987.

Baker, D. H., Odle, J., Funk, M. A., and Wieland, T. M., Bioavailability of copper in cupric oxide, cuprous oxide and in a copper-lysine complex, *Poultry Sci.,* 70, 177, 1991.

Behne, D., Kyriakopoulos, A., Scheid, S., and Gessner, H., Effects of chemical form and dosage on the incorporation of selenium into tissue proteins in rats, *J. Nutr.,* 121, 806, 1991.

Block, E., Manipulation of dietary cation-anion difference on nutritionally related production diseases, productivity and metabolic responses of dairy cows, *J. Dairy Sci.,* 77, 1437, 1994.

Cantor, A. H., Scott, M. L., and Noguchi, T., Biological availability of selenium in feedstuffs and selenium compounds for prevention of exudative diathesis in chicks, *J. Nutr.,* 105, 96, 1975a.

Cantor, A. H., Langevin, M. L., Noguchi, T., and Scott, M. L., Efficacy of selenium in selenium compounds and feedstuffs for prevention of pancreatic fibrosis in chicks, *J. Nutr.,* 105, 106, 1975b.

Chen, X. and Moran, E. T., Jr., Response of broilers to omitting dicalcium phosphate from the withdrawal feed: live performance, carcass downgrading and further-processing yields, *J. Appl. Poultry Res.,* 3, 74, 1994.

Cheng, J., Schell, T. C., and Kornegay, E. T., Influence of dietary lysine on the utilization of zinc from zinc sulfate and a zinc lysine complex by young pigs, *J. Anim. Sci.,* 73 (Suppl. 1), 17, 1995.

Chirase, N. K., Hutcheson, D. P., and Thompson, G. B., Feed intake, rectal temperature and serum mineral concentrations of feedlot cattle fed zinc oxide or zinc methionine and challenged with infectious bovine rhinotracheitis virus, *J. Anim. Sci.,* 69, 4137, 1991.

Coffey, R. D, Cromwell, G. L., and Monegue, H. J., Efficiency of a copper-lysine complex as a growth promotant for weanling pigs, *J. Anim. Sci.*, 72, 2880, 1994.

Cromwell, G. L., Stahly, T. S., and Monegue, H. J., Effects of source and level of copper on performance and liver copper stores in weanling pigs, *J. Anim. Sci.,* 67, 2996, 1989.

Dove, C. R. and Haydon, K. D., The effect of various diet nutrient densities and electrolyte balances on sow and litter performance during two seasons of the year, *J. Anim. Sci.,* 72, 1101, 1994.

Ferket, P. R. and Qureshi, M. A., Effect of level of inorganic and organic zinc and manganese on the immune function of turkey toms, *Poultry Sci.,* 71 (Suppl. 1), 60, 1992.

Ferket, P. R., Nicholson, L., Roberson, K. D., and Yoong, C. K., Effect of level of inorganic and organic zinc and manganese on the performance and leg abnormalities of turkey toms, *Poultry Sci.,* 71 (Suppl. 1), 18, 1992.

Fly, A. D., Izquierdo, O. A., Lowry, K. R., and Baker, D. H., Manganese bioavailability in a Mn-methionine chelate, *Nutr. Res.,* 9, 901, 1989.

Greene, L. W., Lunt, D. K., Byers, F. M. Chirase, N. K., Richmond, C. E., Knutson, R. E., and Schelling, G. T., Performance and carcass quality of steers supplemented with zinc oxide or zinc methionine, *J. Anim. Sci.,* 66, 1818, 1988.

Hahn, J. D. and Baker, D. H., Growth and plasma zinc responses of young pigs fed pharmacologic levels of zinc, *J. Anim. Sci.,* 71, 3020, 1993.

Haydon, K. D., West, J. W., and McCarter, M. N., Effect of dietary electrolyte balance on performance and blood parameters of growing-finishing swine fed in high ambient temperatures, *J. Anim. Sci.,* 68, 2400, 1990.

Henry, P. R., Ammerman, C. B., and Miles, R. D., Bioavailability of manganese monoxide and manganese dioxide for broiler chicks, *Nutr. Rep. Int.,* 36, 425, 1987.

Henry, P. R., Echevarria, M. G., Ammerman, C. B., and Rao, P. V., Estimation of the relative biological availability of inorganic selenium sources for ruminants using tissue uptake of selenium, *J. Anim. Sci.,* 66, 2306, 1988.

Henry, P. R., Ammerman, C. B., and Miles, R. D., Relative bioavailability of manganese in a manganese-methionine complex for broiler chicks, *Poultry Sci.,* 68, 107, 1989.

Henry, P. R., Ammerman, C. B., and Littell, R. C., Relative bioavailability of manganese from a manganese-methionine complex and inorganic sources for ruminants, *J. Dairy Sci.,* 75, 3473, 1992a.

Henry, P. R., Ammerman, C. B., Miles, R. D., and Littell, R. C., Relative bioavailability of iron in feed grade phosphates for chicks, *J. Anim. Sci.,* 70 (Suppl. 1), 228, 1992b.

Hill, D. A., Peo, E. R., Jr., Lewis, A. J., and Crenshaw, J. D., Zinc-amino acid complexes for swine, *J. Anim. Sci.,* 63, 121, 1986.

Jackson, J. A., Jr., Langer, D. L., and Hemken, R. W., Evaluation of content and source of phosphorus fed to dairy calves, *J. Dairy Sci.,* 71, 2187, 1988.

Jongbloed, A. W. and Lenis, N. P., Alteration of nutrition as a means to reduce environmental pollution by pigs, *Livestock Prod. Sci.,* 31, 75, 1992.

Kegley, E. B. and Spears, J. W., Performance and mineral metabolism of lambs as affected by source (oxide, sulfate, or methionine) and level of zinc, *J. Anim. Sci.,* 70 (Suppl. 1), 302, 1992.

Kegley, E. B. and Spears, J. W., Bioavailability of feed-grade copper sources (oxide, sulfate, or lysine) in growing cattle, *J. Anim. Sci.,* 72, 2728, 1994.

Kidd, M. T., Anthony, N. B., Newberry, L. A., and Lee, S. R., Effect of supplemental zinc in either a corn-soybean or a milo and corn-soybean meal diet on the performance of young broiler breeders and their progeny, *Poultry Sci.,* 72, 1492, 1993.

Kincaid, R. L., Biological availability of zinc from inorganic sources with excess calcium, *J. Dairy Sci.,* 62, 1081, 1979.

Kincaid, R. L., Hodgson, A. S., Riley, R. E., Jr., and Cronrath, J. D., Supplementation of diets for lactating cows with zinc oxide and zinc methionine, *J. Dairy Sci.,* 67 (Suppl. 1), 103, 1984.

Kincaid, R. L., Blauwiekel, R. M., and Cornrath, J. D., Supplementation of copper as copper sulfate or copper proteinate for growing calves fed forages containing molybdenum, *J. Dairy Sci.,* 69, 160, 1986.

Kornegay, E. T., Availability of iron contained in defluorinated phosphate, *J. Anim. Sci.,* 34, 569, 1972.

Lardy, G. P., Kerley, M. S., and Paterson, J. A., Retention of chelated metal proteinates by lambs, *J. Anim. Sci.,* 70 (Suppl. 1), 314, 1992.

Ledoux, D. R., Henry, P. R., Ammerman, C. B., Rao, P. V., and Miles, R. D., Estimation of the relative bioavailability of inorganic copper sources for chicks using tissue uptake of copper, *J. Anim. Sci.,* 69, 215, 1991.

Lei, X., Ku, P. K., Miller, E. R., Ullrey, D. E., and Yokoyama, M. T., Supplemental microbial phytase improves bioavailability of dietary zinc to weanling pigs, *J. Nutr.,* 123, 1117, 1993.

Miller, E. R., Bioavailability of minerals, in *Proceedings of the Minnesota Nutrition Conference,* University of Minnesota Press, St. Paul, 1980, 144.

Miller, W. J., Neathery, M. W., Gentry, R. P., Blackmon, D. M., Crowe, C. T., Sare, G. O., and Fielding, A. S., Bioavailability of phosphorus from defluorinated and dicalcium phosphates and phosphorus requirement of calves, *J. Dairy Sci.,* 70, 1885, 1987.

Mongin, P., Recent advances in dietary anion-cation balance: Applications in poultry, *Proc. Nutr. Soc.,* 40, 285, 1981.

National Research Council (NRC), *Nutrient Requirements of Swine,* National Academy Press, Washington, D.C., 1988, 28.

Nelson, J. D., Tracing quality, *Feed Manage.,* May 1982.

Nelson, J. D., Bioavailability of trace mineral ingredients, in *Proceedings of the 3rd Carolina Swine Nutrition Conference,* Raleigh, NC, 1987, 68.

Nockels, C. F., DeBonis, J., and Torrent, J., Stress induction affects copper and zinc balance in calves fed organic and inorganic copper and zinc sources, *J. Anim. Sci.,* 71, 2539, 1993.

Osman, M. and Latshaw, J. D., Biological potency of selenium from sodium selenite, selenomethionine and selenocystine in the chick, *Poultry Sci.,* 55, 987, 1976.

Pehrson, B., Knutsson, M., and Gyllensward, M., Glutathione peroxidase activity in heifers fed diets supplemented with organic and inorganic selenium compounds, *Swed. J. Agric. Res.,* 19, 53, 1989.

Pimentel, J. L., Cook, R. E., and Greger, J. L, Bioavailability of zinc-methionine for chicks, *Poultry Sci.,* 70, 1637, 1991.

Pitzen, D., The trouble with iron, *Feed Manage.,* 44, 9, 1993.

Podoll, K. L., Bernard, J. B., Ullrey, D. E., DeBar, S. R., Ku, P. K., and Magee, W. T., Dietary selenate versus selenite for cattle, sheep, and horses, *J. Anim. Sci.,* 70, 1965, 1992.

Pott, E. B., Henry, P. R., Ammerman, C. B., Merritt, A. M., Madison, J. B., and Miles, R. D., Relative bioavailability of copper in a copper-lysine complex for chicks and lambs, *Anim. Feed Sci. Technol.,* 45, 193, 1994.

Ross, J. G., Spears, J. W., and Garlich, J. D., Dietary electrolyte balance effects on performance and metabolic characteristics in finishing steers, *J. Anim. Sci.,* 72, 1600, 1994a.

Ross, J. G., Spears, J. W., and Garlich, J. D., Dietary electrolyte balance effects on performance and metabolic characteristics in growing steers, *J. Anim. Sci.,* 72, 1842, 1994b.

Smith, M. O., Sherman, I. L., Miller, L. C., and Robbins, K. R., Bioavailability of manganese from different sources in heat-distressed broilers, *Poultry Sci.,* 73 (Suppl. 1), 163, 1994.

Spears, J. W., Zinc methionine for ruminants: Relative bioavailability of zinc in lambs and effects on growth and performance of growing heifers, *J. Anim. Sci.,* 67, 835, 1989.

Spears, J. W., Organic trace minerals in ruminant nutrition, in *Proceedings of the 14th Canadian Western Nutrition Conference,* Calgary, Alberta, 1993, 269.

Spears, J. W., Minerals in forages, in *Forage Quality, Evaluation and Utilization,* Fahey, G. C., Collins, M., Mertins, D. R., and Moser, L. E., Eds., American Society of Agronomy, Madison, WI, 1994, chapter 7.

Spears, J. W., Harvey, R. W., and Brown, T. T., Effects of zinc methionine and zinc oxide on performance, blood characteristics and antibody titer response to viral vaccination in stressed feeder calves, *J. Am. Vet. Med. Assoc.*, 199, 1731, 1991.

Spears, J. W. and Kegley, E. B., Influence of zinc proteinate on performance and carcass characteristics of steers, *J. Anim. Sci.,* 72 (Suppl. 2), 4, 1994.

Stansbury, W. F., Tribble, L. F., and Orr, D. E., Effect of chelated copper sources on performance of nursery and growing pigs, *J. Anim. Sci.,* 68, 1318, 1990.

Swinkels, J. M. G. M., Kornegy, E. T., Webb, K. E., Jr., and Lindermann, M. D., Comparison of inorganic and organic zinc chelate in zinc depleted and repleted pigs, *J. Anim. Sci.,* 69 (Suppl. 1), 358, 1991.

Underwood, E. J., *The Mineral Nutrition of Livestock,* Commonwealth Agricultural Bureaux, Slough, England, 1981, 29.

Ward, J. D. and Spears, J. W., Bioavailability of copper proteinate and copper carbonate relative to copper sulfate in cattle, *J. Anim. Sci.,* 72 (Suppl. 1), 95, 1994.

Ward, J. D., Spears, J. W., and Kegley, E. B., Effect of copper level and source (copper lysine vs copper sulfate) on copper status, performance and immune response in growing steers fed diets with or without supplemental molybdenum and sulfur, *J. Anim. Sci.,* 71, 2748, 1993.

Ward, J. D, Spears, J. W., and Gengelbach, G. P., Differences in copper status and copper metabolism among Angus, Simmental and Charolais cattle, *J. Anim. Sci.,* 73, 571, 1995.

Wedekind, K. J. and Baker, D. H., Zinc bioavailability in feed-grade sources of zinc, *J. Anim. Sci.,* 68, 684, 1990.

Wedekind, K. J., Hortin, A. E., and Baker, D. H., Methodology for assessing zinc bioavailability: Efficacy estimates for zinc-methionine, zinc sulfate, and zinc oxide, *J. Anim. Sci.,* 70, 178, 1992.

Wedekind, K. J., Lewis, A. J., Giesemann, M. A., and Miller, P. S., Bioavailability of zinc from inorganic and organic sources for pigs fed corn-soybean meal diets, *J. Anim. Sci.,* 72, 2681, 1994.

Wiener, G., Suttle, N. F., Field, A. C., Herbert, J. G., and Woolliams, J. A., Breed differences in copper metabolism in sheep, *J. Agric. Sci.,* 91, 433, 1978.

Wittenberg, K. M., Boila, R. J., and Shariff, M. A., Comparison of copper sulfate and copper proteinate as copper sources for copper-depleted steers fed high molybdenum diets, *Can. J. Anim. Sci.,* 70, 895, 1990.

Wong-Valle, J., Henry, P. R., Ammerman, C. B., and Rao, P. V., Estimation of the relative bioavailability of manganese sources for sheep, *J. Anim. Sci.,* 67, 2409, 1989a.

Wong-Valle, J., Jr., Ammerman, C. B., Henry, P. R., Rao, P. V., and Miles, R. D., Bioavailability of manganese from feed grade manganese oxides for broiler chicks, *Poultry Sci.,* 68, 1368, 1989b.

Wright, P. L. and Bell, M. C., Comparative metabolism of selenium and tellurium in sheep and swine, *Am. J. Physiol.,* 211, 6, 1966.

Zhou, W., Kornegay, E. T., van Laar, H., Swinkels, J. W. G. M., Wong, E. A., and Lindemann, M. D., The role of feed consumption and feed efficiency in copper-stimulated growth, *J. Anim. Sci.,* 72, 2385, 1994.

CHAPTER 18

Nutritional, Environmental, and Economic Considerations for Using Phytase in Pig and Poultry Diets

E. T. Kornegay

INTRODUCTION

The growth and development of modern commercial swine and poultry production, essential components of our food supply, are being restricted in some countries and will be restricted in other countries if solutions to the problem of manure disposal are not developed and implemented. Because of the high nutrient content of manure, and thus fertilizing value, land application has been the major means of disposal. However, the amount of manure that can be applied to the land is limited because of buildup of nutrients in and on the soil. The potential environmental impact of nutrient contamination of surface runoff and groundwater is perceived as a major issue facing livestock producers (Coffey, 1992; Van Horn, 1992). The impact of very large amounts of manure being produced in a relatively small land area has led to the development of strict legislation to control pollution caused by animal manure in a few countries. The Dutch case was presented by de Haan (1990) and will be described in Chapter 20 by Jongbloed and Henkens (1996). Also, in a recent review of livestock pollution and politics, Hacker and Du (1993) reviewed legislation and guidelines that have been developed in some countries and regions of the world. The pig and poultry industries have generally received the most attention.

Of the nutrients present in manure, phosphorus (P), nitrogen (N), and trace minerals (probably copper [Cu] and zinc [Zn]) are of greatest concern (Table 1). de Haan (1990) suggests that cadmium (Cd) may also be of concern. Iron (Fe) is usually not a problem because the amount of Fe added in manure would be insignificant since the Fe concentration in most soils is very high. Phosphorus and N are currently the two elements in manure that limit the rate of land application, but there is disagreement as to which one is of greatest concern. Soil analyses (Mueller et al., 1994) of a Sampson County, NC bermudagrass pasture that was fertilized with swine lagoon effluent to satisfy N requirements indicated a buildup of other nutrients during the 3-year period (Table 2). Phosphorus may well be the nutrient that regulates the amount of swine and poultry waste that can be applied to the land.

1-56670-199-6/96/$0.00+$.50

Table 1 Fourteen-Year Average Mineral Concentration of Swine Feed and Manure (Primarily Feces) at Virginia Tech (Dry Matter Basis)[a]

Element[b]	Feed	Manure
	% of DM	
N	2.7	4.5
P	0.62	2.3
Ca	0.94	4.1
Na	0.16	0.46
K	0.83	1.7
Mg	0.17	0.82
	mg/kg of DM	
Cu	257	1329
Zn	93	465
Fe	184	1194
Mn	55	313
B	10	21

a Data from Kornegay, 1992.

[b] Each value is the average of triplicate analyses for two samples taken annually at the same time.

Major ingredients in pig and poultry diets are seeds (cereal grains) or products from seeds (oilseed meal and grain byproducts). Hence a large portion of the P occurs as phytate, since 60 to 80% of the P is present in the form of phytic acid (Table 3). However, pigs and poultry do not utilize phytate P well (Table 4). Bioavailability estimates of P in corn and soybean meal for pigs and poultry range from 10 to 30% (Nelson, 1967; Calvert et al., 1978; Jongbloed and Kemme 1990; Cromwell and Coffey, 1991). This low availability of phytate P poses two problems for producers: (1) the need to add inorganic P supplements to diets and (2) the excretion of large amounts of P in the manure.

Technology does exist for reducing the amount of nutrients excreted and thus the amounts that must be disposed. Nutrient management techniques used to reduce nutrient excretion have been discussed previously (Lenis, 1989; Crenshaw et al., 1994; Cromwell and Coffey, 1994; Kornegay, 1994; de Lange, 1994). This

Table 2 Initial (6-28-90) and Final (12-2-92) Soil Analyses for a Sampson County Bermudagrass Pasture Fertilized with Swine Lagoon Effluent[a]

	P_2O_5		K_2O		Zn		Ca	
Depth	1990	1992	1990	1992	1990	1992	1990	1992
(cm)	kg/ha							
0–15	540	973+	355	460	2.6	10.6	0.9	5.3
15–30	180	869	442	440	0.8	4.8	1.0	3.3
30–61	20	209	855	3347	0.4	2.8	0	3.5
61–91	16	64	719	1919	0.5	2.0	0	2.4

[a] Adapted from Mueller et al. (1994). Swine lagoon effluent was added at a rate to meet the N needs of the bermudagrass pasture.

Table 3 Phytate Phosphorus Content and Phytase Activity of Some Common Feed Ingredients[a]

Ingredient	Phytate P (g/kg)	Phytate P (as % of total P)[b]	Phytase activity (units/kg)[c]
Cereals and byproducts			
Wheat	2.4 (1.9–2.9)[d]	68 (61–78)	1190
Maize	2.0 (1.6–2.6)	73 (61–85)	15
Sorghum	2.2 (1.9–2.9)	68 (61–76)	25
Barley	2.1 (1.9–2.4)	58 (55–62)	580
Oats	2.8 (1.6–3.5)	69 (48–78)	40
Wheat bran	8.8 (6.0–12.7)	76 (68–93)	2960
Grain legumes			
Lupins	3.0 (2.9–3.0)	55 (54–55)	0
Peas	1.7 (1.3–2.1)	45 (36–53)	115
Chick peas	2.1 (2.0–2.3)	51 (49–53)	—
Oilseed meals			
Soyabean meal	3.7 (2.8–4.0)	57 (46–61)	40
Canola meal	6.5 (4.6–7.8)	58 (36–70)	15
Sunflower meal	4.4 (3.2–5.1)	44 (35–47)	60

[a] Data from Ravindran, 1995.

[b] Data adapted from the following sources: Eeckhout and de Paepe, 1994; Kirby and Nelson, 1988; Nelson et al., 1968; Ravindran et al., 1994.

[c] Data from Eeckhout and de Paepe, 1994. One unit is defined as that amount of phytase which liberates inorganic phosphorus from a 5.1-m*M* Na phytate solution at a rate of 1 mol/min at pH 5.5 and 37°C.

[d] Values within parentheses refer to ranges reported in the literature.

Table 4 Bioavailability of P for Pigs and Nonphytate P for Poultry

Feedstuff	Bioavailability of P for pigs[a,b] (%)	Nonphytate P for poultry[c] (%)
Cereal grains		
Corn	12	28
Oats	23	33
Barley	31	36
Triticale	46	33
Wheat	50	31
Corn, high moisture	53	—
High protein meals - plant origin		
Peanut meal	12	21
Canola meal	21	26
Soybean meal, dehulled	25	35
Soybean meal, 44% protein	35	40

[a] Adopted from Cromwell, 1992.

[b] Relative to the availability of P in monosodium phosphate, which is given a value of 100.

[c] NRC, 1994 - poultry.

chapter deals primarily with nutritional, environmental, and economical considerations for reducing excretion of P by using the enzyme microbial phytase, which is known to release phytate P.

OCCURRENCE AND STRUCTURE OF PHYTATE

Phytic acid is an essential component of all seeds and is generally associated with fibers for its nutritional implications. The molecule has a high P content (28.2%), and its six phosphoric acid residues have various affinities for several cations. One mole of phytic acid can bind an average of 3 to 6 mol of calcium (Ca) to form insoluble phytates at the pH of the intestine. Formation of insoluble phytate makes both Ca and P unavailable. Phytic acid has chelating potential, forming a wide variety of insoluble salts with di- and trivalent cations at neutral pH (Vohra et al., 1965; Oberleas, 1973). Zinc, Cu, cobalt (Co), manganese (Mn), Fe, and magnesium (Mg) can also be complexed, but Zn and Cu have the strongest binding affinity (Maddaiah et al., 1964; Vohra et al., 1965). This binding potentially renders these minerals unavailable for intestinal absorption. Phytate can also bind with protein (amino groups of some amino acids) at low and neutral pH (Anderson, 1985). Phytate-protein or phytate-mineral-protein complexes may reduce the utilization of protein. Phytate concentration may vary, depending on the stage of maturity, degree of processing, cultivar, climatic factors, water availability, soil factors, location, and the year of growth (Reddy et al., 1982; Ravindran et al., 1995a).

IMPROVED UTILIZATION OF PHYTATE P BY PIGS AND POULTRY USING SUPPLEMENTAL PHYTASE

Phytate P must be hydrolyzed into inorganic P before it can be utilized by pigs and poultry. Phytase is a special kind of phosphatase that catalyzes the stepwise removal of inorganic orthophosphate from phytate. According to Gibson and Ullah (1990), two classes of phytases are recognized by the International Union of Pure and Applied Chemistry and the International Union of Biochemistry; 3-phytase (E.C.3.1.3.8) first removes the orthophosphate from the 3-position of phytic acid, whereas 6-phytase first removes the orthophosphate from the 6-position. Both classes of phytase then successively remove the remaining orthophosphates, which results in intermediates ranging from free myo-inositol to mono- to tetra-phosphates of inositol. Microorganisms and filamentous fungi normally produce 3-phytases, and the seeds of higher plants typically contain 6-phytases (Gibson and Ullah, 1990). Phytases are known to occur widely in microorganisms, plants, and certain animal tissues (Cosgrove, 1980; Reddy et al., 1982; Nayini and Markakis, 1986).

Four possible sources of phytases could degrade phytate within the digestive tract of pigs and poultry: (1) intestinal phytase in digestive secretions, (2) endogenous phytase present in some feed ingredients, (3) phytase originating from resident bacteria, or (4) phytase produced by exogenous microorganisms.

Contents of the stomach and intestine of pigs (Jongbloed et al., 1992; Yi and Kornegay, 1995b) and crop, stomach, and small intestine of chickens (Leibert et al., 1993) have negligible phytase activity. Also, the endogenous phytase activity of intestinal mucosa was negligible (Pointillart, 1993). Phytase activity has been detected in some animal tissue (McCollum and Hart, 1908; Bitar and Reinhold, 1972; Pointillart et al., 1987); however, this activity, if present, is also negligible for improving the availability of phytate P in nonruminant animals. In addition, the significance of phytase produced by resident bacteria in nonruminants has not been demonstrated and is probably negligible.

Phytase activity has been reported in a wide range of seeds, such as rice, wheat, barley, corn, rye, soybean, and other leguminous and oil seeds (Reddy et al., 1982; Gibson and Ullah, 1990). Phytase activity of seeds varies greatly among species of plants (Table 3). With the exception of wheat and rye and their hybrid triticale, most dormant seeds contain very low phytase activity. Phytase activity in corn and soybean meal is so low that it is not of practical importance. Barley may or may not contain phytase activity (Pointillart, 1993). The majority of the phytase activity in wheat, rye, and triticale is in the bran. Diets formulated using ingredients having high phytase activity, such as wheat bran, wheat, triticale, rye bran, and wheat middling, promote greater absorption of phytate P (Pointillart, 1993). Plant phytase may be about 10% less efficient than microbial phytase when dietary phytase activity is 500 U/kg. The use of plant phytase, however, dictates that these ingredients be used in the diet generally at high levels, which may not be economically feasible.

Nelson et al. (1971) reported that preparations of *Aspergillus ficuum* containing phytase added to corn-soybean meal diets were effectively hydrolyzed by phytate P in the alimentary tract of the chick. Because of their phytase activity, many bacteria and yeast are capable of releasing the P from phytate (Schwarz, 1992). Ruminants utilize phytate P very well because of microbes in the rumen. Microbial phytase was reported to effectively improve the availability of phytate P in corn and soybean meal diets for pigs and poultry (Beers and Jongbloed, 1992; Cromwell et al., 1993). The advantage of microbial phytase is that it is easier to measure the phytase activity and hence to incorporate the optimal amount compared with selecting feed ingredients containing good phytase activity. Most studies have used one of two commercial products, Natuphos® (BASF, Ludwigshafen, Germany) or Finase® (Alko Ltd., Rajamaki, Finland). According to a report by Dunn (1994), the BASF product is now being marketed in the Netherlands, Canada, Germany, Switzerland, Austria, Brazil, Norway, and Finland. Natuphos® was cleared by the Food and Drug Administration for use in the United States on November 17, 1995.

IMPORTANT CONSIDERATIONS FOR USING PHYTASE IN PIG AND POULTRY DIETS

The goal of our research at Virginia Tech has been to develop a knowledge base that would allow the formulation of diets that would minimize the excretion of minerals, primarily P, through the use of microbial phytase. Major objectives are (1) to develop response surfaces (mathematical equations) of various levels of phytase and P added to corn and soybean meal diets fed to young pigs and birds (broilers and turkeys) and to assess the sensitivity of several response measurements; (2) to characterize the influence of the Ca to P ratio on microbial phytase activity; (3) to determine the effects of phytase on the availability of other nutrients, such as Ca, Zn, and N (amino acids); (4) to compare the histological characteristics of proximal tibia of broilers and turkeys fed various combinations of P and phytase; (5) to calculate equivalency values of phytase for inorganic P, Ca, and Zn; (6) to determine the influence of phytase or P excretion; and (7) to estimate the economics of using phytase. In all of our research, Natuphos® phytase was used.

Development of Response Surfaces and Assessment of Sensitivity Measurements and Effectiveness of Phytase

The influence of dietary P and microbial phytase levels was evaluated in two trials with broilers (Denbow et al., 1995; Kornegay et al., 1995) and one trial with turkey poults (Ravindran et al., 1995b) using a 3×7 factorial arrangement of P levels and phytase levels. Second-order translog functions were derived. The model was

$$\text{LnY} = \alpha_0 + \alpha_1\text{LnX}_1 + \alpha_2\text{LnX}_2 + \alpha_3(\text{LnX}_1)^2 + \alpha_4(\text{LnX}_2)^2 + \alpha_5\text{LnX}_1\text{LnX}_2$$

where Y = response measurements, X_1 = nonphytate P (%), and X_2 = phytase added (U/kg of diet). Although not equal, available and nonphytate P are often used interchangeably. An example of a response surface plotted from the second-order translog coefficients is shown in Figure 1 for broiler weight gain. A similar response surface was obtained for toe ash percent.

Second-order translog functions were derived in two pig trials (Yi et al., 1995a; Kornegay and Qian, 1995) using a 2×5 factorial arrangement of P levels and phytase levels. The model was

$$\text{LnY} = \alpha_0 + \alpha_1 D_1 + \alpha_2\text{LnX} + \alpha_3(\text{LnX})^2 + \alpha_4 D_1\text{LnX}$$

where Y = response measurements; X = phytase added (U/kg of diet); D_1 = available P, when available P is level 1 in the diet, $D_1 = 0$, and when available P is level 2, $D_1 = 1$ in the model. Nonlinear and linear functions were also derived for phytase levels at each P level. An example for body weight gain and P digestibility is shown in Figure 2.

Our findings with pigs and poultry using measurements of performance, apparent absorption and retention of P, and bone characteristics confirm a number of previous reports, indicating that microbial phytase is very effective for improving the availability of phytate P in corn and soybean meal diets. Furthermore, we have clearly demonstrated that the magnitude of the response

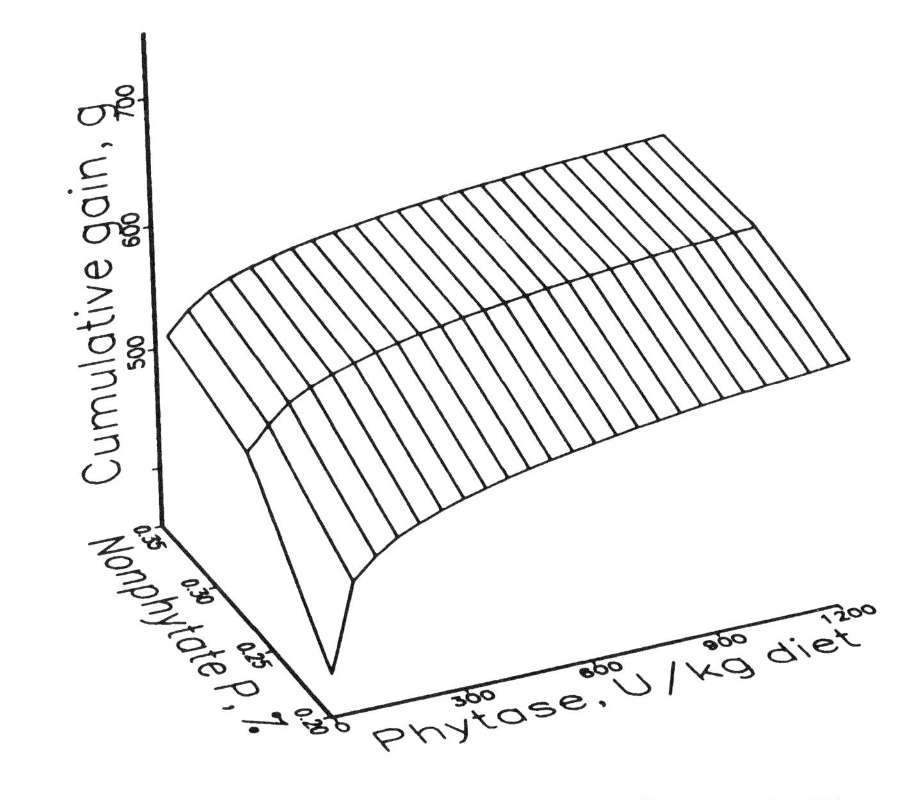

Figure 1 Response surface of weight gain in a broiler study (Denbow et al., 1995).

of several measurements to added phytase is inversely related to the level of available P (and total P to include phytate P) and to the level or amount of supplemental phytase added (Figures 1 and 2). Our results also indicate that adding phytase to the positive control diets containing near NRC (1988) recommended levels of available (a) P (percent) further improved P availability in pigs (Kornegay and Qian, 1995; Yi et al., 1995a). Similar results were also found by Veum et al. (1994).

For poultry, body weight (BW) gain, apparent retention and excretion of P, percentage of ash in dried toes, and percentage of ash in tibia were the most-sensitive measurements, followed by feed intake and shear force of tibia. Apparent retention of Ca, N, and dry matter (DM) and shear stress of tibia were the least-sensitive measurements. For pigs, the apparent absorption of P and percentage of ash of the tenth rib and metacarpals were the most sensitive measurements, followed by BW gain, feed intake, P excretion, and shear force of metacarpal and tenth rib. Apparent absorption of Ca, N, and DM; gain to feed ratio; and shear stress and energy of metacarpals and tenth rib were the least-sensitive measurements.

Growth performance data for pigs and poultry would be the most efficient measurements from the standpoint of being nondestructive and ease and cost of

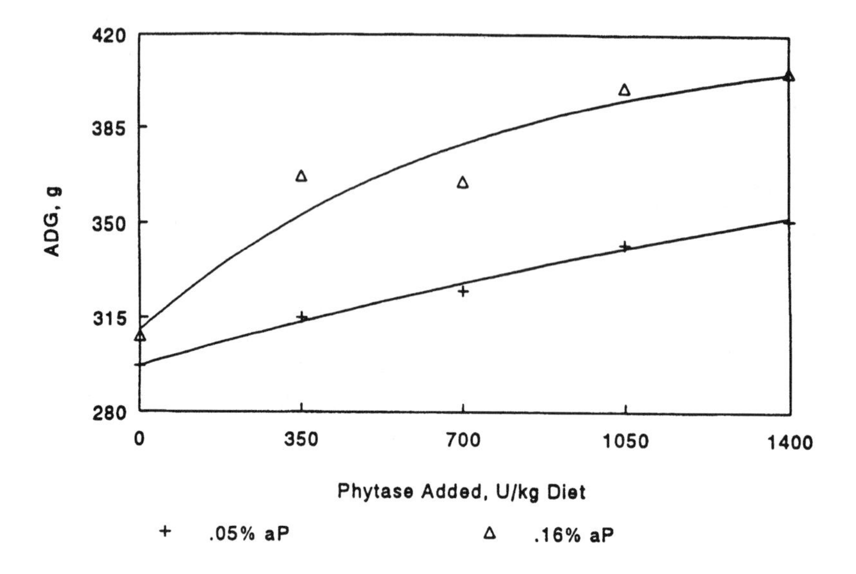

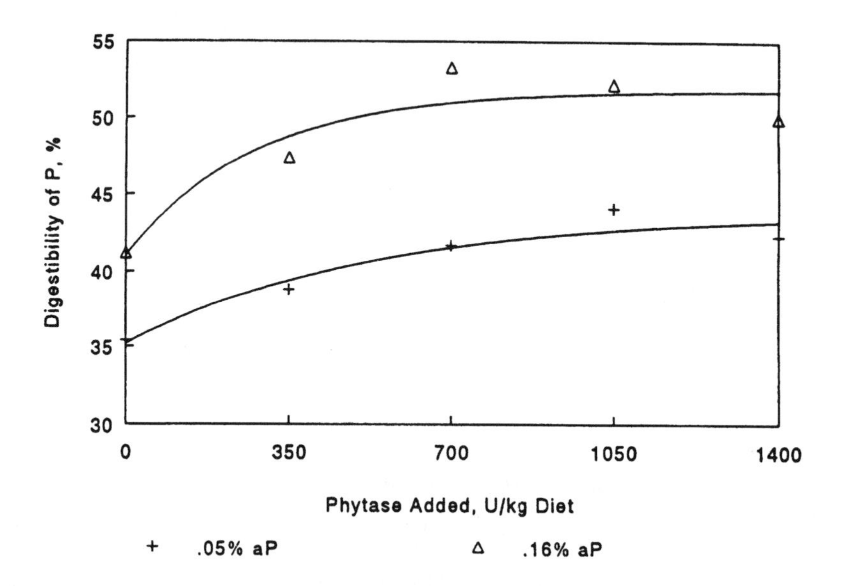

Figure 2 Nonlinear asymptotic curves of cumulative average daily gain (ADG) and digestibility of P of weanling pigs fed soybean meal-based, semipurified diets containing two levels of P and five levels of phytase.

collection. Apparent absorption and retention of P data would also be nondestructive, but would require more labor and more intensive facilities to collect, and there would also be cost associated with analyses of feed, fecal, or excreta samples. Collection of bone data would require that the animals be killed, and there would be laboratory costs associated with bone measurements. For birds, percentage ash of dried toes would require the least amount of laboratory labor. In summary, for nondestructive measurements, BW gain and absorption of P seem to be the most desirable for pigs from the standpoint of efficiency and sensitivity. For birds, BW gain and toe ash would be the most desirable if birds could be killed.

Influence of Ca to P Ratio on Phytase Effectiveness

It is well known that P utilization by animals is influenced by dietary Ca and vitamin D. An excess of dietary Ca will interfere with the availability of P as well as Mg, Mn, and Zn (NRC, 1988, 1994). The dietary level of Ca and the Ca to phytate ratio may be important factors that determine the extent of phytate hydrolysis (Wise, 1983; Ballam et al., 1984) through formation of an insoluble calcium phytate complex (Nelson, 1967). Harms et al. (1962) reported that widening the Ca to total (t) P ratio in broiler diets from 1:1 to 2:1 lowered the availability of the P in phytic acid to a greater extent than in inorganic P supplements. On the other hand, Sanders et al. (1992) confirmed that a very narrow Ca to tP ratio (usually below 1.1:1) would increase the incidence of Ca deficiency rickets, with the magnitude of the effect greater at higher P levels. The NRC (1994) for poultry suggests a Ca to nonphytate (n)P (weight/weight) ratio of 2:1 with the exception of laying birds. Dietary Ca has adverse effects on the availability of phytate P, whereas vitamin D has positive effects (Mohammed et al., 1991; Edwards et al., 1992; Lei et al., 1994).

Only a few studies have investigated the influence of dietary Ca level (Ca to P ratio) on phytase effectiveness. In our study with broilers (1 to 21 d), four Ca to tP ratios (1.1, 1.4, 1.7, and 2.0:1), four microbial phytase levels (0, 300, 600, and 900 U/kg of diet), and two vitamin D_3 levels (66 and 660 μg/kg of diet) were evaluated in a factorial design of treatments (Qian et al., 1995c). The dietary P level in the 23% crude protein (CP) corn-soybean meal diet was formulated at 2.7 g/kg of nP (5.1 g/kg of tP). Added phytase resulted in a linear increase of BW gain, feed intake, toe ash content, and P and Ca retention. These responses were negatively influenced by widening the dietary Ca to tP ratio beyond 1.4:1 and were synergistically improved by the addition of vitamin D_3. In this study, the adverse effect of a wide Ca to tP ratio seemed to be independent of added phytase and vitamin D_3 because all two- and three-way interactions were not significant. Widening the Ca to tP ratio from 1.4 to 2.0:1 reduced the efficacy of phytase, probably by reducing phytase activity, by 13.4 and 14.9%, respectively, for the diets with 66 and 660 μg of vitamin D_3 per kilogram of diet. In agreement, Schoner et al. (1993) reported for broilers (1 to 14 d and 1 to 40 d) that feeding high levels of Ca with a constant level of P (3.5 g/kg) reduced the increase in BW gain, feed intake, and P and Ca retention that was observed when phytase was added. From their lowest (6 g/kg) to highest (9 g/kg) levels of Ca, the Ca to tP ratio varied from 1.7:1 to 2.57:1.

In our study with turkey poults (1 to 21 d), four Ca to tP ratios (1.1, 1.4, 1.7 and 2.0:1), four phytase levels (0, 300, 600, and 900 U/kg of diet), and two P levels

(2.7 and 3.6 g/kg of nP or 5.4 and 6.3 g/kg of tP) were evaluated in a 4 × 4 × 2 factorial design of treatments (Qian et al., 1995b). The 28% corn-soybean meal diet was formulated to supply 66 μg of vitamin D_3 per kilogram of diet. Phytase additions linearly increased BW gain, feed intake, gain to feed ratio, toe ash content, and retention of Ca and P at each Ca to tP ratio and nP level, but the response was influenced by dietary Ca to tP ratio and P level. A negative effect of widening the Ca to tP ratio was observed for all measurements at each phytase and P level, and the magnitude of the effect was greatest at lower phytase and P levels. As with broilers (Qian et al., 1995c), there was very little difference between the Ca to tP ratio of 1.1:1 and 1.4:1. Widening the Ca to tP ratio from 1.4 to 2.0 decreased the phytase efficacy by 7.4 and 4.9%, respectively, for 2.7 and 3.6 g/kg of nP diets. Contrary to the findings with broilers (Qian et al., 1995c), two- and three-way interactions for most measurements were significant, which suggested that the Ca to tP ratio effects were not independent of dietary P and phytase levels.

In pigs, a wide Ca to tP ratio lowers P absorption, which results in decreased growth and bone calcification (Koch et al., 1984; Reinhart and Mahan, 1986; and Pointillart et al., 1989). Detrimental effects of the Ca to tP ratios on performance, bone characteristics, and serum parameters were generally observed when the Ca to tP ratio exceeded 2.0:1; however, no significant effects were observed when the Ca to tP ratio was under 2.0:1, especially at the range of 1.0 to 1.6:1 (Koch et al., 1984; Reinhart and Mahan, 1986; Pointillart et al., 1989; Ketaren et al., 1993). Some studies have not observed detrimental effects on performance, bone measurements, and serum parameters with Ca to tP ratios greater than 2.0:1, such as 3:1 used by Koch et al. (1984). In our study with pigs, three Ca to tP ratios (1.2:1, 1.6:1, and 2:1), two levels of P (0.7 and 1.6 g/kg of aP and 3.6 and 4.5 g/kg of tP), and two levels of phytase (700 and 1050 U/kg of diet) were fed using a corn-soybean meal diet (Qian et al., 1995a). Growth performance, P absorption, and bone measurements were linearly decreased as the Ca to tP ratio became wider. This influence of the Ca to tP ratio appeared independent of P and phytase levels, since most two-way interactions were nonsignificant. Increasing the P level generally did not affect growth performance, but metacarpal and tenth-rib measurements and P absorption were increased. Increasing the phytase level from 700 to 1050 U/kg of diet only increased P absorption.

Lei et al. (1994) reported for pigs fed a corn-soybean meal diet with no added inorganic P (3.1 g/kg of tP) that the ability of phytase to improve phytate P availability was greatly reduced at a normal level (9.2 g/kg) of dietary Ca (Ca to tP ratio was 3:1) compared with the low level (5.0 g/kg) of dietary Ca (Ca to tP ratio was 1.6:1). The depressive effect of a normal level of dietary Ca on performance was greater at the normal vitamin D_3 level or at the optimal phytase level than at the other levels of these two factors. Raising the level of vitamin D_3 in the diet (16.5 vs. 166.5 μg/kg) partially offset this adverse effect, but did not produce further improvement when the Ca level was lower. Jongbloed et al. (1993) investigated three levels of dietary Ca (4, 6, and 8 g/kg) using a basal diet (tapioca, corn, hominy feed, barley, soybean meal, and sunflower meal) containing 4.3 g/kg of P without any added inorganic P. Apparent Ca and P absorption generally decreased as the level of Ca increased, but increased as the level of phytase increased. The magnitude of the increase from phytase was reduced as dietary Ca increased.

Reduced responses to microbial phytase, which are observed when the Ca to tP ratio is widened (greater than 1.1 to 1.4:1), may be a result of decreased activity of dietary microbial phytase. Activity of the microbial phytase added in the mixed diets was decreased by widening the Ca to tP ratio (Qian et al., 1995a, 1995b, 1995c). The magnitude of the reduction in phytase activity was generally of the magnitude of the reduction observed for several other measurements when the Ca to tP ratio was widened. The decrease in the phytase activity as the Ca to tP ratio became wider could be explained as follows: (1) phytate P utilization in corn-soybean meal diets fed to broilers may be influenced by Ca and P levels in the diet (Edwards and Veltmann 1983; Ballam et al., 1984), (2) the extra Ca may bind with phytate to form an insoluble complex that is less accessible to phytase, or (3) the extra Ca may directly repress phytase activity by competing for the active sites of the enzyme (Wise, 1983; Pointillart et al., 1985). The negative effect of wide Ca to tP ratios is stronger at lower levels of supplemental phytase and at lower levels of aP or nP because a smaller amount of P would be released from phytate caused by a lower phytase activity that would contribute to a P-deficient environment.

Effects of Phytase on Other Nutrients

Phytate has impaired the bioavailability of minerals other than P in studies with humans, pigs, poultry, and rats. Minerals that may be bound by phytic acid include Zn, Cu, Mn, Fe, Mg, Ca, and Cr (Maddaiah et al., 1964; Vohra et al., 1965; Oberleas, 1973). Hydrolysis of phytate by phytase should release the minerals that are bound. Supplemental microbial phytase has improved absorption or retention of Ca, Mg, Zn, and Fe in pigs (Pallauf et al., 1992a, 1994a, 1994b; Lei et al., 1993; Lawrence et al., 1995). The magnitude and consistency of improvement of mineral utilization by addition of microbial phytase is influenced by the level of the mineral in the diet. In situations where the dietary requirement is being met or exceeded, a response may not always be observed.

Calcium

Calcium retention and DM digestion were improved when phytase was added to broiler diets (Kornegay et al., 1995; Yi et al., 1995b). Schoner et al. (1991, 1993, 1994) also reported improved Ca retention of broilers fed supplemental phytase. In a broiler study designed to measure the effect of phytase on Ca availability, Schoner et al. (1994) reported that 500 U of microbial phytase was equivalent to 0.35 g of Ca as measured by BW gain and 0.56 g of Ca as measured by phalanx ash. In our study with turkey poults (Kornegay and Denbow, 1995), preliminary estimates based on BW gain and gain to feed ratio within a Ca range of 5.3 to 7.4 g/kg and a phytase level up to 500 U/kg suggest that 500 U of phytase is equivalent to about 0.85 g of Ca. Within a dietary Ca range of 5.3 to 8.8 g/kg and a phytase range of 0 to 625 U/kg, toe ash content was not influenced by dietary treatments; all toe ash content values appeared normal (11 to 12%). Body weight gain appeared to be depressed above 500 U of phytase per kilogram and above 7.4 g/kg of Ca. The dietary tP level was 8.5 g/kg for all diets.

In our pig study (Radcliffe et al., 1995) designed to measure the effect of phytase on Ca availability, preliminary results based on BW gain and digested Ca suggest that 500 U of phytase is equivalent to 1.1 and 0.65 g of Ca, respectively. Calcium digestibility coefficients were found to be inappropriate for estimating Ca equivalency. Mroz et al. (1993) reported enhanced Ca and P digestibility in 30-kg pigs when 300 and 600 U of phytase per kilogram were added to a diet containing suboptimal levels of Ca (4.3 g/kg) and tP (4.3 g/kg). Hypophosphaturia and hypercalciuria developed in those pigs fed the basal diet, which was prevented by adding 0.5 g/kg of P from KH_2PO_4 or 300 U of phytase per kilogram. Body weight gain and gain to feed ratios were similar among treatments. Pointillart (1993) indicated for their studies, primarily with cereal phytase, that improved P utilization was generally accompanied by improved Ca retention. In our studies, the apparent absorption of Ca and N (Yi et al., 1995a; Kornegay and Qian, 1995) in pigs was improved when phytase was added to the diet.

Zinc

Based on enhanced growth, increased plasma zinc concentration and alkaline phosphatase activity, the bioavailability of Zn for pigs was improved when phytase was added to a low P and Zn (no added inorganic P-3 g/kg or Zn-30 mg/kg) corn-soybean meal diet with 0, 30, or 60 mg/kg of Zn added (Lei et al., 1993); however, neither supplemental phytase nor Zn affected Zn retention. Adding microbial phytase to the diets of young pigs significantly improved apparent absorption of Zn and Mg (Pallauf et al., 1992a; Nasi and Helander, 1994). Using a low Zn diet (23 mg/kg) containing 8 g/kg of Ca and 6.2 g/kg of P, Adeola et al. (1995) reported that the growth rate of 9.4-kg pigs was increased, and the retention of Zn, Cu, Ca, and P was increased when 1500 U of phytase per kilogram of diet was fed.

The addition of 800 U of phytase per kilogram to a diet for chicks containing 27 mg/kg of Zn increased the retention of Zn and decreased Zn excretion (Thiel and Weigand, 1992). Thiel et al. (1993) reported that the femoral Zn content of chicks fed a diet containing 30 mg of Zn per kilogram plus 700 U of phytase per kilogram of diet was equal to that of chicks fed a diet containing 39 mg of Zn per kilogram without phytase. Biehl et al. (1995) fed chicks a glucose-soy concentrate diet containing 13 mg of Zn per kilogram of diet with 0, 5, and 10 mg of Zn/ per kilogram or with 1200 U of phytase or 10 μg/kg of 1,25-dihydroxycholecalciferol (di-OH-D_3) for 12 d. Addition of 10 mg of Zn per kilogram increased growth rate by 40% and total tibial Zn by 108%. Phytase or di-OH-D_3 supplementation increased growth rate and tibial Zn to a similar extent. The combination of phytase and di-OH-D_3 increased growth by 43% and tibia Zn by 159%. The addition of 600 to 750 U of phytase per kilogram to a corn-soybean meal diet containing 32 mg of Zn per kilogram diet did not consistently improve Zn absorption or retention in broilers (Roberson and Edwards, 1994). However, in two of the three studies, they reported an increase in tibial Zn. In these studies, when di-OH-D_3 was added at a level of 5 μg/kg, tibial Zn content was increased similarly to that of phytase. Data from our laboratory (Yi et al., 1996c) suggest that about 1 mg of Zn is released per 100 U of microbial phytase over the range of 150 to 600 U of phytase.

Rimbach and Pallauf (1993) reported that Zn absorption of growing rats increased from 33 to 63% when 1000 U of microbial phytase was added per kilogram to a soy protein isolate-corn starch diet containing 15 to 16 mg of Zn per kilogram and 4.0 g of phytic acid per kilogram. Similar positive effects of phytase supplementation were observed for Zn retention, serum Zn concentration activity of alkaline phosphatase, as well as Zn deposition in femur and testes. Comparable results were reported for rats fed semipurified diets based on egg white and corn starch supplemented with sodium phytate and microbial phytase (Rimbach and Pallauf, 1992).

Protein (Amino Acids)

Phytate has the potential of binding with protein at low and neutral pH (Cosgrove, 1980; Anderson, 1985; Thompson, 1986). Phytate-protein complexes may reduce the utilization of proteins and amino acids. In general, at low pH, proteins are positively charged and can form insoluble complexes with negatively charged phytate because of strong electrostatic interactions (Cheryan, 1980; Reddy et al., 1982). Positive charges of proteins could be the terminal α-amino group, ε-amino group of lysine, guanidyl group of arginine, and histidine residues (Barre and van Huot, 1965a, 1965b). When pH is raised, proteins bind with phytate mediated by multivalent cations such as Ca, Mg, and Zn. The common binding site for the ternary complex appears to be ionized carboxyl groups and an unprotonated imidazole group of histidine. Complexing of phytate with proteins may change protein structure, which in turn may lead to decreased solubility, digestibility, and functionality. Phytate may also form complexes with proteases such as trypsin and pepsin (Camus and Laporte 1976; Singh and Krikorian, 1982) in the gastrointestinal tract. The complexes may decrease the activity of digestive enzymes and then decrease the digestibility of dietary proteins.

Ketaren et al. (1993) found that addition of phytase increased protein deposition and retention, but had no effect on CP digestibility. Mroz et al. (1994) reported that supplemental microbial phytase improved the apparent digestibility of protein and amino acids in pigs. The apparent N absorption of pigs and N retention of broilers were improved when phytase was added to the diet (Yi et al., 1996b, 1994b; Kornegay and Qian, 1995). Phytase also improved N absorption in laying hens (Van der Klis and Verstaagh, 1991). Kemme et al. (1995) reported in pigs that the addition of microbial phytase at levels of 900 U/kg exerted a positive effect on the apparent ileal digestibility of CP as well as lysine, tryptophan, isoleucine, and threonine.

Turkey poults fed low P and low CP levels in a corn-soybean meal diet with 750 U of microbial phytase added per kilogram, had enhanced growth performance, toe ash, and N and P retention, especially at the lower P and CP levels (Yi et al., 1996b). Phytase also improved apparent and true ileal digestibility of all essential amino acids up to 2% units. The magnitude of the response seemed to be greater for poults fed the NRC recommended P level and a low CP level (Yi et al., 1996b).

Apparent absorption of Cu tended to be improved in pigs when microbial phytase was added to a corn-soybean meal diet (Pallauf et al., 1992b). Lawrence et al. (1995) reported for young pigs that 1500 U of phytase per kilogram added to a corn-soybean meal diet containing 14 mg of Cu per kilogram increased the retention of Cu, Zn, Ca, P, and Mg. Growth rate was also increased for pigs fed phytase. Aoyagi and Baker (1995) reported that 600 U of phytase per kilogram added to a semipurified diet fed to chicks decreased Cu bioavailability in soybean meal and did not affect Cu bioavailability in cottonseed meal. Biehl et al. (1995) reported that phytase and di-OH-D_3 also were synergistic in stimulating gut absorption of Mn in chicks fed phytate-bound Mn.

Histological Characteristics of Tibias from Broilers Supplemented with Phytase and Phosphorus

Histological, mechanical, and chemical properties of tibia from broilers and turkeys fed P-deficient diets, P-adequate diets, and P-deficient diets supplemented with phytase were examined (Qian et al., 1995d, 1995e). The histological hallmark of a P deficiency in poultry is widening of the hypertrophy zone, which resulted from defective cartilage and bone differentiation. Microbial phytase supplementation improves bone ash and mineral contents; shear force; trabecular bone density; and the orderliness of development, mineralization and arrangement of cartilage and bone cells. Added phytase alleviated effects of P deficiency on the histological structure of tibias. Improvements were similar for either supplemental inorganic P or added phytase.

Estimation of Phosphorus Equivalency of Microbial Phytase

Nonlinear and linear response equations of the effects of various P (no added phytase) and varying phytase levels on the different measurements were derived and evaluated (see section on Evaluation of Sensitive Measurements). For pigs, BW gain and P digestibility were selected, and, for poultry, BW gain and toe ash were selected. A P-equivalency equation was obtained for each of the measurements by setting the equation for added P and the equation for added phytase, at each of the two lower levels of P, equal and then solving. The response for BW gain (pigs and poultry) and P digestibility (pigs) or toe ash percent (poultry) were averaged at each P level for the final calculations. These equations then were used to estimate the equivalent amount of inorganic P released over a range of 250 to 1000 U of phytase per kilogram of diet. The P-equivalency equations of inorganic P for microbial phytase for pigs and poultry fed soybean meal-based semipurified diets and corn-soybean meal-based diets were calculated (Table 5). Only the two lower levels of P for broilers (Denbow et al., 1995; Kornegay et al., 1995) and

Table 5 Phosphorus Equivalency of Microbial Phytase for Pigs and Poultry

Author	Type of diet	Minerals in diet: aP(tP) (g/kg)	Minerals in diet: Ca to tP ratio	Phytase addition (U/kg)	Equivalency equation[a]	P equivalency[b]
Pigs						
Yi et al. (1996a)	SBM based	0.5(2.2)	2:1	0–1400	$Y = 1.546–1.504e^{-0.0015X}$	676
		1.6(3.2)	2:1	(5 levels)		
Kornegay and Qian (1996)	Corn SBM	0.7(3.6)	2:1	0–1400	$Y = 2.622–2.559e^{-0.00185X}$	246
		1.6(4.5)	2:1	(5 levels)		
Broilers						
Denbow et al. (1995)	SBM based	2.0(3.8)	2:1	0–1200	$Y = 1.126–1.065e^{-0.0026X}$	822
		2.7(4.5)	2:1	(7 levels)		
		3.4(5.2)	2:1			
Kornegay et al. (1996)	Corn SBM	2.0(4.0)	2:1	0–1200	$Y = 1.849–1.799e^{-0.0008X}$	939
		2.7(5.1)	2:1	(7 levels)		
		3.4(5.8)	2:1			
Yi et al. (1996b)	SBM based	2.7–5.4 (4 levels)	2:1	0–1050 (4 levels)	$Y = -0.17 + 0.001268X$	922
	Corn SBM	2.7–5.4 (4 levels)	2:1	0–1050 (4 levels)	$Y = 2.451–2.233e^{-0.006X}$	760
Qian et al. (1996c)	Corn SBM	2.7(5.1)	2:0	0–900	$Y = 2.330–2.288e^{-0.00074X}$	735
			1:7	(4 levels)		
			1:4		$Y = 2.767 - 2.703e^{-0.00067X}$	635
			1:1			
Turkey poults						
Ravindran et al. (1995)	SBM based	2.7(4.9)	2:1	0–1200	$Y = 2.000–1.876e^{-0.009X}$	699
		3.6(5.8)	2:1	(7 levels)		
		4.5(6.7)	2:1			
Qian et al. (1996b)	Corn SBM	2.7(5.4)	2:0	0–900	$Y = 1.279–2.426e^{-0.0037X}$	585
		3.6(6.3)	1:7	(4 levels)		
		at both	1.4		$Y = 1.448–2.768e^{-0.0039X}$	467
		P levels	1.1			

[a] Based on BW gain during 5 weeks and P absorption during weeks 4 and 5 for pigs; based on BW gain and toe ash percent during 1 to 21 d for broilers and turkey poults. Y = P released (g/kg) and X = phytase (U/kg of diet).

[b] Equivalent to 1 g P as defluorinated phosphate.

[c] SBM = soybean meal.

turkeys (Ravindran et al., 1995b) were used in the calculation. At the higher level in each of these studies, the response to phytase was reduced and the fit of equations was poor. The P equivalency of 1 g of P ranged from 246 to 922 U of phytase.

Results of these studies using multiple levels of phytase and P indicate that for pigs (Yi et al., 1995a; Kornegay and Qian, 1995) the efficacy of microbial phytase was 20 to 30% greater for the diet containing 1.6 g of aP per kilogram vs. the diets containing 0.5 or 0.7 g of aP per kilogram. In both studies, the lower P diet had no added inorganic P. In the broiler trial (Denbow et al., 1995) where the soybean meal-based diet (supplied 1.8 g/kg of phytate P) was fed, the efficacy of microbial phytase was much greater for the diet containing 2.0 g of nP per kilogram vs. the diet containing 2.7 g of nP per kilogram. However, in the broiler trial (Kornegay et al., 1995) where a corn-soybean meal diet (supplied 2.4 g/kg of phytate P) was fed, the efficacy of microbial phytase was very similar (only 3% difference in favor of the 2.7 vs. 2.0 g of aP per kilogram). In the two turkey trials (Ravindran et al., 1995b; Qian et al., 1995b), the efficacy of microbial phytase was higher for the diet containing 2.7 g of aP per kilogram vs. the diet containing 3.6 g of aP per kilogram; 16% higher for the soybean meal-based diet (2.2 g/kg of phytate P) and 34% higher for the corn-soybean meal diet (2.7 g/kg of phytate P).

Hoppe and Schwarz (1993) presented a similar summary of available data for pigs. They calculate P equivalency of microbial phytase for 1 g of P from monocalcium phosphate (MCP) based on apparent P digestibility. Their P equivalency of 1 g of P ranged from 176 to 930 U of phytase. When only studies using treatments with phytase activities up to 500 U/kg were used, the mean P equivalency was 432 ± 169 U/kg = 1 g of P. Hoppe et al. (1993) reported 1 g of P from MCP was equivalent to 380 U of phytase when based on P retention and 403 U of phytase when based on phalanx crude ash, when pigs were fed a corn, oat, and soybean meal-based diet.

The amount of inorganic P released per 100 U of phytase over the range of phytase levels indicated that the magnitude of the response per unit of phytase decreased as the phytase level of the diet increased. When the amount of P released was also expressed as a percentage of phytate, the total amount of P released as a percentage of phytate increased at a decreasing rate as total amount of phytase added increased. More than 1000 U of phytase would be required to totally release phytate P in corn and soybean meal diets. This confirms the *in vitro* results (Simons et al., 1990). Based on current research, phytase levels of 350 to 800 U/kg of diet appear reasonable to add to diets of pigs and poultry. However, phytase data, now available for various species, should be combined into common useful data sets that could be used to derive more equations.

PHOSPHORUS EXCRETION

Supplementing microbial phytase provides a means of reducing the potential of P pollution from pig and poultry manure. Phosphorus excretion (in feces of pigs and in excreta for poultry) could be reduced by 25 to 50% with the addition of 200 to 1000 U of phytase. Based on available information, a 109-kg pig consuming 318 kg of feed from 18 kg to market would consume 1.43 kg of P (if fed NRC recommended levels, average of 4.5 g of P per kilogram of diet) and would excrete about 0.71 kg of P. When phytase is added to the diet, the amount of P fed would be reduced to at least 1.11 kg per pig and the amount of P excreted would be reduced to about 0.56 kg or less. If higher than NRC recommended levels of P are fed, as is often the case (average of 5.5 g/kg), the amount of P excreted per pig would be at least 0.88 kg. Based on 80 million pigs (average P intake of 1.59 kg per pig) marketed annually in the United States, a 30% reduction in P excretion would mean about 38.2 million kg (38,200 metric tons) less P excreted annually in the United States. A broiler model of P excretion has been proposed in which one manure unit corresponded to 350 broilers with a P discharge of 55 kg P_2O_5 per year (Schoner et al., 1990). Adding phytase to broiler diets could decrease P discharge from 55 to 28 kg of P_2O_5 per year and increase the number of broilers per manure unit from 350 to 525 for the same amount of P excretion.

ECONOMICS OF USING PHYTASE

Hoppe and Schwarz (1993) reviewed published data for pigs. Based on the increase in apparent digestibility averaged across all data, they calculated the increase in digestible P to be 0.86 g/kg. By assuming a linear relationship between phytase activity and the increase in digestible P, a phytase activity of 930 U was calculated as equivalent to 1 g of P from MCP. However, several studies used phytase levels near and above 1000 U where the response to addition of phytase is minimal. If only data for phytase levels of 500 U/kg and below were used, Hoppe and Schwarz (1993) reported the mean P equivalency to be 432 ± 169 U of phytase = 1 g of P. For practical feeding purposes, BASF (1992) suggested an equivalency of 500 U = 1 g of P from MCP. Hoppe and Schwarz (1993) also reported a study in which 0, 200, 400, and 600 U of phytase per kilogram were added to a barley-wheat-soybean meal diet containing 500 U of intrinsic phytase activity. By regression analysis of bone ash and using theoretical phytase and P supplementation rates, a mean equivalency of 690 U = 1 g of P was calculated.

Based on the data from nine broiler studies in which corn-soybean meal diets were fed, Yi and Kornegay (1995b) derived the following response equation

$$Y = 1.554(1 - 0.9964e^{-.0012X})$$

where Y = inorganic P (g/kg) and X = phytase (U/kg). Using this equation, 1 g of P would be equivalent to 856 U of phytase. This value is a little higher than a value of 800 reported previously by Yi et al. (1994b).

Equivalency values reported in the literature are quite variable, which can be expected because of a number of known factors already discussed and probably

Table 6 Example of Cost Comparison Between Microbial Phytase and Inorganic Phosphorus

Equivalency of phytase (U/1 g of P)	Cost of phytase ($/kg[a])	Cost of P ($/kg)	Cost of phytase ($)	Cost of P ($)	Ratio phytase:P
350	20	1.5	0.0014	0.0015	0.93
		2.0	0.0014	0.0020	0.70
		2.5	0.0014	0.0025	0.56
	30	1.5	0.0021	0.0015	1.40
		2.0	0.0021	0.0020	1.05
		2.5	0.0021	0.0025	0.84
500	20	1.5	0.0020	0.0015	1.33
		2.0	0.0020	0.0020	1.00
		2.5	0.0020	0.0025	0.80
	30	1.5	0.0030	0.0015	2.00
		2.0	0.0030	0.0020	1.50
		2.5	0.0030	0.0025	1.20
650	20	1.5	0.0026	0.0015	1.70
		2.0	0.0026	0.0020	1.73
		2.5	0.0026	0.0025	1.04
	30	1.5	0.0039	0.0015	2.60
		2.0	0.0039	0.0020	1.95
		2.5	0.0039	0.0025	1.56
800	20	1.5	0.0032	0.0015	2.13
		2.0	0.0032	0.0020	1.60
		2.5	0.0032	0.0025	1.28
	30	1.5	0.0048	0.0015	3.20
		2.0	0.0048	0.0020	2.40
		2.5	0.0048	0.0025	1.92

[a] Each kilogram of Natuphos® phytase contained 5,000,000 U of phytase activity.

unknown factors. Nevertheless, phytase is very effective for pigs, broilers, turkeys, and layers.

Sample calculations are shown in Table 6. Whether or not microbial phytase is as economical as inorganic P depends on the response (P-equivalency value assumed), the cost of phytase, and the cost of inorganic P. The cost of adding 350 to 800 U of phytase per metric ton (2200 lb) can vary from $0.56 to $3.20, depending upon the cost and P-equivalency value chosen. However, these cost estimates do not consider P disposal cost, which would be much less if P excretion was reduced 30 to 50%. If the cost of phytase is discounted for the reduced cost associated with P disposal, the economics can be more favorable for the use of the enzyme. It is interesting to note that Han (1989) reported that the cost of treating soybean meal with microbial phytase was about 17 times greater than using inorganic phosphate.

OTHER NEEDED RESEARCH

The effect of microbial phytase on Ca, trace mineral, and amino acid utilization need to be further investigated. The negative effect of a wide Ca to tP ratio and the beneficial effects of vitamin D_3 need to be further characterized. More accurate and consistent P-equivalency equations need to be developed and tested in *growing-finishing* pigs and in birds from *hatch to market.* Information on P disposal costs for different farm and environmental situations must be obtained.

SUMMARY

Supplemental microbial phytase is well known for its effectiveness in improving P availability of plant ingredients. Supplemental phytase has also been reported to improve the availability of Ca, Zn, and protein (amino acids, AA) in diets high in phytate. Using multiple levels of phytase and P and several response measurements, we have demonstrated that the magnitude of the response of selected measurements to microbial phytase is inversely related to the dietary level of a P (and t P to include phytate P) and to the amount of phytase activity added. Nonlinear functions usually best described the response to phytase. The amount of P released per unit of phytase definitely decreased as the total units of phytase increased. Equivalency values of microbial phytase for P have been calculated, and estimates for Ca and Zn are being determined. The P equivalency of 1 g of P as defluorinated or dicalcium phosphate has been shown to range from 467 to 939 U of microbial phytase. Current research clearly show that adding microbial phytase to diets containing high levels of phytate P and low endogenous levels of phytase will enhance Ca, P, and Zn availability and can result in lower amounts of these minerals being excreted. The dietary level of Ca or the Ca to phytate ratio and the Ca to P ratio are important factors that influence the release and utilization of P. Ratios of Ca to P above 1.4:1 have an adverse effect on the release of P from phytase; whereas, added vitamin D_3 has a positive effect. Decreasing the Ca to P ratio from 2.0:1 to 1.4:1 or 1.1:1 has been shown to improve phytase efficiency in the range of 5 to 12%. The Ca to P ratio is more critical as the level of P is lowered. The use of microbial phytase in pig and poultry diets will depend upon a number of factors, including the environmental need to reduce the excretion of P and other minerals (disposal costs), costs of P, and phytase. In any case, optimizing the effectiveness of phytase will be necessary.

ACKNOWLEDGMENTS

Appreciation is expressed to the United Soybean Board and the John Lee Pratt Animal Nutrition Program for financial support; to BASF for supplying phytase; to Zhixong Yi and Hao Qian for their advice, assistance, and help in conducting the research; to Dave Notter for advice with statistical treatments of data; to Don Conner, Jr., Barbara Self, and Lisa Flory for laboratory support; to Drs. D. M. Denbow and V. Ravindran for their help and advice; and to Cindy Hixon for manuscript preparation. This material is based on work supported in part by the Cooperative State Research Service, USDA, under project number 6129880.

REFERENCES

Adeola, O., Lawrence, B. V., Sutton, A. L., and Cline, T. R., Mineral utilization in phytase-and-zinc supplemented diets for pigs, *J. Anim. Sci.,* 73 (Suppl. 1), 73, 1995.

Anderson, P. A., Interactions between proteins and constituents that affect protein quality, in *Digestibility and Amino Acid Availability in Cereals and Oilseeds,* Finley, G. W. and Hopkins, D. T., Eds., American Association of Cereal Chemists, St. Paul, MN, 1985, 31.

Aoyagi, S. and Baker, D. H., Effect of microbial phytase and 1,25-dihydroxycholecalciferol on dietary copper utilization in chicks, *Poultry Sci.,* 74, 121, 1995.

Ballam, G. C., Nelson, T. S., and Kirby, L. K., Effect of fiber and phytate source and of calcium and P level on phytate hydrolysis in the chick, *Poultry Sci.,* 63, 333, 1984.

Barre, R. and van Huot, N., Etude de la combinason de l'acide phytique avec la serum albumin humane native, acetylee et des amine, *Bull. Soc. Chim. Biol.,* 47, 1399, 1965a.

Barre, R. and van Huot, N., Etude de la combinason de l'ovalbumine avec les phosphorique, B-glycerophosphorique et phytique, *Bull. Soc. Chim. Biol.,* 47, 1419, 1965b.

BASF, Natuphos 5,000, Technical leaflet, BASF AG, Ludwigshafen, 1992.

Beers, S. and Jongbloed, A. W., Effect of supplementary Aspergillus niger phytase in diets for piglets on their performance and apparent digestibility of phosphorus, *Anim. Prod.,* 55, 425, 1992.

Biehl, R., Emmert, J., and Baker, D., 1,25-Dihydroxycholecalciferol acts additively with microbial phytase to release phosphorus and zinc from phytate present in soy-based diets, *FASEB J.,* (Abstract), 1995.

Bitar, K. and Reinhold, J. G., Phytase and alkaline phosphatase activities in intestinal mucosal of rat, chicken, calf and man, *Biochim. Biophys. Acta,* 268, 442, 1972.

Calvert, C. C., Besecker, R. J., Plumlee, M. P., Cline, T. R., and Forsyth, D. M., Apparent digestibility of phosphorus in barley and corn for growing swine, *J. Anim. Sci.,* 47, 420, 1978.

Camus, M. C. and Laporte, J. C., Inhibition de la proteolyse pepsique en vitro par de ble. Role de l'acide phytique des issues, *Ann. Biol. Biochem. Biophys.,* 16, 719, 1976.

Cheryan, M., Phytic acid interactions in food systems, *CRC Crit. Rev. Food Sci. Nutr.,* 13, 297, 1980.

Coffey, M. T., An industry perspective on environmental and waste management issues: Challenge for the feed industry, *GA Nutr. Conf.,* 144, 1992.

Cosgrove, D. J., *Inositol Phosphates: Their Chemistry, Biochemistry and Physiology,* Elsevier Scientific Publishing, New York, 1980.

Crenshaw, T. D., Gahl, M. J., Blemings, K. P., and Benevenga, N. J., Swine feeding programs optimum performance and economic considerations, *10th Annual Carolina Swine Nutrition Conference,* November 10, 1994, 21.

Cromwell, G. L., The biological availability of phosphorus in feedstuffs for pigs, *Pig News Inf.,* 13, 75N, 1992.

Cromwell, G. L. and Coffey, R. D., Nutritional Methods to Reduce Nitrogen and Phosphorus Levels in Swine Waste, Invited Paper, Southern Section ASAS, February 8, 1994.

Cromwell, G. L. and Coffey, R. D., Phosphorus-a key essential nutrient, yet a possible major pollutant-its central role in animal nutrition, in *Biotechnology in the Feed Industry,* Lyons, T. P., Ed., Alltech Technical Publications, Nicholasville, KY, 1991, 133.

Cromwell, G. L., Stahly, T. S., Coffey, R. D., Monegue, H. J., and Randolph, J. H., Efficacy of phytase in improving the bioavailability of phosphorus in soybean meal and corn-soybean meal diets for pigs, *J. Anim. Sci.,* 71, 1831, 1993.

de Haan, F. A. M., Agronomic and environmental implications of soil enrichment with mineral elements originating from animal slurries, in *Animal Agriculture for the 90's,* 1990, 6.

de Lange, C. F. M., Formulation of Diets to Minimize the Contribution of Livestock to Environmental Pollution, American Feed Industry Association Nutrition Council Symposium, St. Louis, MO, November 10–11, 1994.

Denbow, D. M., Ravindran, V., Kornegay, E. T., Yi, Z., and Hulet, R. M., Improving phosphorus availability in soybean meal for broilers by supplemental phytase, *Poultry Sci.,* 74, 1831, 1995.

Dunn, N., A feed additive to fight pollution, *Pigs-Misset,* March, 1994, 21 and 23.

Edwards, H. M., Jr. and Veltmann, J. R., Jr., The role of calcium and phosphorus in the etiology of tibial dyschondroplasia in young chicks, *J. Nutr.,* 113, 1568, 1983.

Edwards, H. M., Jr., Elliot, M. A., and Sooncharernying, S., Effect of dietary calcium on tibial dyschondroplasia. Interaction with light, cholecalciferol, 1,25-dihydroxycholecalciferol, protein and synthetic zeolite, *Poultry Sci.,* 71, 2041, 1992.

Eeckhout, W. and de Paepe, M., Total phosphorus, phytate-phosphorus and phytase activity in plant feedstuffs, *Anim. Feed Sci. Technol.,* 47, 19–29, 1994.

Gibson, D. M. and Ullah, A. B. J., Phytase and their actions on phytic acid, in *Inositol Metabolism in Plants,* Morre, D. J., Boss, W. F., and Loewus, F. A., Eds. Wiley-Liss, New York, 1990, 77.

Hacker, R. R. and Du., Z., Livestock pollution and politics, in *Nitrogen Flow in Pig Production and Environmental Consequences,* Verstegen, M. W. A., den Hartog, L. A., van Kempen, G. J. M., and Metz, J. H. M., Eds., Pudoc Scientific Publishers, Wageningen, The Netherlands, 1993, 3.

Han, Y. W., Use of microbial phytase in improving the feed quality of soyba bean meal, *Anim. Feed Sci. Technol.,* 24, 345, 1989.

Harms, R. W., Waldroup, P. W., Shirley, R. L., and Anderson, C. B., Availability of phytic acid P for chicks, *Poultry Sci.,* 41, 1189, 1962.

Hoppe, P. P. and Schwarz, G., Experimental approaches to establish the phosphorus equivalency of Aspergillus-niger-phytase in pigs, in *Enzymes in Animal Nutrition,* Wenk, C. and Boessinger, M., Eds., Proc. 1st Symp. Kartause Ittingen, Switzerland, Oct. 13–16, 1993, 187.

Hoppe, P. P., Schoner, F. J., Wiesche, H., Schwarz, G., and Safer, A., Phosphorus equivalency of *Aspergillus-niger*-phytase for piglets fed a grain-soybean-meal diet, *J. Anim. Physiol. Anim. Nutr.,* 69, 225, 1993.

Jongbloed, A. W. and Henkens, C. H., Environmental Concerns of Using Animal Manure — he Dutch Case, International Symposium on Nutrient Management of Food Animals to Enhance the Environment, June 4–7, Virginia Tech, Blacksburg, 1995.

Jongbloed, A. W. and Kemme, P. A., Apparent digestible phosphorus in the feeding of pigs in relation to availability, requirement and environment. 1. Digestible phosphorus in feedstuffs from plant and animal origin, *Neth. J. Agric. Sci.,* 38, 567, 1990.

Jongbloed, A. W., Mroz, Z., and Kemme, P. A., The effect of supplementary *Aspergillus niger* phytase in diets for pigs on concentration and apparent digestibility of dry matter, total phosphorus, and phytic acid in different sections of the alimentary tract, *J. Anim. Sci.,* 70, 1159, 1992.

Jongbloed, A. W., Mroz, Z., Kemme, P. A., Geerse, C., and Van Der Honing, Y., The effect of dietary calcium levels on microbial phytase efficacy in growing pigs, *J. Anim. Sci.,* 71 (Suppl. 1) , 166, 1993.

Kemme, P. A., Jongbloed, A. W., Mroz, Z., and Makinen, M., Apparent Illeal Digestibility of Protein and Amino Acids From a Maize-Soybean Meal Diet With or Without Extrinsic Phytate and Phytase in Pigs, International Symposium on Nutrient Management of Food Animals to Enhance the Environment, June 4–7, Virginia Tech, Blacksburg, 1995.

Ketaren, P. P., Batterham, E. S., Dettmann, E. B., and Farrell, D. J., Phosphorus studies in pigs. 3. Effect of phytase supplementation on the digestibility and availability of phosphorus in soy-bean meal for grower pigs, *Br. J. Nutr.,* 70, 289, 1993.

Kirby, L. K. and Nelson, T. S., Total and phytate phosphorus content of some feed ingredients derived from grains, *Nutr. Rep. Int.,* 37, 277, 1988.

Koch, M. E., Mahan, D. C., and Corley, J. R., An evaluation of varying biological characteristics in assessing low phosphorus intake in weaning swine, *J. Anim. Sci.,* 59, 1546, 1984.

Kornegay, E. T., Unpublished data, 1992.

Kornegay, E. T. and Denbow, D. M., Unpublished data, 1995.

Kornegay, E. T., Nutrient Management of Swine to Enhance and Protect the Environment, 3rd International Symposium on Animal Nutrition, Kaposvar, Hungary, October 18, 1994.

Kornegay, E. T. and Qian, H., Replacement of inorganic phosphorus by microbial phytase for young pigs fed a corn soybean meal diet, *Br. J. Nutr.,* in press, 1996.

Kornegay, E. T., Denbow, D. M., Yi, Z., and Ravindran, V., Response of broilers to graded levels of Natuphos® phytase added to corn-soybean meal-based diets containing three levels of nonphytate phosphorus, *Br. J. Nutr.,* 75, 839, 1996.

Lawrence, B. V., Adeola, O., Sutton, A. L., and Cline, T. R., Phytase-Induced Increase in Mineral Absorption and Utilization in Diets for Young Pigs, International Symposium on Nutrient Management of Food Animals to Enhance the Environment, June 4–7, Virginia Tech, Blacksburg, 1995.

Lei, X. G., Ku, P. K., Miller, E. R., Ullrey, D. E., and Yokoyama, M. T., Supplemental microbial phytase improves bioavailability of dietary zinc to weanling pigs, *J. Nutr.,* 123, 1117, 1993.

Lei, X. G., Ku, P. K., Miller, E. R., Yokoyama, M. T., and Ullrey, D. E., Calcium level affects the efficacy of supplemental microbial phytase in corn-soybean meal diets of weanling pigs, *J. Anim. Sci.,* 72, 139, 1994.

Lenis, N. P., Lower nitrogen excretion in pig husbandry by feeding: Current and future possibilities, *Neth. J. Agric. Sci.,* 37, 61, 1989.

Liebert, F., Wecke, C., and Schoner, F. J., Phytase activities in different gut contents of chickens are dependent on level of phosphorus and phytase supplementations, in *Enzymes in Animal Nutrition,* Wenk, C. and Boessinger, M., Eds., Kartause Ittingen, Switzerland, 1993, 202.

Maddaiah, V. T., Kurnick, A. A., and Reid, B. L., Phytic acid studies, *Proc. Soc. Exp. Biol. Med.,* 115, 391, 1964.

McCollum, E. V. and Hart, E. B., On the occurrence of a phytin-splitting enzyme in animal tissue, *J. Biol. Chem.,* 4, 497, 1908.

Mohammed, A., Gibney, M. J., and Taylor, T. G., The effects of dietary levels of inorganic phosphorus, calcium and cholecalciferol on the digestibility of phytate-P by the chick, *Br. J. Nutr.,* 66, 251, 1991.

Mroz, Z., Jongbloed, A. W., and Kemme, P. A., Apparent digestibility and retention of nutrients bound to phytate complexes as influenced by microbial phytase and feeding regimen in pigs, *J. Anim. Sci.,* 72, 126, 1994.

Mroz, Z., Jongbloed, A. W., Kemme, P. A., and Geerse, K., Digestibility and urinary losses of calcium and phosphorus in pigs fed a diet with suboptimal levels of both elements and graded doses of microbial phytase (Natuphos®), in *Enzymes in Animal Nutrition,* Kartause Ittingen, Switzerland, 1993.

Mueller, J. P., Zublena, J. P., Poore, M. H., Barker, J. C., and Green, J. T., Managing Pasture and Hay Fields Receiving Nutrients From Anaerobic Swine Waste Lagoons, AG-506, N.C. Coop. Ext. Service, 1994.

Nasi, M. and Helander, E., Effects of microbial phytase supplementation and soaking of barley-soybean meal on availability of plant phosphorus for growing pigs, *Acta Agric. Scand.,* 44, 79, 1994.

National Research Council (NRC), *Nutrient Requirements of Swine,* 9th edition, National Academy Press, Washington, DC, 1988.

National Research Council (NRC), *Nutrient Requirements of Poultry,* 9th edition, National Academy Press, Washington, DC, 1994.

Nayini, N. R. and Markakis, P., Phytases, in *Phytic Acid: Chemistry and Applications,* Graf, E., Ed., Pilatus Press, Minneapolis, 1986, 101.

Nelson, T. S., The utilization of phytate P by poultry–a review, *Poultry Sci.,* 46, 862, 1967.

Nelson, T. S., Ferrara, L. W., and Storer, N. L., Phytate P content of feed ingredients derived from plants, *Poultry Sci.,* 47, 1372, 1968.

Nelson, T. S., Shieh, T. R., Wodzinski, R. J., and Ware, J. H., Effect of supplemental phytase on the utilization of phytate phosphorus by chicks, *J. Nutr.,* 101, 1289, 1971.

Oberleas, D., Phytates, in *Toxicants Occurring Naturally in Foods,* National Academy of Sciences, Washington, DC, 1973, 363.

Pallauf, V. J., Hohler, D., and Rimbach, G., Effect of microbial phytase supplementation to a maize-soya-diet on the apparent absorption on Mg, Fe, Cu, Mn and Zn and parameters of Zn-status in piglets, *J. Anim. Physiol. Anim. Nutr.,* 68, 1, 1992a.

Pallauf, V. J., Hohler, D., Rimbach, G., and Neusser, H., Effect of microbial phytase supplementation to a maize-soy-diet on the apparent absorption of phosphorus and calcium in piglets, *J. Anim. Physiol. Anim. Nutr.,* 67, 30, 1992b.

Pallauf, J., Rimbach, G., Pippig, S., Schindler, B., and Most, E., Effect of phytase supplementation to a phytate-rich diet based on wheat, barley and soya on the bioavailability of dietary phosphorus, calcium, magnesium, zinc and protein in piglets, *Agribiol. Res.,* 47, 39, 1994a.

Pallauf, J., Rimbach, G., Pippig, S., Schindler, B., Hohler, D., and Most, E., Dietary effect of phytogenic phytase and an addition of microbial phytase to a diet based on field beans, wheat, peas and barley on the utilization of phosphorus, calcium, magnesium, zinc and protein in piglets, *Z. Ernahrungswiss,* 33, 128, 1994b.

Pointillart, A., Importance of phytates and cereal phytases in the feeding of pigs, in *Enzymes in Animal Nutrition,* Wenk, C. and Boessinger, M., Eds., Kartause Ittingen, Switzerland, 1993, 192.

Pointillart, A., Fontaine, N., Thomasset, M., and Jay, M. E., Phosphorus utilization, intestinal phosphatases and hormonal control of calcium metabolism in pigs fed phytic phosphorus: Soyabean or rapeseed diets, *Nutr. Rep. Int.,* 32, 155, 1985.

Pointillart, A., Fourdin, A., Bourdeau, A., and Thomasset, M., Phosphorus utilization and hormonal control of calcium metabolism in pigs fed phytic phosphorus diets containing normal or high calcium levels, *Nutr. Rep. Int.,* 40, 517, 1989.

Pointillart, A., Fourdin, A., and Fontaine, N., Importance of cereal phytase activity for phytate phosphorus utilization by growing pigs fed diets containing triticale or corn, *J. Nutr.,* 117, 907, 1987.

Qian, H., Kornegay, E. T., and Conner, D. E., Jr., Adverse effects of wide calcium:phosphorus ratios on supplemental phytase efficacy for weanling pigs fed two dietary phosphorus levels, *J. Anim. Sci.,* 74, 1288, 1996a.

Qian, H., Kornegay, E. T., and Denbow, D. M., Phosphorus equivalence of microbial phytase in turkey diets as influenced by Ca:P ratios and P levels, *Poultry Sci.,* 75, 69, 1996b.

Qian, H., Kornegay, E. T., and Denbow, D. M., Utilization of phytate phosphorus and calcium as influenced by microbial phytase, cholecalciferol and calcium:total phosphorus ratios in broiler diets, *Poultry Sci.,* 1996c (in press).

Qian, H., Kornegay, E. T., and Veit, H. P., Effects of supplemental phytase and phosphorus on histological, mechanical and chemical traits of tibia and performance of turkeys fed soybean meal-based semi-purified diets high in phytate phosphorus, *Br. J. Nutr.,* 76, 263, 1996d.

Qian, H., Veit, H. P., Kornegay, E. T., Ravindran, V., and Denbow, D. M., Effects of supplemental phytase and phosphorus on histological and other tibial bone characteristics and performances of broilers fed semi-purified diets, *Poultry Sci.,* in press, 1996e.

Radcliffe, J. S., Kornegay, E. T., and Conner, D. E., Jr., The effect of phytase on calcium release in weanling pigs fed corn-soybean meal diets, *J. Anim. Sci.,* 73, 173, 1995.

Ravindran, V., Phytases in poultry nutrition, an overview, *Proc. Aust. Poultry Sci. Symp.,* 7, 135, 1995.

Ravindran, V., Bryden, W. L., and Kornegay, E. T., Phytates: Occurrence, bioavailability and implications in poultry nutrition, *Poultry and Avian Biol. Rev.,* 6, 125, 1995.

Ravindran, V., Kornegay, E. T., Denbow, D. M., Yi, Z., and Hulet, R. M., Response of turkey poults to tiered levels of Natuphos® phytase added to soybean meal-based semi-purified diets containing three levels of nonphytate phosphorus, *Poultry Sci.,* 74, 1820, 1996.

Ravindran, V., Ravindran, G., and Sivalogan, S., Total and phytate phosphorus contents of various foods and feedstuffs of plant origin, *Food Chem.,* 50, 133, 1995.

Reddy, N. R., Sathe, S. K., and Salunkhe, D. K., Phytates in legumes and cereals, in *Advances in Food Research,* Chichester, C. O., Mrak, E. M., and Stewart, G. F., Eds., Academic Press, New York, 1982, 1.

Reinhart, G. A. and Mahan, D. C., Effect of various calcium:phosphorus ratios at low and high dietary phosphorus for starter, grower and finishing swine, *J. Anim. Sci.,* 63, 457, 1986.

Rimbach, G. and Pallauf, J., Effekt einer zulage mikrobieller phytase auf die zink-verfugbarkeit, *Z. Ernahrungswiss,* 31, 269, 1992.

Rimbach, G. and Pallauf, J., Enhancement of zinc utilization from phytate-rich soy protein isolate by microbial phytase, *Z. Ernahrungswiss,* 32, 308, 1993.

Roberson, K. D. and Edwards, H. M., Jr., Effects of 1,25-dihydroxycholecalciferol and phytase on zinc utilization in broiler chicks, *Poultry Sci.,* 73, 1312, 1994.

Sanders, A. M., Edwards, H. M., Jr., and Rowland, G. N., III, Calcium and P requirements of the very young turkey as determined by response surface analysis, *Br. J. Nutr.,* 67, 421, 1992.

Schoner, F. J., Schwarz, G., and Hoppe, P. P., Influence of the P compound on the P and Ca retention and the P discharge in broilers, *102nd VDLUFA Congr. Berlin,* 32, 437, 1990.

Schoner, F. J., Schwarz, G., Hoppe, P. P., and Wiesche, H., Effect of Microbial Phytase on Ca-Availability in Broilers, Third Conference of Pig and Poultry Nutrition in Halle, November 29–December 1, 1994.

Schoner, F. J., Hoppe, P. P., and Schwarz, G., Comparative effects of microbial phytase and inorganic phosphorus on performance and on retention of phosphorus, calcium and crude ash in broilers, *J. Anim. Physiol. Anim. Nutr.*, 66, 248, 1991.

Schoner, F. J., Hoppe, P. P., Schwarz, G., and Wiesche, H., Effects of microbial phytase and inorganic phosphate in broiler chickens: Performance and mineral retention at various calcium levels, *J. Anim. Physiol. Anim. Nutr.*, 69, 235, 1993.

Schwarz, G., *Benefit of Microbial Phytase on Phosphorus Availability and Excretion in Poultry and Swine Feeding,* NFIA Nutrition Institute, Chicago, 1992, 1.

Simons, P. C. M., Versteegh, H. A. J., Jongbloed, A. W., Kemme, P. A., Slump, P., Bos, K. D., Wolters, M. G. E., Beudeker, R. F., and Verschoor, G. J., Improvement of phosphorus availability by microbial phytase in broilers and pigs, *Br. J. Nutr.*, 64, 525, 1990.

Singh, M. and Krikorian, A. D., Inhibition of trypsin activity in vitro by phytate, *J. Agric. Food Chem.*, 30, 799, 1982.

Thiel, U. and Weigand, E., Influence of dietary zinc and microbial phytase supplementation on Zn retention and Zn excretion in broiler chicks, *Worlds Poultry Congr.*, 19, 20–24.9, 1992.

Thiel, U., Weigand, E., Schoner, F. J., and Hoppe, P. P., Zinc retention of broiler chicken as affected by dietary supplementation of zinc and microbial phytase, in *Trace Elements in Man and Animals (TEMA 8),* Anke, M., Meissner, D., and Mills, C. F., Eds., Gersdorf, Dresden, 1993, 658.

Thompson, L. U., Phytic acid: A factor influencing starch digestibility and blood glucose response, in *Phytic Acid: Chemistry and Applications,* Graf, E., Ed., Pilatus Press, Minneapolis, 1986, 173.

Van der Klis, J. D. and Verstaagh, H. A., Ileal Absorption of P in Lightweight White Laying Hens Using Microbial Phytase and Various Calcium Contents in Laying Hen Feed, Spelderholt Publ. No. 563, 1991.

Van Horn, H. H., Achieving environmental balance with manure and cropping systems, *GA Nutr. Conf.*, 1992, 110.

Veum, T. L., Liu, J., Bollinger, D. W., Zyla, K., and Ledoux, D. R., Use of a microbial phytase in a canola-grain sorghum diet to increase phosphorus utilization by growing pigs, *J. Anim. Sci.*, in press (Abstr.), 1994.

Vohra, P., Gray, G. A., and Kratzer, F. H., Phytic acid-metal complexes, *Proc. Soc. Exp. Biol. Med.*, 120, 447, 1965.

Wise, A., Dietary factors determining the biological activities of phytase, *Nutr. Abstr. Rev.*, 53B, 791, 1983.

Yi, Z. and Kornegay, E. T., Sensitive indicators to evaluate efficacy of supplemental phytase on phosphorus availability in diets fed to pigs and poultry, *J. Anim. Sci.*, 73 (Suppl. 1), 174, 1995a.

Yi, Z. and Kornegay, E. T., Unpublished data, 1995b.

Yi, Z. and Kornegay, E. T., Sites of phytase activity in the gastrointestinal tract of young pigs, *Anim. Feed Sci. Technol.*, in press, 1996a.

Yi, Z., Kornegay, E. T., and Denbow, D. M., Supplemental microbiol phytase improves zinc utilization in broilers, *Poultry Sci.*, 75, 540, 1996c.

Yi, Z., Kornegay, E. T., and Denbow, D. M., Effect of microbial phytase on nitrogen and amino acid digestibility and nitrogen retention of turkey poults fed corn-soybean meal diets, *Poultry Sci.*, 75, 979, 1996b.

Yi, Z., Kornegay, E. T., Lindemann, M. D., Ravindran, V., and Wilson, J. H., Effectiveness of Natuphos® phytase in improving the bioavailabilities of phosphorus and other nutrients in soybean meal-based semi-purified diets for young pigs, *J. Anim. Sci.*, 74, 1601, 1996a.

Yi, Z., Kornegay, E. T., and McGuirk, A., Replacement values of inorganic phosphorus by microbial phytase for pigs and poultry, *J. Anim. Sci.,* 72 (Suppl. 1), 330, 1994.

Yi, Z., Kornegay, E. T., Ravindran, V., and Denbow, D. M., Improving phytate phosphorus availability in corn and soybean meal for broilers using microbial phytase and calculation of phosphorus equivalency values for phytase, *Poultry Sci.,* 75, 240, 1996b.

CHAPTER 19

Chromium Picolinate for the Enhancement of Muscle Development and Nutrient Management

M. D. Lindemann

INTRODUCTION

Animal-derived food products are excellent and readily available sources of micronutrients (vitamins and minerals) and high-quality proteins not easily duplicated with plant-derived food products. However, a shift in consumer acceptability of animal-derived food products has occurred during the last 25 years. Consumer concern is expressed about the nutritional value of red meat with a specific concern being the contribution of animal fat to total dietary fat intake. In response to concerns about animal fats, the swine industry has implemented several approaches to increase lean, and lower fat, content. These approaches have ranged from increased trimming of cuts at the consumer level to altered payment schemes at the packer level to altered genetics and nutrition at the producer level. Despite these current efforts, new technologies or nutritional strategies that are more efficacious, timely, and cost effective are needed to continue to provide lean, low-fat meat products that are, both in fact and perception, safe and nutritious.

Concern about the condition of the environment is also increasing throughout the world. Of particular importance to the livestock industry in many countries are potential regulations relating to waste management and disposition. The increasing size of animal production units in many areas has resulted in greater manure loading rates on given quantities of land. Animal waste has traditionally been an excellent source of nutrients for crops. However, when manure application rates exceed the need of the growing crops, the excess becomes a potential pollutant. The nitrogen in animal waste is of particular interest with regard to possible pollution. Nitrogen is found in animal waste largely in the inorganic form as the ammonium (NH_4^+) cation, which is immobile in the soil because it binds to negative charges on the surface of the clay particles. If it accumulates sufficiently, some may undergo the process of nitrification wherein it is converted to nitrate (NO_3^-). Nitrate does not bind to clay particles, so it is free to percolate down through the soil until it reaches the groundwater; soil nitrate is also subject to denitrification, during which the nitrogen can escape to the atmosphere. Excess nitrogen applications have the potential, therefore, to result in contamination of

1-56670-199-6/96/$0.00+$.50

either the air or the groundwater, which consequently necessitates attention from the livestock industry.

At times increasing carcass lean content to meet consumer demands can be in conflict with efforts to reduce the excretion of certain nutrients. An exciting development in recent years is the understanding of the nutritional need for the mineral chromium. Proper supplementation of this mineral is demonstrating benefits in both carcass lean content and in improved nutrient utilization, which results in reduced nutrient excretion.

REDUCING NITROGEN EXCRETION

Reductions in animal nitrogen excretion in order to reduce soil-applied nitrogen are possible via both dietary and management means. Some of the means may add to production costs, while others simply require attention to practices of which producers are already aware. One of the most obvious means of reducing excretion of all nutrients is to reduce the amount of feed required for each unit of gain (that is to improve feed efficiency). This can be accomplished by physical methods such as proper feeder selection and adjustment to biological methods such as the selection of more efficient breeding stock.

A management decision to change diets more frequently across weight ranges will result in less excretion. This is because the excretion of nitrogen is related to the biological availability of the forms of nitrogen and its supply relative to the need of the pig; the more closely the supply can be matched to the need of the pig the less there will be excreted in the manure. The utilization of split-sex feeding is another means whereby nutrient supply is more closely matched to the needs of the pig. Additionally, changes in feed intake due to changes in environmental conditions will necessitate dietary changes to match nutrient supply to nutrient need; lack of dietary reformulation in instances of thermal stress results in waste of nutrients and increased excretion.

With regard to dietary alterations as a means of reducing nitrogen in swine waste, one of the easiest means is by the use of crystalline amino acids to effect an improved amino acid balance of the diet. When typical grain-soybean meal diets are formulated based on the first-limiting amino acid, which is lysine, relatively large excesses can occur of several other amino acids. These excess amino acids are of no use to the pig for protein synthesis, and the nitrogen portion of those amino acids is excreted in the urine and contributes to total nitrogen excretion. Formulation based on the second-, third-, or fourth-limiting amino acid with supplementation of the others that are more limiting will markedly reduce nitrogen excretion.

CONFLICTING INTERESTS

A factor that now contributes to increased nitrogen excretion is the current emphasis on leanness of market swine. Market payment programs reward the sale of lean market swine and penalize swine with less lean and more fat. This type of payment schedule is understandable from the standpoint of producing what the consuming public demands, but it does create added problems from the production standpoint with regard to the effort to minimize nitrogen excretion.

It is well recognized that to maximize carcass leanness, a greater concentration of protein is required in the diet than is required to maximize growth rate. Feeding of swine to maximize carcass leanness to provide a product desired by the consumer and to benefit from the current payment standards will require that more protein be fed. Additionally, the continued supplementation of greater amounts of protein to maximize carcass leanness is subject, as many biological and mechanical systems, to the law of diminishing returns — that is, that with each subsequent unit of input, the output is diminished from that of the previous input. This means that the closer to the genetic carcass lean maximum, the less efficient the use of nitrogen for lean deposition. For example (Table 1), taking a portion of the data from a recently reported study by Lindemann et al. (1995b) in which pigs were fed an increase in protein content of the diet from 100% to 120% of the NRC (1988) requirement estimates, the carcass effects of that increase in protein, although not of a large magnitude, were as expected to increase loin muscle area and to decrease backfat. The increased protein fed in this trial would have resulted in an increase in nitrogen intake over the entire trial period of 516 g per pig. However, the increase in nitrogen retained in the carcass is only about 25 g per pig, a minimal change given the input of 516 g per pig. Not all of the remaining nitrogen would have been excreted, but given that dressing percentage in pigs is in excess of 70% and if one assumes increases in nitrogen content of other parts of the body similar to that of the carcass, the amount of nitrogen accounted for by the whole body would not be in excess of 150% of that of the carcass, or no more than 38 g in this example.

It is also demonstrated in Table 1 that increasing only the loin muscle area (LMA) by 55% (from 29.0 to 45.0 cm^2, equal to an absolute increase of 6.8% in percentage lean) with no decrease in backfat accounts for just 239 g of nitrogen; decreasing backfat alone by 58% (from 33.6 to 14 mm, equal to an absolute increase of 7.5% in percentage lean) with no increase in LMA accounts for just 262 g of nitrogen. An absolute increase of greater than 14% in percentage carcass lean is required to account for the entire 500 g of nitrogen input in the carcass. It is impossible to achieve this level of increase in carcass lean on a given production unit without changing the genetic base of production. It should be evident that the attempt to increase carcass leanness through increased dietary protein levels will result in increased nitrogen excretion for that production unit. This leads to the conclusion that dietary alterations directed at improving carcass leanness can only decrease nitrogen excretion if they are of a nonintact-protein nature.

RECENT ADVANCES

One of the exciting areas of recent mineral nutrition research is the discovery of marked improvements in human and animal health as a result of organic

Table 1 Relationship of Change in Carcass Measurements to Changes in Retained Nitrogen in the Carcass[a]

Item	LMA (cm^2)	10th rib BF (mm)	Carcass lean (kg)	Carcass lean (%)	Carcass Nitrogen (g)
Example baseline[b]	29.0	33.6	33.8	43.8	1535
1. Increased protein 20%[b]	30.1	33.0	34.3	44.5	1560
2. Increased LMA, constant BF	33.0	33.6	35.1	45.5	1595
	37.0	33.6	36.4	47.2	1654
	41.0	33.6	37.7	48.9	1714
	45.0	33.6	39.0	50.6	1774
3. Constant LMA, decreased BF	29.0	30.0	34.8	45.2	1584
	29.0	26.0	36.0	46.7	1637
	29.0	22.0	37.2	48.2	1690
	29.0	18.0	38.4	49.8	1744
	29.0	14.0	39.6	51.3	1797
4. Increased LMA, decreased BF	33.0	30.0	36.2	46.9	1643
	37.0	26.0	38.6	50.1	1756
	41.0	22.0	41.1	53.3	1869
	45.0	18.0	43.6	56.6	1982
	47.0	16.0	44.9	58.2	2038

[a] Absolute lean and percent lean were calculated using the formulae in Procedures to Evaluate Market Hogs (NPPC, 1991). The nitrogen content of the carcass was calculated based on the report (Dickerson and Widdowson, 1960) that adult pig skeletal muscle is 28.4% protein coupled with the assumption that nitrogen is 16% of that protein. LMA = loin muscle area; BF = backfat depth.

[b] Data from Lindemann et al., 1995.

chromium (Cr) supplementation to the diet. Some of the effects observed have been improved glucose tolerance, positive alteration of serum lipids, and, in some studies, the increase of lean body mass (Evans, 1989). The National Research Council (1989) recommends a Cr intake of 50 to 200 μg/d for adult humans. Estimates exist (Kumpulainen, 1992) that over 50% of the human population have a Cr intake that falls at or below the lower limit of the present estimated safe and adequate daily dietary intake as established by the NRC (1989).

Since the glucose tolerance factor (GTF) was identified (Schwartz and Mertz, 1957) and the active ingredient was determined to be Cr (Schwartz and Mertz, 1959), the role of Cr has been the subject of extensive investigation in humans and in laboratory animals. The best-known function of the GTF is stimulation of the action of insulin in Cr-deficient tissue (Mertz, 1969). Insulin, a polypeptide hormone, promotes anabolic processes and inhibits catabolic ones in muscle, liver, and adipose tissue. Insulin stimulates the active transport of glucose and amino acids into muscle cells and enhances protein synthesis.

The nutritional status and function of Cr in livestock species, however, has not been extensively investigated. Samsell and Spears (1989) indicated that Cr supplementation lowered fasting plasma glucose concentrations in lambs fed a low-fiber diet. Chang et al. (1991) suggested that calves fed corn silage following market-transit stress may be deficient in Cr, and supplemental GTF-Cr did decrease serum cortisol and improve immune status. Chromium has also been shown to promote growth of turkey poults (Steele and Rosebrough, 1979). It had not been reported that pigs responded to dietary Cr supplementation until recently when Page (1991) reported increased LMA and decreased carcass backfat when the specific chelate Cr picolinate was fed.

EFFECTS OF CHROMIUM ON RETAINED CARCASS NITROGEN

The fact that a nonintact-protein nutrient such as Cr has been demonstrated to favorably affect carcass parameters (Page, 1991; Lindemann et al., 1995) means that this dietary nutrient will decrease nitrogen excretion whenever it elicits the carcass improvements (provided that there is not an increase in feed intake, or an increase in dietary protein, required to effect the response). By calculating the protein intake in relation to the carcass protein, it is possible to determine whether there is a net increase in nitrogen utilization. The carcass results of recent trials at Virginia Tech in Blacksburg, and their resultant effects on carcass nitrogen, are given in Table 2. In the initial trial, the effects on the carcass were not large, but, as a result of reductions in protein intake associated with the Cr supplementation, there was a 9 to 12% nitrogen savings when expressed as a percentage of noncarcass nitrogen in the control pigs. The second trial was conducted with two protein/lysine levels, and, depending on the protein level used, the nitrogen savings varied from –2 to 10%.

A wide variety of trials were conducted at Louisiana State University by Page (1991). Calculations of the effects of Cr on retained carcass nitrogen from those trials are given in Table 3. The nitrogen savings as a percentage of noncarcass nitrogen in the control animals ranged from –2 to 13%. Disregarding all factors in the trials at these two locations that may have affected magnitude of response, the mean nitrogen savings of the six trials in Tables 2 and 3, which involved 16 estimates of nitrogen savings, was 5.6% with 5 of the 16 estimates being slightly negative or negligible positive responses, whereas the other 11 responses ranged from 2.6 to 13.5%. Direct determinations of the effect of Cr supplementation on nitrogen retention are few. Wang et al. (1995) reported on two digestibility trials in which the mean absolute increase in retained nitrogen with Cr supplementation was 2.35% (48.5% for control-fed pigs and 50.85% for Cr-supplemented pigs), which equates to a relative increase of 4.85%.

EFFECTS OF CHROMIUM ON REPRODUCTIVE EFFICIENCY

Reproductive efficiency is integrally related to the profitability of swine production units and to nutrient excretion charges made to the offspring. Pigs weaned per sow annually are one of the most widely used indicators of breeding herd productivity and are an excellent indicator of the overall efficiency of the breeding female. This productivity measurement is a function of litter size born, preweaning mortality, lactation length, days to estrus, and herd conception rate.

Table 2 Effect of Chromium From Chromium Picolinate on Carcass Characteristics and Calculated Nitrogen Retention (Virginia Tech Trials)[a]

	Virginia Tech I[b]			Virginia Tech II[c]				
Chromium (ppb):Lysine:	0	250	500	0 100	200 100	0 120	100 120	200 120
Loin muscle area (cm^2)	39.9	41.2	41.4	29.0	33.6	30.1	31.1	32.1
Tenth-rib fat depth (mm)	28.8	29.5	31.4	33.6	27.8	33.0	31.2	29.6
Lean percentage	50.3	50.6	49.9	44.0	47.9	44.7	45.5	46.8
Total protein consumption[d] (g)	28637	26489	26910	36478	34441	39703	39897	40895
Total nitrogen consumption[e], (g)	4582	4238	4306	5836	5511	6352	6384	6543
Difference from unsupplemented		–344	–276		–325		31	191
Lean in 77.1 kg carcass[f] (kg)	38.8	39.0	38.5	33.9	36.9	34.4	35.1	36.1
Nitrogen in 77.1 kg carcass[f] (g)	1762	1772	1749	1542	1678	1566	1594	1640
Difference from control		10	–13		136		28	74
Nitrogen savings (carcass increase - diet increase) (g)		354	263		461		–3	–117
Noncarcass nitrogen	2820			4295		4786		
Savings as % of control noncarcass nitrogen		12.6	9.3		10.7		–0.1	–2.4

[a] Taken from Lindemann et al., 1995.

[b] Initial weight of 40.9 kg, final weight of 98.8 kg; diet formulated to ca. 105 to 116% of NRC (1988) for the respective growth phases.

[c] Initial weight of 14.5 kg, final weight of 102.2 kg; diet formulated to 100 or 120% of NRC (1988) lysine requirement with intact protein for the respective growth phases.

[d] Computed directly from performance values and diet compositions relative to the portion of total feed consumed during each growth phase.

[e] As 16% of protein consumption.

[f] As footnote a of Table 1.

Table 3 Effect of Chromium From Chromium Picolinate on Carcass Characteristics and Calculated Nitrogen Retention (Louisiana State University Trials)[a]

	LSU I[b]					LSU IV[c]				
Chromium (ppb):	**0**	**25**	**50**	**100**	**200**	**0**	**100**	**200**	**400**	**800**
Loin muscle area	34.9	35.9	35.3	34.2	37.2	34.0	40.4	39.9	41.7	40.3
Tenth-rib fat depth	28.3	24.2	23.3	26.6	24.4	31.5	23.4	26.3	22.0	24.6
Percentage of muscling	52.9	54.7	54.5	53.6	54.3	51.7	56.1	54.7	57.4	56.2
Total protein consumption[d] (g)	30763	28785	29448	30869	29160	33010	32330	33109	31317	33310
Total nitrogen consumption[e] (g)	4922	4606	4712	4939	4666	5282	5173	5297	5011	5330
Difference from unsupplemented		–316	–210	17	–256		–109	15	–271	48
Lean in 77.1 kg carcass[f] (kg)	40.8	42.2	42.0	41.3	41.9	39.9	43.3	42.2	44.3	43.3
Nitrogen in 77.1 kg carcass[f] (g)	1854	1917	1910	1878	1903	1812	1966	1917	2011	1969
Difference from control		63	56	24	49		154	105	199	157
Nitrogen savings (carcass increase — diet increase) (g)		379	266	7	305		263	90	470	109
Noncarcass nitrogen	3068					3470				
Savings as % of control noncarcass nitrogen		12.4	8.7	0.2	9.9		7.6	2.6	13.5	3.1

	LSU III[g]			LSU IV[h]	
Chromium (ppb):	**0**	**100**	**200**	**0**	**200**
Loin muscle area	31.5	38.1	38.4	31.8	36.6
Tenth-rib fat depth	30.7	25.4	23.9	33.4	26.6
Percentage of muscling	52.3	54.8	55.7	50.0	53.4
Total protein consumption[d] (g)	33451	34612	34180	38544	38068
Total nitrogen consumption[e] (g)	5352	5538	5469	6167	6091
Difference from unsupplemented		186	117		–76
Lean in 77.1 kg carcass[f] (kg)	40.3	42.3	43.0	38.6	41.2
Nitrogen in 77.1 kg carcass[f] (g)	1833	1920	1952	1752	1871
Difference from control		87	119		119

Table 3 Effect of Chromium From Chromium Picolinate on Carcass Characteristics and Calculated Nitrogen Retention (Louisiana State University Trials)[a] (Continued)

	LSU III[g]			LSU IV[h]	
Chromium (ppb):	**0**	**100**	**200**	**0**	**200**
Nitrogen savings (carcass increase - diet increase) (g)		–99	2		195
Noncarcass nitrogen	3519			4415	
Savings as % of control noncarcass nitrogen		–2.8	0.1		4.4

[a] From Page, 1991; diets formulated to 120% of NRC (1988) for the respective growth phases. All intake and carcass calculations are as in Table 2.

[b] Initial weight of 37.8 kg, final weight of 97.6 kg.

[c] Initial weight of 30.5 kg, final weight of 103.2 kg.

[d] Computed from performance values and diet compositions with assumptions of the portion of total feed consumed during each growth phase.

[e] As 16% of protein consumption.

[f] As footnote a of Table 1.

[g] Initial weight of 22.4 kg, final weight of 93.6 kg.

[h] Values are the mean of two genotypes used in the study; initial weight of 21.9 kg, final weight of 102.2 kg.

Table 4 Effects of Dietary Chromium From Chromium Picolinate on Sow Fecundity and Pig Weight[a]

Reproductive phase	Chromium (ppb): 0	200	P value
Litter size			
Total born	9.58	11.82	0.03
Born live	8.93	11.25	0.02
Day 21	8.15	10.30	0.03
Litter weight, kg			
Total born	13.79	17.04	0.02
Born live	12.86	16.27	0.01
Day 21	46.51	54.57	0.07
Survival to d 21 (%)[b]	93.3	93.2	0.89

[a] From Lindemann et al., 1995b.

[b] Means are adjusted by covariance analysis for litter size.

Since all costs associated with each sow must be paid for by receipts from the pigs produced, any increase in the number of pigs per sow will have marked effects on the amount of cost (monetary or nutrient charges) that each pig must bear and, consequently, on total enterprise profitability and on total enterprise nutrient efficiency.

Proper nutrition of the sow is an essential part of optimizing reproductive efficiency. The modern high-producing sow loses relatively large amounts of weight in lactation due to high milk output. Large changes in body weight and composition (body condition) have deleterious effects on reproduction. Increases in body weight loss during lactation are associated with increased loss of backfat and muscle, increased interval to rebreeding, decreased number bred in each sow group, decreased conception rate, decreased embryo survival that results in smaller litters at the next farrowing, and premature culling from the herd. Given the large negative nutrient balance in lactating sows and the recent evidence that Cr is related to utilization of carbohydrates, proteins, and fat, it would seem that Cr might markedly improve sow performance and condition.

Additionally, because the negative nutrient balance of lactational catabolism is associated with depressed insulin, it would seem that insulin, which is an important signal related to the metabolic state of the sow, may be an important signal related to the reinitiation of the estrus cycle. Indeed, exogenous insulin has been demonstrated to increase the frequency of LH pulses in restricted-fed gilts and has been reported to increase the ovulation rate of gilts (Britt et al., 1988). Research at Mississippi State University has demonstrated increases in the ovulation rate in gilts in response to preestrus administration of insulin (Cox et al., 1987) and in litter size at second farrowing in response to insulin administration from weaning of the first litter to remating (Ramirez et al., 1994). Due to the established relationship of Cr to insulin function, it seemed logical that supplementation of the reproducing female would elicit positive responses. This was confirmcd (Lindcmann ct al., 1995b; Table 4) by research at Virginia Tech that demonstrated a 2.2-pig per litter increase when gilts were fed Cr from Cr picolinate during growth and development and then continued on a 200-ppb

Table 5 Calculation of Pigs Per Sow Per Year and Nitrogen Charge Per Pig[a]

Conception rate	Total born	Litters/ sow/year	Pigs/ sow/year	Lactation sow feed (kg)	Lactation N[b] (g)	Nonlactation sow feed (kg)	Nonlactation N (g)	Total N (g)	N charge/ pig (g)
Altered conception rate									
70	10.0	2.26	22.59	380	10327	603	13517	23844	1056
75	10.0	2.29	22.88	385	10464	602	13479	23943	1046
80	10.0	2.32	23.19	390	10601	600	13442	24043	1037
85	10.0	2.35	23.50	395	10739	598	13404	24143	1027
88	10.0	2.37	23.70	398	10830	597	13379	24209	1021
90	10.0	2.38	23.83	400	10876	597	13367	24243	1017
Altered litter size									
80	10.0	2.32	23.19	390	10601	600	13442	24043	1037
80	10.5	2.32	24.35	390	10601	600	13442	24043	987
80	11.0	2.32	25.51	390	10601	600	13442	24043	942
80	11.5	2.32	26.67	390	10601	600	13442	24043	901
80	12.0	2.32	27.83	390	10601	600	13442	24043	864

[a] Based on a lactation length of 28 d, 7 d to rebreed, and those sows failing to conceive being rebred in 42 d. It is assumed that the sow eats 6 kg/d in lactation and 2 kg/d in nonlactation states, that the lactation diet is 17% crude protein, and that the gestation diet is 14% crude protein.

[b] N = nitrogen

level of supplementation throughout two parities. Recently completed research (Lindemann et al. 1995a) with gilts that began receiving Cr for the first time at breeding revealed small responses in litter size during the initial two litters, but again a 2-pig per litter advantage in the third parity. Improvements in the conception rate were also observed in both trials, ranging from 5 to 20%. Clearly, Cr supplementation is a part of proper sow nutrition, and future research should help to establish both the level of supplementation that is optimal and the length of time that supplementation needs to occur to see optimal responses.

The effects of these changes in sow productivity on nitrogen charges per pig are illustrated in Table 5. Improvements in either conception rate or litter size are associated with lower nitrogen charges per pig with the largest reductions obviously occurring with the increased litter size. Improvements in both parameters will accelerate the reduction in nitrogen charge per pig.

CONCLUSION

Current levels of dietary formulation knowledge and changes in management already being implemented demonstrate that swine production enterprises are already moving in the direction of reduced nutrient excretion and that further reductions are achievable. The field of Cr research is in its infancy, and much remains to be done, but results already generated demonstrate the positive effect of supplementation of this mineral on improved carcass quality and on improved nutrient management.

REFERENCES

Britt, J. H., Armstrong, J. D., and Cox, N. M., Metabolic interfaces between nutrition and reproduction, *Proceedings of the 11th International Congress on Animal Reproduction,* Artificial Insemination, Dublin, Ireland, 1988, 117.

Chang, X., Mowat, D. N., and Bateman, K. G., GTF chromium benefits stressed feeder calves, *J. Anim. Sci.,* 69 (Suppl. 1), 158, 1991.

Cox, N. M., Stuart, M. J., Althen, T. G., Bennett, W. A., and Miller, H. W., Enhancement of ovulation rate in gilts by increasing dietary energy and administering insulin during follicular growth, *J. Anim. Sci.,* 64, 507, 1987.

Dickerson, J. W. T. and Widdowson, E. M., Chemical changes in skeletal muscle during development, *Biochem J.,* 74, 247, 1960.

Evans, G. W., The effect of chromium picolinate on insulin controlled parameters in humans, *Int. J. Biosoc. Med. Res.,* 11, 163, 1989.

Kumpulainen, J. T., Chromium content of foods and diets, *Biol. Trace Element Res.,* 32, 9, 1992.

Lindemann, M. D., Harper, A. F., and Kornegay, E. T., Further assessments of the effects of supplementation of chromium from chromium picolinate on fecundity in swine, *J. Anim. Sci.,* 73 (Suppl. 1), 185, 1995a.

Lindemann, M. D., Wood, C. M., Harper, A. F., Kornegay, E. T., and Anderson, R. A., Dietary chromium additions improve gain/feed and carcass characteristics in growing/finishing pigs and increase litter size in reproducing sows, *J. Anim. Sci.,* 73, 457, 1995b.

Mertz, W., Chromium occurrence and function in biological systems, *Physiol. Rev.,* 49, 163, 1969.

National Research Council (NRC), *Nutrient Requirements of Swine,* 9th ed., National Academy Press, Washington, DC, 1988.

National Research Council (NRC), *Recommended Dietary Allowances,* 10th ed., National Academy Press, Washington, DC, 1989.

NPPC, *Procedures to Evaluate Market Hogs,* 3rd edition, National Pork Producers Council, Des Moines, IA, 1991.

Page, T. G., Chromium, tryptophan, and picolinate in diets for pigs and poultry, Ph.D. Thesis, Louisiana State University, Baton Rouge, 1991.

Ramirez, J. L., Cox, N. M., and Moore, A. B., Enhancement of litter size in sows by insulin administration prior to breeding, *J. Anim. Sci.,* 72 (Suppl. 1), 79, 1994.

Samsell, L. J. and Spears, J. W., Chromium supplementation effects on blood constituents in lambs fed high or low fiber diets, *Nutr. Res.,* 9, 889, 1989.

Schwartz, K. and Mertz, W., A glucose tolerance factor and its differentiation from factor 3, *Arch. Biochem. Biophys.,* 72, 515, 1957.

Schwartz, K. and Mertz, W., Chromium (III) and the glucose tolerance factor, *Arch. Biochem. Biophys.,* 85, 292, 1959.

Steele, N. C. and Rosebrough, R. W., Trivalent chromium and nicotinic acid supplementation for the turkey poult, *Poultry Sci.,* 58, 983, 1979.

Wang, Z., Kornegay, E. T., Wood, C. M., and Lindemann, M. D., Effect of supplemental chromium picolinate on dry matter digestibility, nitrogen retention, and leanness in growing-finishing pigs, *J. Anim. Sci.,* 73 (Suppl. 1), 18, 1995.

CHAPTER 20

Environmental Concerns of Using Animal Manure — The Dutch Case

Age W. Jongbloed and Chris H. Henkens

INTRODUCTION

In the past, animals were fed on farm-produced feeds, and the manure produced was regarded as a scarce and valuable commodity for maintaining soil fertility. As long as the manure was used on the farm, nutrients remained within the cycle, except for some losses associated with storage, transport, and nutrients deposited in milk and meat.

During the last decades, livestock production changed greatly in several countries to minimize production costs. The production level per animal and the production of animal product per hectare of land increased considerably. This was possible by improvement in genetic potential, better housing systems and mechanization, the use of mineral fertilizers, and the use of complete feeds, which may partly originate from imported feedstuffs or may be transported from other regions for a long distance. In addition, agricultural enterprises expanded to provide a sufficient income per worker. The solution for getting a sufficient income on farms with a small acreage was to turn to activities independent of soil. Thus large confinement systems for livestock have been developed on holdings with limited acreage, sometimes in the neighborhood of big cities.

In the last decades, an increasing interest in, and respect for, the environment can be noticed. This also has consequences for livestock production. Air became polluted by noxious odors from animal husbandry while this sector causes almost 90% of the total ammonia (NH_3) emission in the Netherlands (Lenis, 1989). Also, the emission of methane (CH_4), from ruminants primarily, is regarded as a threat in relation to the greenhouse effect. Furthermore, some minerals like phosphorus (P), copper (Cu), and zinc (Zn) accumulate in the soil and contribute via leaching and runoff to eutrophication of groundwater and freshwater sources. Eutrophication may cause excessive growth of algae, sometimes resulting in massive fish mortality (Roland et al., 1993). Besides, Cu, Zn, and other heavy metals accumulate in the top layer with consequences for plant growth and human and animal health. This is one of the hazards of using sewage sludge too. Because of excessive application of manure and fertilizers per hectare of land and a surplus of precipitation, nitrate (NO_3^-) was leached and often exceeded tolerated values in fresh water (50 mg of NO_3^-/kg). Similarly,

1-56670-199-6/96/$0.00+$.50

the tolerated level of 12 mg of potassium (K) per liter of fresh water is exceeded, although it is not clear at this moment what are its environmental consequences (Van Boheemen et al., 1991).

Negative aspects of livestock production on the environment already have led to legislation in some countries or states that limit the use of animal manure. The amount of manure applied per hectare of land is related to its nitrogen (N) or P content (Hacker and Du, 1993). Another possibility is to restrict the number of animals (or animal units) per hectare of cultivated land. The goal of this chapter is to describe the developments in livestock production in the Netherlands and the policy to restrain environmental pollution on the short and the long terms. The discussion still continues. Also, insight is given in the approach to reduce environmental pollution by means of nutrition of pigs.

GENERAL INFORMATION OF THE NETHERLANDS

The Netherlands is a small and flat country with a total area of 40,000 km^2 of which 59% is used as cultivated land and 9% as wood (C.B.S., 1995). About 27% of the area is below sea level and has a population of 15.2×10^6. It has a moderate sea climate with cool summers (July, 17°C) and soft winters (January, 2°C). Precipitation during the year averages 76 cm, but most of it falls in autumn. In the northern and western part, the soil types are clay or peat, whereas in the eastern and southern part, soils are mostly sandy and farm size is small. On these sandy soils, pig and poultry production is very intensive, whereas cattle production is spread all over the country. Arable farming is predominantly on the clay soils.

DEVELOPMENT OF THE LIVESTOCK PRODUCTION

Farms with small acreage usually have intensified production of swine and poultry based mainly on imported feedstuffs. The development of the use of concentrates in the Dutch livestock production is shown in Table 1 (C.B.S., 1995). The use of concentrates has tremendously increased (250%) from 1964 to 1980, after which concentrate use has slightly increased for pigs and poultry, but has declined for cattle. The latter is mainly because of the introduction of the quota system for milk production. As almost all feedstuffs for the concentrates are imported, a large proportion of minerals is imported. In addition, most of the concentrates are used in the southern and eastern regions on the sandy soils, leading to a high production of animal manure per hectare of cultivated land.

RECOGNITION OF A DISTURBED MINERAL BALANCE

As a result of the warnings by experts on soil fertility in 1969, an official working group was established by the National Council for Enterprise Development

Table 1 Progressive Usage of Concentrates (10^6 Tons) in the Netherlands by Animals During Last 30 Years

	1964/1965	1971/1972	1979/1980	1986/1987	1993
Cattle	1.43	2.37	4.84	4.64	4.12
Swine	2.27	4.08	6.11	7.46	8.01
Poultry	1.73	2.16	2.79	3.31	3.47
Other	0.05	0.51	0.20	0.40	1.22
Total	5.48	9.12	13.94	15.81	16.81

to give advice on transport and elimination of manure surpluses. This resulted in a subsidy on transport of manure to arable farms in 1970. Henkens (1972) pointed out that the mineral content of concentrates should be minimized. In 1974, this working group advised limiting overdosing of P by manure. This advice was supported in 1980 by the Board Agricultural Emission and established by the Minister of Agriculture.

In 1973, another official working group (Working Group on Minerals in Concentrates in Relation to Fertilization and Environment) was established by the Dutch National Committee for Agricultural Research to survey the current status of the mineral balances on livestock farms. An extensive sampling programme was set up to get insight in the contents of some minerals (N, calcium [Ca], P, K, and Cu) in feeds for cattle, swine, poultry, and veal calves. Based on the intake and retention of minerals for animals, excretion was estimated. Subsequently, the mineral balance per hectare of cultivated land was calculated for several regions in the Netherlands. In areas with a high concentration of swine and poultry, P and Cu imput far exceeded the mineral removed by the crop, leading to saturation of the P-holding capacity of soils in the future. First, it was recommended to reduce the high Cu levels from 250 to 175 ppm in the pig feeds as soon as possible. Additionally, it was suggested to investigate a possibility of lowering the P content in swine and poultry feeds by 25%. Some research was initiated to study the mechanism of the growth-promoting effect of Cu and studies into the P requirements of swine and poultry. Also research was started to reduce the noxious odors from livestock farms. Despite the warnings of experts, nothing changed and the number of animals increased dramatically.

In 1979, a second report of the working group was published (Anonymous, 1979). Apart from N, minerals such as Ca, P, K, and Cu and also Zn, cadmium (Cd), and lead (Pb) were considered. Despite a somewhat lowered P content in feeds for pigs and poultry, P and Cu surpluses per hectare of cultivated land were considerably increased. As a result of this report, attempts were made to reduce Cd content in the feeds by selected choice of feedstuffs. Furthermore, in 1984, a maximum of 175 ppm Cu in pig feeds was only allowed up to 16 weeks of age, after which the maximum was reduced to 100 ppm.

In 1984, a third report of the working group published analyses of Ca, P, Cu, Zn, Cd, and Pb in several roughages and mixed feeds. The Cd content in most feeds was about 25% of the value observed earlier. The P content in the feeds had slightly decreased. In this year, the environmental issue became politically interesting as well. Table 2 lists the contributions of phosphate (P_2O_5) from animal

Table 2 Amount of P_2O_5 in Animal Manure and Fertilizers (kg/ha Cultivated Land)

Province/country	1970	1975	1980	1987	1990
Noord Brabant	110	155	195	245	200
Gelderland	115	150	170	200	175
Limburg	110	145	165	215	160
The Netherlands (manure)	78	95	115	130	110
The Netherlands (fertilizer)	49	39	41	40	37
The Netherlands (total)	127	144	156	170	147

manure and fertilizers for some animal dense provinces and for the whole Netherlands (C.B.S., 1992).

Considering that crops withdraw on average 50 kg of P_2O_5 per hectare, it is apparent that accumulation of P takes place. It was concluded that such a high accumulation of P and some other minerals (like Cu) per hectare of cultivated land was not acceptable. Besides, it was shown in 1984 that of the total deposition of acid or acidifying components in the Netherlands one third originated from ammonia emission for which 90% was from animal husbandry.

Furthermore, because of excessive application of manure per hectare of land, heavy metals accumulated and also leached, as illustrated in Table 3 (Van Erp and Van Lune, 1991). Especially, Cd and Cu are regarded as the most threatening metals in the Netherlands. High levels of Cd in manure result in a higher Cd content in the crop, which may harm man and animal. Copper intoxication of sheep may occur with large applications of pig manure on fields (Henkens, 1975).

LEGISLATION IN THE NETHERLANDS

In 1984, new legislation was enforced in the Netherlands. Global aims of governmental policy were (1) equilibrium fertilization, (2) reduction of acid deposition, and (3) protect surface water and groundwater quality. With respect to soil protection, the policy stated that the soil has four essential functions and that, in the long term, no harm for plant growth or health risk for humans and animals was permitted. In the same year, pig and poultry farmers were not allowed to expand their number anymore.

The following criteria were formulated for N. The concentration of NO_3^- in groundwater should not exceed 50 mg/l, whereas surface water should not contain more than 2.2 mg of N per liter (about 10 mg of NO_3^- per liter). Moreover, surface water should not contain more than 0.02 mg of NH_3 N per liter. Furthermore, NH_3

Table 3 Leaching of Heavy Metals from Arable Fields (g/ha/year)

Soil type	Cd	Cr	Cu	Pb	Ni	Zn
Sand	2.6	47	78	13.7	47	207
Clay	2.8	55	87	13.9	29	88

Table 4 Amount of P_2O_5 from Animal Manure That Can Be Applied (kg/ha) From 1987 Onwards in the Netherlands

Phase	Year	Grass land	Arable land	Corn silage
1	≤1990	250	125	350
2	<1994	200	125	200
2	1994	200	125	150
3	1995	150	110	110
3	1996	135	100	100
4	1998	120	100	100
4	2000	85	85	85
5	≥2002	80	80	80

emission should be reduced by 50% in the year 2000 as compared with 1980. From 1992 onwards, all manure pits should be covered. For P, the groundwater should not exceed 0.10 mg of *ortho*-P per liter (about 0.15 mg of P_t per liter). Surface water should also contain not more than 0.15 mg of P_t per liter.

For the minerals of interest, the aim was to achieve an equilibrium in fertilization, which means a good balance between input and output, taking into account obligatory losses. Local authorities can ask for additional measurements, especially when the farms are located close to woods or natural parks.

Three phases were implemented to achieve these goals gradually. The amount of animal manure that could be applied per hectare of land was based on its P content (Table 4). In order to reduce leaching of NO_3^-, application of manure on the field was restricted more and more during the autumn and winter. Application also depended on the soil type and crop (Table 5). Additionally, restrictions were made with regard to the method of application. A greater use of the injection method was recommended to prevent ammonia emission (Table 5). For arable and maize land on clay and peat soils, animal manure can be applied throughout the year, provided that emission is minimized, which means ploughing immediately after application. If a farmer has a higher production of P_2O_5 per hectare of land than allowed, he has to pay a fine for the surplus and it has to be transported to other regions. Manure transport can be facilitated by the manure bank. If the P_2O_5 or NH_3 goals cannot be still achieved, finally the number of animals must be reduced.

Implications for Research

In 1985, a large research program (Programme Manure and Ammonia Emission) was set up to find solutions for the so-called manure problem. Three main solutions were proposed. The first one was a reduction of input of minerals via the feed. The second one was a stimulation of practical solutions at the farm level, such as distribution and application of manure. The third solution was to upgrade manure by processing of manure on a large scale for export purposes. Costs of this additional research program was $\$120 \times 10^6$, of which about 25% was financed by animal-related commodities and the remainder by the government. The various nutrition-related projects were supervised both by members of research institutes and the feed industry. The advantage of such supervision is that application of the results was not delayed.

Table 5 Restrictions for Period and Method of Application of Manure on Grass, Arable, and Maize Land

	Jan	Feb	Mar	Apr	May	Jun	Jun	Jul	Aug	Sep	Oct	Nov	Dec
Grassland													
1991	+	+	+	+	+	+	+	+	+	+	–	–	–
1992	–	++	++	++	++	++	+	+	+	+	–	–	–
1993	–	++	++	++	++	++	+	+	+	+	–	–	–
1994	–	++	++	++	++	++	+	+	+	+	–	–	–
1995	–	++	++	++	++	++	++	++	++	–	–	–	–
Arable and Corn Silage Land on Sandy Soils													
1991	+	+	+	+	+	+	+	+	–	–	–	–	—
1992	++	++	++	++	++	++	++	++	–	–	–	–	—
1993	–	++	++	++	++	++	++	++	–	–	–	–	—
1994	–	++	++	++	++	++	++	++	–	–	–	–	—
1995	–	++	++	++	++	++	++	++	–	–	–	–	—

Note: Legend: – = no application; ++ = low-emission techniques; + = application no restrictions.

In the first 5 years, a large amount of the research was carried out monodisciplinarily, but later on it was recognized that multidisciplinarily research was required. Interactions between feeding and housing and feeding and manure processing became more evident. So far, the approach of feeding and nutrition has been most successful, as will be summarized for pigs later in this chapter. The same approach has been successful for poultry. Ruminants are not discussed in this series. At the end of 1994, it became clear that processing of manure on a large scale was too expensive to be implemented as a tool for reducing surpluses of manure by export.

REDUCTION OF EXCRETION OF N AND P BY PIGS BY ALTERING NUTRITION AND FEEDING

Supply Nutrients for Pigs in Better Agreement with Their Requirement

Better Knowledge of the Supply of N and P in the Feed

It has been recognized for one or two decades that total digestible protein or total P in a feedstuff does not indicate its nutritive value correctly. Therefore, many efforts have been undertaken to base the nutritive value on the apparent digestibility of amino acids (AAs) and total P over the total digestive tract. Quality of dietary protein was evaluated by using the apparent ileal digestible AA content (at the end of the small intestine). This method was considered to be a further improvement, as compared with the apparent total tract digestible AAs (Sauer and Ozimek, 1986; Lenis, 1992; Batterham, 1994a). Based on many ileal digestibility experiments of van Leeuwen et al. (1989), a protein evaluation system in the Netherlands was introduced on the basis of apparent ileal digestibility of lysine, methionine, cystine, threonine, and tryptophan. This data set has been incorporated in the Dutch Feedstuff Table (C.V.B., 1994; Lenis, 1992). In Table 6, data are presented on ileal digestibility of these AAs. Large differences exist in digestibility among feedstuffs. Tanksley and Knabe (1984) and Southern (1990) indicated a rather wide range in ileal AA digestibility of grain byproducts and protein-rich feeds.

In addition, new evaluation systems were introduced, or are in preparation, based on the apparent digestibility or availability of P (D.L.G., 1987; NRC, 1988; C.V.B., 1994). The most advanced is that of C.V.B. (1994), with data obtained from more than 120 experiments with about 40 feedstuffs. As noted by Jongbloed and Kemme (1990) the digestibility of P in feed of vegetable origin also varies substantially (Table 6).

Better Knowledge of the Animal's Requirement

Much work was done in several countries to establish the animal's requirement for protein and P. Apart from the protein evaluation system, current

Table 6 Ileal Digestibility of N and Some AAs (%)[a] and Apparent Total Tract Digestibility of P (%) in Some Feedstuffs[b]

Feedstuff	N	Lys	Met	Cys	Thr	Try	P
Corn	70	56	82	70	62	48	17
Barley	73	70	80	71	66	73	39
Peas	75	83	74	63	69	67	45
Wheat middlings	69	71	76	69	63	74	28
Soybean extract	81	86	86	76	79	83	38
Sunflower ext	75	74	86	73	72	79	16
Meat meal fat	74	83	85	55	78	74	80

[a] Data from C.V.B., 1990.

[b] Data from Jongbloed and Kemme, 1990.

recommendations for AA and P requirements vary widely in different countries. Requirements may vary because of differences in housing conditions, genotype of animals, level of feeding, major ingredients used in the diets, and response criteria.

It is generally accepted that a starting point for AA requirement of pigs should be the concept of ideal protein. The first attempt to specify "ideal protein" for pigs was made by the ARC (1981). Reassessments of ideal protein have been reported by Moughan and Smith (1984), Wang and Fuller (1989), Fuller et al. (1989), Baker and Chung (1992), Lenis (1992), Batterham (1994b), and Baker (1996). Recently, Lenis (1992) concluded from his experiments and from calculations on literature data that, for maximum growth performance of growing-finishing pigs, the requirements for ileal digestible methionine + cystine, threonine and tryptophan, relative to ileal digestible lysine, should be 59, 63, and 19%, respectively. With regard to the methionine + cystine requirement, there are indications that the methionine to cystine ratio in the diet should not be below 50% (Roth and Kirchgessner, 1989; Lenis et al., 1990).

Most recent estimations of the requirement for digestible P of pigs were based on the factorial method (Jongbloed and Everts, 1992; Jongbloed et al., 1994).

Better Agreement of Supply and Requirement

Both for AAs and P, the concentration per kilogram of feed decreases as the liveweight of the pig increases from 30 to 110 kg (Table 7). Therefore, introduction of one additional feed for growing-finishing pigs will help balance AAs and digestible P in the diet to the requirements of the animals so less N and P is excreted (phase feeding).

A slightly larger reduction in N and P excretion by growing pigs can be achieved by mixing a feed rich in protein and minerals with a feed having a low concentration of protein and minerals in a changing ratio during the fattening period (multiphase feeding). This mixing system can be achieved with a computerized mechanical feeding system (Henry and Dourmad, 1993). A feeding strategy can be developed with a good fit of energy, protein, and mineral supply based

Table 7 Estimated Requirement for Ileal Digestible AAs and Digestible P (g/kg of Feed) of Pigs From 10 to 110 kg[a]

Weight (kg)	Lys	Met + Cys	Thr	Try	Dig P
10–25	9.5	5.6	5.7	1.7	3.7
25–45	8.8	5.2	5.3	1.6	3.0
40–70	7.7	4.5	4.6	1.4	2.2
70–110	6.5	3.8	3.9	1.2	1.7

[a] Data from C.V.B., 1990 and Jongbloed et al., 1994.

on pig potential, stage of production, production objective, and environmental constraints.

Required concentrations of N and P per kilogram of feed for breeding sows are much lower during pregnancy than during lactation. The use of separate diets for pregnancy and lactation compared with one diet for both reduced the excretion of N and P by 20% (Everts and Dekker, 1994).

Enhancement of Digestibility of Phosphorus and Protein

Phytase

Two thirds of total P in most feed of vegetable origin is present as phytic acid P, which is almost indigestible for monogastrics. However, in the presence of the enzyme phytase, phytic acid can be hydrolyzed to inositol and orthophosphate and subsequently absorbed. Phytases, as well as being present in feedstuffs such as wheat, barley, and rye, are produced by the microbial flora in the intestine and are probably also secreted into the lumen of the intestinal tract (alkaline phosphatase) from the brush border of the small intestine (Davies and Flett, 1978; Cooper and Gowing, 1983). Digestibility of P from wheat or wheat by-products is quite high, which is attributed to their high phytase activity (Cromwell et al., 1985; Jongbloed and Kemme, 1990; Pointillart, 1993). The digestibility of P in corn, sorghum, sunflower seed meal, rice bran, and some other plant ingredients, however, is very low.

The first promising results by using microbial phytase in pig feeds were reported by Simons et al. (1990) who obtained the phytase from a strain of *Aspergillus ficuum.* When 1000 units of phytase per kilogram (1 unit = release of 1 μmol inorganic P per minute from an excess of sodium phytate at 37°C and pH 5.5) was used, the digestibility of P increased from 27 to 51%. Studies with cannulated pigs indicated that the hydrolysis of phytic acid by microbial phytase occurred mainly in the stomach (Jongbloed et al., 1992). Now several types of microbial phytase have substantially enhanced P digestibility (Jongbloed et al., 1993). The prospect of using microbial phytase in pig feeds appears very promising. Phytase-supplemented feeds for growing-finishing pigs and for pregnant sows may need little or no supplementary feed phosphate. Microbial phytase is commercially available now and has been incorporated in pig feeds in the Netherlands since 1991.

Enzymes for Nonstarch Polysaccharides

The carbohydrate fraction of feed ingredients from vegetable origin consists, apart from starch and sugar, of nonstarch polysaccharides (NSP) like cellulose, hemicellulose, pectins, and oligosaccharides. These components are resistant to digestive enzymes (Low, 1985). Digestibility of these fibrous feeds can be improved by treatment with enzymes that can hydrolyze the NSP to monosaccharides (Inborr, 1994). Although several reports have shown positive responses for performance when enzymes for NSP are supplemented (Chesson, 1987), data showing significant improvement of AA and P digestibility in pigs are very scarce. These enzymes could further reduce N and P excretion by pigs.

Reduction of Antinutritional Factors and Processing

Most legume seeds contain different antinutritional factors (ANFs) like protease inhibitors, lectins, tannins, and α-amylase inhibitors. ANFs can affect ileal digestibility and absorption of protein, as was clearly demonstrated by Sauer and Ozimek (1986), Van der Poel (1990), and Huisman (1990). Elimination of ANFs from the feed and better processing conditions can improve N utilization in pigs which reduces N excretion.

Changes in Feedstuff Composition

The wide variety of feedstuffs available for pig diets show considerable variation not only in AA and P content, but also in their digestibilities. Therefore, in order to decrease excretion of N and P, those feedstuffs in the mixed feed are chosen which have a high digestibility of AAs and P.

Phosphorus

The data shown in Table 7 indicate great differences in P digestibility among feed of vegetable origin, ranging from 16 to 45%. Differences also exist in P digestibility among feedstuffs of animal origin (Jongbloed and Kemme, 1990), ranging from 68 to 91% and between feed phosphates ranging from 63 to 90% (Dellaert et al., 1990). Because feed phosphates are only used for their P supply, one can easily choose those with a high digestibility of P. In practice in the Netherlands, this has already led to a total shift to monocalcium phosphates at the expense of dicalcium phosphates.

Protein

Feedstuffs not only vary in N and AA content, but also in AA content expressed per kilogram of N. For economic reasons, it is impossible to formulate diets without oversupply of certain AAs, particularly in those countries where many by-products are used. Lowering the proportion of some by-products and other feedstuffs with a low ileal protein digestibility in the diet in favor of cereals and other feedstuffs with higher protein digestibilities will result in a better balancing of dietary protein.

The N excretion can be substantially lowered by reducing the protein level by more than 2%. In doing so, feeds need to be supplemented with lysine, methionine, and in most cases also with threonine and tryptophan. Lenis et al. (1990) showed in their studies on fast-growing boars and gilts that dietary protein levels can be reduced by 2% without any disadvantageous effect on growth performance when limiting AAs are supplemented sufficiently. The same was concluded by Fremaut and De Schrijver (1990) and Schutte et al. (1990) with young pigs. Lowering dietary crude protein level for growing-finishing pigs by 2% units will reduce N excretion by approximately 20%. This is more than double the reduction observed for a 1% unit reduction in dietary protein because of changes from less digestible feedstuffs to better digestible ones (Lenis, 1989).

Improvement of the Feed Conversion Ratio

In general, a better feed conversion ratio leads to a lower excretion of N and minerals by the animal. Therefore, attempts to improve the feed conversion ratio are beneficial for the environment (Jongbloed and Lenis, 1993). First, the feed should cover all nutritional needs of the animal; a suboptimal level may lead to decreased performance. Optimal management with regard to housing, choice of breeds, and feeding strategy is necessary to obtain better performance of the pigs. Animals with a higher daily protein deposition rate (Pd_{max}) have a higher lean meat percentage and, consequently, a better feed conversion ratio.

Supplementation of feed additives (growth promoters) may also reduce excretion of N and P as a result of a better feed conversion ratio as compared to nonsupplemented feeds. Jongbloed (1992) estimated that the excretions of N and P per weaned piglet and growing pigs were 7 and 3%, respectively, higher when no feed additives were used.

Unfavorable feed conversion ratios in castrates compared with boars can be improved with β-agonists which shift from fat to lean meat production (Berschauer, 1990). Both N and P excretion can be further reduced with recombinant porcine somatotropin (rPST), as shown by Noblet et al. (1993). Permission to use rPST or β-agonists is doubtful in the Netherlands because of consumer's attitude.

Reduction of Ammonia Emission from Piggeries and of the Amount of Manure

Ammonia emission from pig manure originates from urea in the urine. Due to the urease activity of fecal microbes, urea is rapidly converted into ammonia, which easily volatilizes into the air. Factors influencing the rate of ammonia emission are concentrations of urea and ammonia/ammonium in the manure, temperature and air velocity, pH, emitting surface, and dry matter content (Aarnink et al., 1993; Van Vuuren and Jongbloed, 1994.)

In order to reduce ammonia emission many measures have been developed to increase utilization of dietary protein: changing the ratio between urinary N and fecal N (bacterially fermentable carbohydrates), reducing urea degradation (separation of urine and feces, urease inhibitors), binding the ammonia (Yucca extract, clay minerals), lowering the slurry pH (acidification), and reducing the emitting surface (flushing, sloped floors, slurry removal (Van der Peet-Schwering et al., 1996), airtight storage, soil injection). Some measures are still speculative and need further research.

Concerning bacterially fermentable carbohydrates, several authors investigated possibilities to reduce the ratio between urinary N and fecal N by including these carbohydrates in the diet. Nitrogen incorporated in bacterial protein in feces is less easily degraded to ammonia than urea-N excreted in urine. Microbial fermentation of organic matter (OM) in the hindgut can increase the excretion of N in feces, while it reduces the N excretion in the urine (Kreuzer and Machmüller, 1993; Mroz et al., 1993; Canh, 1994). They showed indeed a lower ammonia emission compared with control diets ranging from 22 to 45%. Recently, Bakker et al. (1996) showed that inclusion of raw potato starch (PS) in diets for growing pigs increased the amounts (g/d) of OM disappearing from the hindgut, while less N disappeared from the hindgut to blood circulation. With the PS diet there was even a net N appearance in the hindgut. The N excretion with urine was lower with more PS in the diets, while the N retention was not different. According to Canh et al. (1996), a lower slurry pH related to a higher VFA content may also affect ammonia emission. Recently, Mroz et al. (1996) measured the effect of dietary acid-base difference (DABD) and acidifying salts on urinary pH, nutrient retention, and indoor ammonia emission by growing pigs. They found that acidifying Ca-salts reduced urinary pH by 1.6 to 1.8 units, thereby diminishing ammonia emission by 26 to 53%. Besides, reducing DABD from 320 to 100 meq/kg DM lowered urinary pH by 0.48 units, and ammonia emission by 11%.

To predict the amount of urine produced, quantitative insight is required with regard to factors that determine the water requirement. Besides, as a surplus of slurry on a farm often has to be transported over a long distance more attention should be paid to increasing DM content of slurry. In this respect, not only the effect of NaCl, but also of K, protein (Brooks and Carpenter, 1990; Pfeiffer, 1991; Mroz et al., 1996b), non-starch polysaccharides, and structural carbohydrates in the diet should be studied to reduce water intake by pigs. (Mroz et al., 1995a,b).

CURRENT STATUS

Recently two studies, one on N (Anonymous, 1995) and one on P (Anonymous, 1994), were completed to determine if an equilibrium in fertilization can be realized. These studies have given insight into possible losses on a farm level. Some results of these studies are reported here. Surveys on N and P excretion by pigs and poultry have shown that N excretion has not changed, but P excretion has been substantially decreased. Data in Table 8 show that excretion of P in growing-finishing pigs has more than halved in 20 years. This lower P excretion has, apart from increased nutritional knowledge, undoubtedly been stimulated by legislation based on P.

Table 8 Mean Excretion of P of a Growing-Finishing Pig From 25 to 108 kg

Year	P in feed (g/kg)	Feed conversion ratio	Excretion of P (kg/pig)
1973	7.4	3.37	1.62
1983	6.2	3.08	1.18
1988	6.0/5.0[a]	2.94	0.85
1992	5.5/4.9	2.86	0.77
1994	5.5/4.8	2.82	0.74
1995	5.3/4.6	2.76	0.68

[a] 6.0 in starter diet, 5.0 in finisher diet.

Table 9 Mean NO_3^- Concentration (mg/l) in Upper Groundwater on Sandy Soil (Anonymous, 1995)

Category	Mean	95% interval
Arable farm	134	103–165
Dairy extensive (<2.8 animals/ha)	135	94–176
Dairy intensive (>2.8 animals/ha)	243	178–308
Mixed farming extensive	189	150–218
Mixed farming intensive	236	186–285

Table 10 N Surplus (kg/ha) in 1992 at Good Agricultural Practice (GAP) and at Acceptable Environmental Level (Anonymous, 1995)

	1992 Practice	2000 (GAP)	Acceptable environmental level
1.5 Dairy cow/ha	300–310	180–310	Grass clay 80–270
2.5 Dairy cow/ha	380–390	240–360	Grass sand dry 50–130
Winter wheat (no manure)	–6	31	Grass sand wet 50–260
Winter wheat (with manure)	—	84	Arable clay 35–200
Potatoes (no manure)	384	79	Arable sand dry 50–130
Potatoes (with manure)	—	165	Arable sand wet 50–260

Nitrogen

The surplus of N (the difference between input and output at the farm) using good agricultural practice has been empirically estimated. Good agricultural practice means that current recommendations with regard to fertilizing and management are followed. Also, surplus of N was calculated that is acceptable from an environmental point of view. One of the criteria was a maximal level of 50 mg of NO_3^- per liter of groundwater. From 1992 to 1994, random samples were collected from the upper groundwater of 93 farms on sandy soils, and nitrate content was measured (Table 9; Anonymous, 1995). The concentration of NO_3^- was 2.5 to 4.5 higher than the target of 50 mg/l, and 93% of the farms exceeded this value. In addition, in 1992, 89% of the surface water samples exceeded the value of 2.2 mg of N_t per liter. No correlation could be detected between surplus of N and NO_3^- content in groundwater.

In Table 10, a comparison is made between N surplus in 1992 at good agricultural practice and from the acceptable environmental level (Anonymous, 1995). Based on these data, the environmental targets for surplus N can only be obtained on grassland with substantial adaptations. However, on dry sandy soils, this cannot be realized. On arable farms, no problems are foreseen except for dry sandy soils. The surplus N goal for sandy dry soils can be obtained only when fertilizer mineral N is used or a suboptimal harvest is accepted.

For pigs and poultry, the surplus of N was defined as the amount of N in emitted NH_3, assuming that there is no land available. Therefore, the emission of NH_3 should be reduced by 50% in the year 2000 or even by 70% in the future to reach an environmentally acceptable level. The target of NH_3 emission can be reached with laying hens, but probably not with other animals. With high investments, it may be possible, but economically this is difficult to exploit at this moment.

Phosphorus

From the environmental point of view, groundwater should not exceed 0.15 mg of P_t per liter. For Dutch conditions with a surplus of precipitation of 300 mm, this means that 1 kg of P_2O_5 per hectare per year can be leached. Furthermore, a P-saturated soil has been defined as a soil in which 25% of the P-binding capacity of the soil to a depth to the highest groundwater level has been used. If a soil is P saturated, then application of P_2O_5 is only allowed equal to the withdrawal by the crop. In a survey of sandy soils in 1992, 70% of the soil samples were P saturated (>25%), of which 21% had more than 50% of the P-binding capacity used. The P saturation of soil will lead to leaching on the long term. The time that leaching occurs depends on P-binding capacity and P status of the soil, P surplus per hectare, and the distance from the surface to the groundwater level.

Results of a recent study indicated that at a good agricultural practice, inevitable losses are between 25 and 50 kg of P_2O_5 per hectare (Anonymous, 1994). This amount seems to be needed to maintain an optimal P level in the soil, not only for grassland, but also for arable land. Inevitable losses of P can be caused by redistribution of P between the top and subsoil by ploughing, fixation, precipitation, and immobilization of P in the top soil or by runoff. The big gap between good agricultural practice and environmentally acceptable losses of P should be evaluated with regard to what will be acceptable for the future, allowing a socioeconomic strong agriculture.

LEGISLATION IN THE NEAR FUTURE

In the autumn of 1995 a preliminary political decision was made as to how to proceed with legislation about the Dutch manure problem from 1998 onwards. The goal of the governmental policy is to reach an equilibrium in fertilization. In addition to P_2O_5, N is included in the new legislation. Final political decisions will be made between 1996 and 1998.

In 1998, farms having more than 2.0 animal units per hectare will be obliged to have a bookkeeping system of the main sources of the input and output (or removal) from the farm for both P (P_2O_5) and N. Main sources of input are, for example, supply of animals, feed, fertilizers, and animal manure from other farms.

Table 11. Accepted Losses of P (kg P_2O_5/ha) and N (kg/ha) in the Netherlands, as Proposed in New Legislation (≥ 1998).

Year	Grassland and arable land P	Grassland N[a]	Arable Land N[a]
1998	40	300	175
2000	35	275	150
2002	30	250	125
2005	25	200	110
2008/2010	20	180	100

[a] Deposition and mineralization not included

Main items of output are discharges of slaughter animals, animal manure, and crops. Two animal units equal 30.4 starter pigs (between 6 weeks of age and 25 kg live weight) or 11.1 growing/finishing pigs in the live weight range 25-110 kg or 4–6 sows. All specialized pig farms, having no land available, will fall under the obligation to have a bookkeeping system. Farmers having less than 2.0 animal units per hectare do not have this obligation. On these farms restrictions are enforced with regard to the storage of manure (covered) and the time and method of application of manure on the land (autumn and winter and low-emission techniques). In the new bookkeeping system the farmer has to prove all entries at the input and output balance. All manure discharged from the farm will be sampled and analyzed for its P and N contents. The farmer has to pay for that. For the specialized pig and poultry farmer, having no or little land, the main costs due the manure legislation will be for discharging, sampling, and analyzing the manure.

The difference between input and output of minerals on a farm is the mineral surplus. From the mineral surplus the farmer may deduct both the so-called "accepted losses" of P_2O_5 and N per hectare and an allowed loss of N per animal (because of ammonia volatilization).

If there is a surplus left, then a fine has to be paid for both a surplus of P_2O_5 and N. The accepted losses, recently formulated as political goals in the new legislation, are rather stringent, as shown in Table 11. The accepted P loss for each hectare of arable land and grassland will gradually decrease from 40 kg P_2O_5 per hectare in 1998 to 20 kg in the year 2008/2010. For grassland the accepted N loss will decrease from 300 to 180 kg per hectare. This probably will affect the "supply and demand" for animal manure and therefore the costs of discharging of animal manure from pig and poultry farms.

Apart from the accepted losses, the allowed amounts of P_2O_5 to be applied per hectare will also be lowered after 1998 (Table 4). Besides, pig and poultry farmers will be obliged to build low ammonia emission animal facilities.

The new legislation, especially the stringent accepted losses, concerns Dutch farmers who have doubts whether under these conditions a socioeconomically strong agriculture is possible. Farmers organizations ask for a more regional approach. If the environmental targets prevail over the possibilities of good agricultural practice, then either high investments are to be made or the number of animals has to be reduced. Maybe the targets formulated in the new legislation will be postponed a bit or a more regional approach will come forward. Possible new solutions will come forward from research. Anyway, under the new legislation farmers will have to pay attention as much as possible to an optimal agreement between supply and requirement of minerals

FUTURE NEEDS

If the environmental targets prevail over the possibilities of good agricultural practice, then either high investments are to be made or the number of animals has to be reduced. Maybe the targets formulated for the year 2000 will be postponed to a later date. Possibly new solutions will come forward from research.

More research is required to supply data on digestibility of AAs at ileal level and total tract digestibility of P. Even more sophisticated evaluation systems may be introduced (true digestible) in the future. Also, more knowledge is required on factors that affect digestibility and utilization of N and P. Many processes in the animal are still unclear, especially with regard to AAs. A better understanding of these processes can lead to better balanced feeding.

Requirements for several categories of animals and types of production can be best estimated by the factorial approach. However, many data are lacking on the requirement of AAs and P for production. Modeling growth can give more insight in the processes in the animals and help to estimate the requirements in different situations.

CONCLUSION

Soil fertility and quality of the groundwater determine the amount of manure (minerals) that can be applied per hectare. Despite early warnings of researchers that the mineral balance on a farm was disturbed, 15 years passed before it became politically interesting. Therefore, a lot of precious time was lost and expansion of large operations took place. Environmental constraints will probably be more severe in the future. If environmental constraints can be anticipated at an early stage, money may be saved in the long term. Money should be invested in research, but also at the farm level. Adequate knowledge is lacking to give the policy of the government a scientific basis. An intensive research program, in which government and business participated, could result in faster solutions. In our opinion, not enough attention has been paid to a more fundamental and strategic approach. By means of a levy on a surplus of P_2O_5 in manure per hectare of land, excretion of P in manure of swine and poultry has decreased considerably. To prevent problems on the long term, the production of minerals per amount of animal product should be at least as possible in animal-dense areas. The surplus has to be transported to other areas or abroad. If this does not succeed, then the number of animals should be reduced.

Nutrition management can substantially contribute to reduction in N and P excretion by pigs. Adequate knowledge is required on the digestibility of AAs and P in the feed used and on the requirement of these nutrients at any stage and type of production. The digestibility of the AAs and P also can be enhanced. Supplementary microbial phytase can enhance the digestibility of P by 20% or more so that feeds for growing-finishing pigs and for pregnant sows may need little or no supplementary feed phosphate. Phosphorus excretion can be lowered by 20 to 30% by using microbial phytase. The use of enzymes for hydrolyzing NSP seems interesting. Elimination of ANFs from feeds by technological treatments can improve N utilization. The incorporation of more free AAs in the feeds and lowering crude protein content in the feed by 2% units can lower N excretion of growing pigs by 20%.

For a further decrease of N and P excretion, we recommended research to study more in-detail physiological processes in the animal with emphasis on amino acid absorption and utilization. Current knowledge concerning the possible reduction of the manure surplus has to be integrated into future feed strategies. A further integration of the nutrition research with other disciplines is necessary. In this respect, both the genetical potential of the animals and hygienic conditions should be evaluated.

REFERENCES

Aarnink, A. J. A., Wagemans, M. J. M., and Keen, A., Factors affecting ammonia emission from housing for weaned piglets, in *Proceedings of the First International Symposium on Nitrogen Flow in Pig Production and Environmental Consequences*, Verstegen, M. W. A., Den Hartog, L. A., Van Kempen, G. J. M., and Metz, J. H. M., Eds., EAAP Publication No. 69, Wageningen, The Netherlands, 1993, 286.

Agricultural Research Council (ARC), Pigs, in *The Nutrient Requirements of Farm Livestock,* Agricultural Research Council, London, 1981.

Anonymous, The Concentration of Some Minerals in Mixed Feeds in Relation to the Requirement of Animals, the Excretion in Faeces and Urine, and Some Consequences for Soil, Plant and Animal II, *Second Report of the Working Group on Minerals in Concentrates in Relation to Fertilization and Environment,* NRLO, The Hague, 1979.

Anonymous, Phosphate Losses and Surplusses of Phosphate in the Dutch Agriculture, Report of the Technical Project Group "P-deskstudy" (in Dutch), Ministry of Agriculture, Nature Management and Fisheries, 1994.

Anonymous, Nitrogen Losses and Nitrogen Surplusses in the Dutch Agriculture, Report of the Technical Working Group Admissible Nitrogen Surplus (in Dutch), Ministry of Agriculture, Nature Management and Fisheries, 1995.

Baker, H., Advancing Our Understanding of Amino Acid Utilization and Metabolism in Swine and Poultry Tissues, International Symposium on Nutrient Management of Food Animals to Enhance the Environment, June 4–7, Virginia Tech, Blacksburg, 1996, 43.

Baker, H. and Chung, T. K., Ileal protein for swine and poultry, *Kyowa Hakko Technical Review-4,* Nutri-Quest Inc., Chesterfield, MO, 1992.

Bakker, G. C. M., Bakker, J. G. M., Dekker, R. A., Jongbloed, R., Everts, H., Van der Meulen, J., Ying, S. C., and Lenis, N. P., The quantitative relationship between abosrption of nitrogen and starch from the hindgut of pigs, *J. Anim. Sci.*, 74 (Suppl. 1), 188, 1996.

Batterham, E. S., Ideal digestibilities of amino acids in feedstuffs for pigs, in *Amino Acids in Farm Animal Nutrition,* D'Mello, J. P. F., Ed., CAB International, Oxford, U.K., 1994a, 113.

Batterham, E. S., Protein and energy relationships for growing pigs, in *Principles of Pig Science,* Cole, D. J. A., Wiseman, J., and Varley, M. A., Eds., University Press, Nottingham, U.K., 1994b, 107.

Berschauer, F., Einfluss von B-Rezeptor-Agonisten auf den Protein- und Fettstoffwechsel, *Übersichten Tierernährung,* 18, 227, 1990.

Brooks, P. H. and Carpenter, J. L., The water requirement of growing-finishing pigs — theoretical and practical considerations, in *Recent Advances in Animal Nutrition,* Haresign, W. and Cole, D. J. A., Eds., Butterworths, London, 1990, 115

Canh, T. T., Influence of dietary composition on the pH, the ammonium content and the ammonia emission of slurry from fattening pigs at different storage temperatures, M.Sc. thesis, Agricultural University Wageningen, The Netherlands, 1994.

Canh, T. T., Aarnink, A. J. A., Bakker, G. C. M., and Verstegen, M. W. A., Effect of dietary fermentable carbohydrates on the pH of and the ammonia emission from slurry of growing-finishing pigs, *J. Anim. Sci.*, 74 (Suppl. 1), 191, 1996.

C.B.S., Algemene milieustatistiek 1992, SDU uitgeverij, s'Gravenhage, 1992, 86.

C.B.S., Landbouwcijfers 1994, Landbouweconomisch Instituut, s'Gravenhage, 1995, 67.

Chesson, A., Supplementary enzymes to improve the utilization of pig and poultry diets, in *Recent Advances in Animal Nutrition,* Haresign, W. and Cole, D. J. A., Eds., Butterworths, London, 1987, chapter 6.

Cooper, J. S. and Gowing, H. S., Mammalian small intestinal phytase (EC 3.1.3.8), *Br. J. Nutr.,* 50, 673, 1983.

Cromwell, G. L., Stahly, T. G., and Moneque, H. J., Bioavailability of the phosphorus in wheat for the pig, *J. Anim. Sci.,* 61, (Suppl. 1), 320, 1985.

C.V.B., *Provisional Table on Apparent Ileal Digestible Amino Acids in Feed for Pigs* (in Dutch), Centraal Veevoederbureau, Lelystad, The Netherlands, 1990.

C.V.B., *Table of Feedstuffs. Information About Chemical Composition, Digestibility and Feeding Value* (in Dutch), Centraal Veevoederbureau, Lelystad, The Netherlands, 1994.

Davies, N. T. and Flett, A. A., The similarity between alkaline phosphatase (EC 3.1.3.1.) and phytase (EC 3.1.3.8.) activities in rat intestine and their importance in phytate-induced zinc deficiency, *Br. J. Nutr.,* 39, 307, 1978.

Dellaert, B. M., Van der Peet, G. F. V., Jongbloed, A. W., and Beers, S., A comparison of different techniques to assess the biological availability of feed phosphates in pig feeding, *Neth. J. Agric. Sci.,* 38, 555, 1990.

D.L.G., *Empfehlungen zur Energie und Nährstofversorgung der Schweine*, nr. 4, DLG Verlag, Frankfurt, 1987.

Everts, H. and Dekker, R., Effect of nitrogen supply on the retention and excretion of nitrogen and on energy utilization of pregnant sows, *Anim. Prod.,* 59, 293, 1994.

Fremaut, D. and De Schrijver, R., Protein reduction in diets for growing-finishing pigs: Effects upon animal performance and nitrogen excretion (in Dutch), *Landbouwtijdschrift,* 43, 1007, 1990.

Fuller, M. F., MacWilliam, R., Wang, T. C., and Giles, L. R., The optimum dietary amino acid pattern for growing pigs. 2. Requirements for maintenance and for tissue protein accretion, *Br. J. Nutr.,* 62, 255, 1989.

Hacker, R. R. and Du, Z., Livestock pollution and politics, in *Proceedings of the First International Symposium on Nitrogen Flow in Pig Production and Environmental Consequences,* Verstegen, M. W. A., Den Hartog, L. A., Van Kempen, G. J. M., and Metz, J. H. M., Eds., EAAP Publication No. 69, Wageningen, The Netherlands, 1993, 3.

Henkens, Ch. H., Production of manure and the environment (in Dutch), *Bedrijfs-ontwikkeling,* 3, 343, 1972.

Henkens, Ch., H., Border between application and dumping of organic manure (in Dutch), *Bedrijfsontwikkeling,* 6, 247, 1975.

Henry, Y. and Dourmad, J. Y., Feeding strategy for minimizing nitrogen output in pigs, in *Proceedings of the First International Symposium on Nitrogen Flow in Pig Production and Environmental Consequences,* Verstegen, M. W. A., Den Hartog, L. A., Van Kempen, G. J. M., and Metz, J. H. M., Eds., EAAP Publication No. 69, Wageningen, The Netherlands, 1993, 137.

Huisman, J., Anti-Nutritional Effects of Legume Seeds in Piglets, Rats and Chickens, Ph.D. thesis, Agricultural University, Wageningen, The Netherlands, 1990.

Inborr, J., Supplementation of pig starter diets with carbohydrate degrading enzymes — stability, activity and mode of action, *Agric. Sci. Finland,* 3, (Suppl. 2), 1994.

Jongbloed, A. W., Quantification of the Efficacy of Feed Additives on the Excretion of Nitrogen and Phosphorus, Amount of Manure and Feed Consumption of Pigs (in Dutch), Mededelingen IVVO-DLO No. 17, Lelystad, The Netherlands, 1992.

Jongbloed, A. W. and Everts, H., The requirement of digestible phosphorus for piglets, growing-finishing pigs and breeding sows, *Neth. J. Agric. Sci.,* 40, 123, 1992.

Jongbloed, A. W., Everts, H., and Kemme, P. A., The Requirement of Digestible Phosphorus for Pigs (in Dutch), CVB-Documentation Report No. 10, Lelystad, 1994.

Jongbloed, A. W. and Kemme, P. A., Apparent digestible phosphorus in the feeding of pigs in relation to availability, requirement and environment. 1. Digestible phosphorus in feedstuffs from plant and animal origin, *Neth. J. Agric. Sci.,* 38, 567, 1990.

Jongbloed, A. W., Kemme, P. A., and Mroz, Z., The role of microbial phytases in pig production, in *Enzymes in Animal Production,* Wenk, C. and Boessinger, M., Eds., Kartause Ittingen, Switzerland, 1993, 173.

Jongbloed, A. W., Kemme, P. A., and Mroz, Z., Phytase in swine rations: impact on nutrition and environment, in *BASF Technical Symposium*, January 29, 1996, Des Moines, Iowa, BASF, Mount Olive, New Jersey, USA, 44, 1996.

Jongbloed, A. W. and Lenis, N. P., Alteration of nutrition as a means to reduce environmental pollution by pigs, *Livestock Prod. Sci.,* 31, 75, 1992.

Jongbloed, A. W. and Lenis, N. P., Excretion of nitrogen and some minerals by livestock, in *Proceedings of the First International Symposium on Nitrogen Flow in Pig Production and Environmental Consequences,* Verstegen, M. W. A., Den Hartog, L. A., Van Kempen, G. J. M., and Metz, J. H. M., Eds., EAAP Publication No. 69, Wageningen, The Netherlands, 1993, 22.

Jongbloed, A. W., Mroz, Z., and Kemme, P. A., The effect of supplementary Aspergillus niger phytase in diets for pigs on concentration and apparent digestibility of dry matter, total phosphorus and phytic acid in different sections of the alimentary tract, *J. Anim. Sci.,* 70, 1159, 1992.

Kreuzer, M. and Machmüller, A., Reduction of gaseous nitrogen emission from pig manure by increasing the level of bacterially fermentable substrates in the ration, in *Proceedings of the First International Symposium on Nitrogen Flow in Pig Production and Environmental Consequences,* Verstegen, M. W. A., Den Hartog, L. A., Van Kempen, G. J. M., and Metz, J. H. M., Eds., EAAP Publication No. 69, Wageningen, The Netherlands, 1993, 151.

Krieger, R., Hartung, J., and Pfeiffer, A., Experiments with a feed additive to reduce ammonia emisssion from pig housing — preliminary results, in *Proceedings of the First International Symposium on Nitrogen Flow in Pig Production and Environmental Consequences,* Verstegen, M. W. A., Den Hartog, L. A., Van Kempen, G. J. M., and Metz, J. H. M., Eds., EAAP Publication No. 69, Wageningen, The Netherlands, 1993, 295.

Leeuwen, P., Slump, P., van Weerden, E. J., Huisman, J., Tolman, G. H., and van Kempen, G. J. M., Table Ileal Digestible Amino Acids for Pigs (in Dutch), ILOB-TNO Report No. I 89-3641, Wageningen, The Netherlands, 1989.

Lenis, N. P., Lower nitrogen excretion in pig husbandry by feeding: Current and future possibilities, *Neth. J. Agric. Sci.,* 37, 61, 1989.

Lenis, N. P., Digestible amino acids for pigs. Assessment of requirements on ileal digestible basis, *Pig News Inf.,* 13, 31N, 1992.

Lenis, N. P., van Diepen, J. Th. M., and Goedhart, P. W., Amino acid requirements of pigs. 1. Requirements for methionine + cystine, threonine and tryptophan of fast-growing boars and gilts, fed ad libitum, *Neth. J. Agric. Sci.,* 38, 577, 1990.

Low, A. G., Role of dietary fibre in pig diets, in *Recent Advances in Animal Nutrition,* Haresign, W. and Cole, D. J. A., Eds., Butterworths, London, 1985, chapter 4.

Moughan, P. J. and Smith, W. C., Assessment of a balance of dietary amino acids required to maximise protein utilisation in the growing pig (20-80 kg live weight), *N. Z. J. Agric. Res.,* 27, 341, 1984.

Mroz, Z., Jongbloed, A. W., and Lenis, N. P., Water in pig nutrition: Physiology, allowances and environmental implications, *Nutr. Res. Rev.,* 8, 137, 1995.

Mroz, Z., Jongbloed, A. W., Beers, S., Kemme, P. A., DeJonge, L., Van Berkum, A. K., and Van der Lee, R. A., Premilinary studies on excretory patterns of nitrogen and anaerobic deterioration of faecal protein from pigs fed various carbohydrates, in *Proceedings of the First International Symposium on Nitrogen Flow in Pig Production and Environmental Consequences,* Verstegen, M. W. A., Den Hartog, L. A., Van Kempen, G. J. M., and Metz, J. H. M., Eds., EAAP Publication No. 69, Wageningen, The Netherlands, 1993, 247.

Mroz, Z., Jongbloed, A. W., Van Diepen, J. Th. M., Vreman, K., Jongbloed, R., Lenis, N. P., and Kogut, J., Excretory and physiological consequences of reducing drinking water supply to non-pregnant sows, *J. Anim. Sci.,* 73 (Suppl. 1), 213, 1995b.

Mroz, Z., Jongbloed, A. W., Canh, T. T., and Lenis, N. P., Lowering ammonia volatilization from pig excreta by manipulating dietary acid-base differnce, *Abstract 8th Animal Science Congress of the Asian-Australasian Association of Animal Production (AAAP),* October, 13–18, 1996, Chiba (Tokyo), Japan.

Noblet, J., Dourmad, J. Y., and Dubois, G., The effect of porcine somatotropin and nitrogen losses in growing pigs, in *Proceedings of the First International Symposium on Nitrogen Flow in Pig Production and Environmental Consequences,* Verstegen, M. W. A., Den Hartog, L. A., Van Kempen, G. J. M., and Metz, J. H. M., Eds., EAAP Publication No. 69, Wageningen, The Netherlands, 1993, 189.

NRC (National Research Council), *Nutrient Requirements of Swine,* 9th revised edition, National Academy Press, Washington D.C., 1988.

Pfeiffer, A. M., Untersuchungen über den Einfluss proteinreduzierter Rationen auf die Sstickstoff- und Wasserbilanzen sowie die Mastleistungen an wachsenden Schweinen, Ph.D. thesis, Christian-Albrechts University, Kiel, Germany, 1991.

Pointillart, A., Importance of phytates and cereal phytases in the feeding of pigs, in *Enzymes in Animal Production,* Wenk, C. and Boessinger, M., Eds., Kartause Ittingen, Switzerland, 1993, 192.

Roland, D. A., Gordon, R. W., and Rao, S. K., Phosphorus solubilization and its effect on the environment, *Proc. Maryland Nutr. Conf.,* 1993, 138.

Roth, F. X. and Kirchgessner, M., Influence of the methionine: Cystine relationship in the feed on the performance of growing pigs, *J. Anim. Physiol. Anim. Nutr.,* 61, 265, 1989.

Sauer, W. C. and Ozimek, L., Digestibility of amino acids in swine: Results and their practical applications. A review, *Livestock Prod. Sci.,* 15, 367, 1986.

Schutte, J. B., Bosch, M. W., Lenis, N. P., De Jong, J., and Van Diepen, J. Th. M., Amino acid requirements of pigs. 2. Requirement for apparent digestible threonine of young pigs, *Neth. J. Agric. Sci.,* 38, 597, 1990.

Simons, P. C. M., Versteegh, H. A. J., Jongbloed, A. W., Kemme, P. A., Slump, P., Bos, K. D., Wolters, M. G. E., Beudeker, R. F., and Verschoor, G. J., Improvement of phosphorus availability by microbial phytase in broilers and pigs, *Br. J. Nutr.*, 64, 525, 1990.

Southern, L. L., Ileal protein for swine and poultry, *Kyowa Hakko Technical Review-4*, Nutri-Quest Inc., Chesterfield, MO, 1990.

Tanksley, T. D., Jr. and Knabe, D. A., Ileal digestibilities of amino acids in pig feeds and their use in formulating diets, in *Recent Advances in Animal Nutrition,* Haresign, W. and Cole, D. J. A., Eds., Butterworths, London, 1984, chapter 6.

Van Boheemen, P. J. M., Berghs, M. E. G., Goosensen, F. R., Hotsma, P., and Meeuwissen, P. C., Environmental pollution in 2000 by application of K, in *Manure and Environment in 2000. Scope from Agricultural Research,* Verkerk, N. A. C., Ed., DLO, Wageningen, The Netherlands, 1991, 61.

Van Erp, P. J. and Van Lune, P., Long-term heavy-metal leaching from soils, sewage sludge and soil/sewage sludge mixtures, in *Treatment and Use of Sewage Sludge and Liquid Agricultural Wastes*, EC-COST-action, 1991, 122.

Van der Peet-Schwering, C. M. C., Verdoes, N., Voermans, M. P., and Beelen, G. M., Effect of feeding and housing on the ammonia emission of growing and finishing pig facilities (in Dutch), Report Experimental Pig Station No. P1.145, Rosmalen, The Netherlands, 1996.

Van der Poel, A. F. B., Effects of processing on bean (Phaseolus vulgaris L) protein quality, Ph.D. thesis, Agricultural University, Wageningen, The Netherlands, 1990.

Van Vuuren, A. M. and Jongbloed, A. W., General Plan for the Role of Feeding Measures to Restrict Ammonia Emission from Livestock Buildings (in Dutch), Report ID-DLO (IVVO) No. 272, 1994.

Wang, T. C. and Fuller, M. F., The optimum dietary amino acid pattern for growing pigs. 1. Experiments by amino acid deletion, *Br. J. Nutr.,* 62, 77, 1989.

Index

A

B

D

N

O

P

R

S

T

U